AF572586

Telescope OPTICS

A Comprehensive Manual for Amateur Astronomers

Rutten & van Venrooij

Published by

P.O. Box 35025 • Richmond, Virginia 23235

☎ (804) 320-7016 • **www.willbell.com**

Fifth printing 2002
Printed in the United States of America

Library of Congress Cataloging-in-Publication Data
Rutten, Harrie G. J., 1950-
Telescope optics : evaluation and design / Harrie G.J. Rutten, Martin A.M. van Venrooij ; edited by Richard Berry ; programs adapted for the IBM-PC by Diane Lucas.
p. cm.
Includes bibliographies and index.
ISBN 0-943396-18-2
1. Telescope. 2. Optics. 3. Telescope-Design and construction - Amateurs' manuals. I. Venrooij, Martin A. M. van, 1934- II. Berry, Richard, 1946- . III. Lucas, Diane. IV. Title.
QB88.R87 1988 88-14243
522'.2–dc 19 CIP

Cover Photograph Credits:
Jupiter by Donald C. Parker
Full Moon by Richard Berry

Preface

The authors gratefully acknowledge the help of Klaas Compaan, who was our advisor during the writing of this book. He guarded us against a number of errors and unclarities and also allowed us to publish the data for some of the new optical systems he has designed.

We received further support from Richard Berry, our editor. He invested a great deal of time in editing the text into a consistent whole, ready for the press.

We also gratefully acknowledge the highly skilled assistance of Diane Lucas, who patiently edited the original Apple programs. She translated these programs into very convenient IBM programs that are easy to use.

The authors are indebted to the publisher, Willmann-Bell, Inc., for the excellent care bestowed on the design and production of the book.

We want also to record our gratitude to Ans Colaris, who typed and patiently retyped our manuscript.

Finally, we wish to thank our wives, Elly and Ans, and our children, Maurice, Marie-Pauline, and Aad, for accepting the restricted social life that accompanied the preparation of this book during a period of more than a year.

HARRIE G. J. RUTTEN
MARTIN A. M. VAN VENROOIJ

Editor's Note

With this book, Martin van Venrooij and Harrie Rutten have placed in the hands of amateur astronomers and telescope makers both knowledge of optics and powerful tools to apply that knowledge.

This book will both arouse your curiosity and answer your questions. Why are there so many different kinds of telescopes? What does each type have to offer? What makes one telescope better than another? Which are best? Why? What are the tradeoffs? As a telescope buyer, you will be better informed; as a telescope maker, you will be able to design custom optics.

Many readers will find the analyses of existing designs the most valuable part of the book. Newtonians, Cassegrains, Maksutovs, Schmidt cameras—name a design and you'll find it in here—are all described and analyzed so that you can easily compare them. What's your dream telescope? This book will help you choose it.

Others will make use of the power they now have to check, test, and analyze new telescope designs. The optional design and raytrace programs give you the tools you need to begin with a basic design and work systematically until you have created an optimized optical system that meets your personal design criteria. You'll be able to try new types of glass, design a telescope around that corrector shell you have parked in the basement, even compare the performance of many different eyepieces on your telescope.

No longer must you, as an amateur astronomer, meekly accept someone else's opinion about a telescope design. You can scrutinize existing designs and improve them to meet your own standards. Is that new astrographic camera all it's cracked up to be? By raytracing it, you'll know the answer.

This book will make you an optical expert. After you've read and studied it, the intricacies of the Maksutov-Cassegrain and apochromatic refractor will make sense to you. Of course you may have to work hard to understand, but the rewards will be great. You'll be privy to the terminology of optics, familiar with the methods of optical design, and able to share the design goals of professional opticians. You'll be able to design your own telescopes.

Many books on optics have come and gone. This book will stay. As it finds its audience, I look forward to seeing a multitude of exciting new developments in telescope design that will surely spring from it.

I want to thank Mary Steinke Tingbald for keying hundreds of thousands of text characters into my computer; Eleanor Berry for proofing the entire text, marking and correcting fuzzy wording and textual inconsistencies; Eugene W. Cross, Ralph K. Dakin, Harold R. Suiter, John Gregory, Paul A. Valleli, Roger W. Sinnott, Robert E. Cox, Robert D. Sigler, Richard A. Buchroeder, and David E. Stoltzmann who so generously read and commented on the book before publication. Their practical experience, design sensitivity, and familiarity with optical practice have made the book more useful and valuable for the amateur telescope maker.

Finally, all who use the accompanying software owe Diane Lucas—a telescope builder and designer herself—their gratitude and appreciation for her work on the programs. Whether you use them for the design of new systems or the evaluation of existing designs, you will benefit from Diane's dedication. It was a real thrill to see this software become fast, efficient, and easy to use.

RICHARD BERRY
Cedar Grove, WI
May 1988

Note Added April 1999: The text for this printing has been completely reset and the cover redone. Willmann-Bell, Inc, wishes to thank Jack Koester, Sandra Lucas, and Christopher Bechtler for their help in maintaining the accuracy of this printing.

Table of Contents

Preface ... **iii**

Editor's Note ... **iii**

Chapter 1 Introduction ... **1**

Chapter 2 Development of the Amateur Telescope ... **3**
2.1 Early Developments ... 3
2.2 Twentieth Century Developments ... 5

Chapter 3 First Order Optics: Lenses and Mirrors ... **9**
3.1 Refraction and Reflection ... 9
3.2 Image Formation ... 10
3.3 The Optical System of the Telescope ... 16
3.4 Flat Plates and Prisms ... 18

Chapter 4 Image Aberrations and Their Presentation ... **21**
4.1 The Spot Diagram ... 21
4.2 Image Aberrations ... 24
4.2.1 Spherical Aberration ... 25
4.2.2 Coma ... 26
4.2.3 Astigmatism ... 29
4.2.4 Curvature of Field ... 31
4.2.5 Distortion ... 31
4.3 Chromatic Aberrations ... 31
4.3.1 Longitudinal Chromatic Aberration ... 35
4.3.2 Lateral Color ... 35
4.4 Presentation of Image Aberrations with Spot Diagrams ... 35
4.5 Scaling Optical Systems ... 41
4.6 Concluding Remarks ... 43

Chapter 5 The Newtonian Telescope ... **45**
5.1 Introduction ... 45
5.2 The Spherical Mirror ... 46
5.3 The Paraboloidal Mirror ... 47
5.4 The Size of the Secondary Mirror ... 49

Chapter 6 The Refractor ... **53**
6.1 Correction of Aberrations ... 53
6.2 Residual Aberrations in Objective Lenses ... 57
6.2.1 Chromatic Aberration ... 57

6.2.2 Spherical Aberration and Spherochromatism 60
6.3 Evaluation of Lens Objectives 60

Chapter 7 The Cassegrain Telescope 65
7.1 Introduction 65
7.2 Curvature of Field 67
7.3 Optical Performance 69
7.4 Baffling 75

Chapter 8 The Schmidt Camera 75
8.1 Introduction 75
8.2 Optical Principles 75
8.3 The Schmidt Corrector 76
8.4 Characteristics of the Schmidt Camera 78
8.5 Results of Optical Ray Tracing 79
8.6 The Field-Flattened Schmidt Camera 79
8.7 The Lensless Schmidt 82

Chapter 9 The Schmidt-Cassegrain Telescope 83
9.1 General Classification 83
9.2 Treatment of Systems 85
9.3 "Visual" Schmidt-Cassegrain Telescope 85
9.4 Close Focusing in the SCT 91
9.5 Flat-Field Schmidt-Cassegrain Systems 92
9.6 Computer-Aided Design 95

Chapter 10 The Maksutov Camera 97
10.1 Introduction 97
10.2 Maksutov Camera Designs 97
10.3 The Optimum Meniscus Corrector 101

Chapter 11 The Maksutov-Cassegrain Telescope 107
11.1 Introduction 107
11.2 Maksutov-Cassegrain Systems 107
11.3 Meniscus Correctors 113
11.4 Curved- and Flat-Field Maksutov-Cassegrain 115

Chapter 12 The Schiefspiegler 117
12.1 Introduction 117
12.2 Optical Principles of Schiefspieglers 118
12.3 Results of Optical Ray Tracing 122

Chapter 13 Other Compound Systems 127
13.1 Introduction 127
13.2 Full-Aperture Correctors: Schmidt Derivatives 127
13.3 Full-Aperture Correctors: Houghton Derivatives 131
13.4 Focal Correctors: Jones, Bird, and Brixner 133
13.5 Unusual Compound Systems 136
13.6 Gregorians, Relay Telescopes, and Wright's Off-Axis Catadioptric 142

Chapter 14 Field Correctors **147**
14.1 Introduction 147
14.2 The Single-Lens Field Flattener 147
14.3 The Distant Field Flattener 150
14.4 Field Correctors for Newtonians 151

Chapter 15 Focal Extenders and Reducers **155**
15.1 Focal Extenders 155
15.2 Focal Reducers 158
15.3 Remarks on Achromatic Combinations 160

Chapter 16 Eyepieces for Telescopes **163**
16.1 Introduction 163
16.2 Eyepiece Types 165
16.3 Aberrations and Other Eyepiece Characteristics 167
16.4 Ray-Tracing Eyepieces 171
16.5 Ray-Trace Results for Eyepieces 173
16.6 Eyepieces Used for Projection 186
16.7 The Performance of Objective-Eyepiece Combinations 187
16.7.1 Introduction 187
16.7.2 Astigmatism and Field Curvature 188
16.7.3 Accommodation of the Eye 189
16.7.4 Analyzing Objective-Eyepiece Combinations 190
16.7.5 Combinations Examined 191
16.7.6 Results of Ray Tracing 191
16.7.7 Discovering Favorable Objective-Eyepiece Combinations 197

Chapter 17 Deviations, Misalignments, and Tolerances **199**
17.1 Introduction 199
17.2 Surface Accuracy 199
17.3 Deviations and Misalignment 201
17.4 Influence of Deviations and Misalignments 204
17.5 Tolerance Analysis 205
17.6 Correcting Manufacturing Deviations 206

Chapter 18 Resolution, Contrast, and Optimum Magnification **211**
18.1 Introduction 211
18.2 Resolving Point Sources 212
18.3 Resolving Power and Contrast for Extended Objects 215
18.4 Contrast Transfer in a Perfect Optical System 216
18.5 Contrast Transfer for Imperfect Optical Systems 218
18.6 Central Obstructions 218
18.7 Obstructed Telescopes for Visual Use 220
18.8 Residual Aberrations 222
18.9 The Value of the Contrast Transfer Function 222
18.10 Optimum Magnification 223

Chapter 19 Opaquing and Vignetting ... 227
19.1 Introduction ... 227
19.2 Baffles for Refractors and Newtonians ... 228
19.3 Baffling for Cassegrain Telescopes ... 230
19.4 Stops and Vignetting ... 234
19.5 Internal Reflections in Catadioptric Systems ... 240
19.6 Lens Coatings ... 241

Chapter 20 Optical Calculations ... 243
20.1 Introductory Remarks to Chapters 20 and 21 ... 243
20.2 Methods of Optical Calculation ... 243
20.3 Optical Surfaces ... 244
20.3.1 Conic Sections ... 245
20.3.2 Higher-Order Surfaces ... 247
20.4 Sign Conventions ... 249
20.5 The Paraxial Calculation ... 249
20.6 The Seidel Calculation ... 254
20.7 The Meridional Calculation ... 261
20.8 The Skew-Ray Trace ... 262
20.8.1 Introduction ... 262
20.8.2 Flat Surfaces ... 264
20.8.3 Spherical Surfaces ... 264
20.8.4 Conic Sections ... 266
20.8.5 Higher-Order Surfaces ... 269
20.9 Calculation of Non-Centered Systems ... 272
20.10 Using Ray-Trace Results ... 274
20.10.1 Magnitude of the Image Aberrations ... 274
20.10.2 Determining the Diameters of Optical Elements ... 275
20.11 Other Optical Calculations ... 276

Chapter 21 Designing Telescope Optical Systems ... 277
21.1 Introduction ... 277
21.2 Designing a Cassegrain ... 278
21.3 Designing a Catadioptric Cassegrain ... 283
21.4 Designing a Schmidt-Cassegrain ... 284
21.5 Designing a Houghton-Cassegrain ... 289
21.6 Designing a Maksutov-Cassegrain ... 292
21.7 Designing Single-Mirror Catadioptrics (Astrocameras) ... 298
21.8 Designing Schmidt and Wright Cameras ... 298
21.9 Designing a Houghton Camera ... 299
21.10 Designing a Maksutov Camera ... 300
21.11 The Shape of the Schmidt Corrector ... 300
21.12 Optimization Techniques ... 303
21.13 Designing a Two-Element Achromatic Refractor Objective ... 305
21.13.1 Introduction ... 305

21.13.2 Doublet Design Procedure ... 306
21.13.3 Achromatizing a Doublet Lens ... 306
21.13.4 Correcting Spherical Aberration ... 314
21.13.5 Correcting Coma ... 318
21.13.6 Reducing Spherochromatism ... 320
21.14 Other Degrees of Freedom ... 322
21.15 An Alternate Method of Designing a Doublet ... 322
21.16 Designing a Three-Element Apochromatic Refractor Objective ... 322
21.16.1 Choosing Glass for a Triplet ... 323
21.16.2 The Powers of the Elements ... 326
21.16.3 Designing a Triplet ... 327
21.16.4 Examples of Triplets ... 329
21.17 Thick Optical Elements ... 330

Chapter 22 How to Use the Telescope Design Programs ... 331
22.1 Capabilities ... 331
22.2 Designing Telescopes with TDESIGN ... 332
22.2.1 Designs Available with TDESIGN ... 332
22.2.2 Using TDESIGN ... 335
22.3 Lens Design with LENSDES ... 340
22.3.1 Designing Lenses ... 341
22.3.2 Using LENSDES ... 341
22.3.3 Doublet Design with LENSDES ... 343
22.3.4 Triplet Design with LENSDES ... 346
22.3.5 Rescaling Doublet and Triplet Designs ... 347
22.4 The Telescope Optics Ray-Tracing Program ... 347
22.4.1 Using RAYTRACE ... 347
22.4.2 Vignetting Calculations ... 360
22.4.3 Tilted and Decentered Surfaces ... 360
22.4.4 Notes on Vignetting Computations ... 361
22.4.5 Data Input Exercises ... 361
22.5 Optimizing Predesigns from TDESIGN ... 362
22.5.1 The Wright Design ... 362
22.5.2 The Schmidt-Cassegrain Telescope ... 363
22.5.3 The Houghton Camera ... 365
22.5.4 The Houghton-Cassegrain Telescope ... 366
22.5.5 The Maksutov Camera ... 367
22.5.6 The Maksutov-Cassegrain Telescope ... 368
22.5.7 Automatic Optimizations ... 370

Appendix A: Schott Optical Glass Specifications ... 371

References ... 381

Index ... 389

Computer Software ... 397

Chapter 1

Introduction

This book has been written for every amateur astronomer who owns a telescope and wants to know more about it. It was not so long ago that we found ourselves in much the same position and set out to learn more about optics. We soon realized, as we asked questions and searched for answers, that we were hardly alone in our ignorance. When we set out to find the answers to our questions about telescopes, this book was born.

We were possibly too ambitious in trying to bridge the gap between amateur astronomer and professional optical designer, but we felt it needed to be done. In the past, when amateur telescopes were relatively simple and correspondingly easy to understand, there was no need for a book such as this. But today's telescopes and astrocameras, with their more sophisticated optics, require an understanding of optical theory considerably beyond anything now available in most amateur-level books. As a result, the modern telescope has become a "black box" to many of its most dedicated users.

Anyone, of course, can consult the professional literature. However, not only is this literature difficult to understand, but it is also not widely available, and, to make matters still worse, little of it deals directly with telescopes. In this book, when we discuss optics, we apply the theory directly to amateur telescopes and astrocameras. We have kept the book practical rather than theoretical, first analyzing a wide range of optical systems for the reader, then showing the reader how to investigate the performance of any optical system himself.

We believe it is quite possible for amateurs to design new optical systems and improve existing designs. Until recently, optical design was carried out only by professionals because optical calculations required the power of a mainframe computer. With the advent of fast and powerful home computers, fast ray tracing and optimization have come within the reach of determined amateurs.

In the process of writing this book, we carried out a comprehensive investigation of many optical systems, analyzing them with our own computer. As we investigated the different telescope designs, we decided we could not ignore astrocameras, field correctors, and eyepieces. These optical systems, we feel, are as important as the telescopes themselves in understanding today's amateur optics.

It is evident that a book this size cannot consider all the optical systems available to the amateur. In our search of the patent literature, we found hundreds of designs for objectives with five or more optical elements. Since we had to restrict ourselves somehow, we chose to include only those designs that amateurs would find practical to build. In the end, and admittedly somewhat arbitrarily, we decided to limit ourselves to systems consisting of four or fewer optical elements. Even with this restriction, the number of possible designs remains far too large to cover exhaustively.

We wish to make it clear that we offer no answer to the often-asked question, "Which telescope is best?" The answer, if there is one, depends largely on the task for which an instrument is to be used. If you wish to build or buy a low-cost telescope, your choice is likely to be a Newtonian reflector. If you demand the highest possible sharpness and contrast in observing planetary surfaces for a given aperture, you may prefer a refractor. For those requiring a compact and portable telescope, the obvious choice is a catadioptric such as the Schmidt-Cassegrain or Maksutov-Cassegrain. For optimum performance in wide-field photography, a modern flat-field camera is the best choice; for smaller fields, there are many catadioptric objectives to chose from. The advantage of optical ray tracing is that it permits an observer to analyze the performance of an optical system before constructing or purchasing it, and thus determine whether the system can do the job.

To aid the reader in comparing different optical systems, we have chosen 200 millimeters (8 inches) as the standard aperture for the systems analyzed in this book. This is a very popular aperture among amateur astronomers, for it is neither too small to take seriously nor too large for practical construction. We discuss conversion of optical specifications and performance data from our standard aperture to other apertures in chapter 4. Because so much is already available in the literature, we do not discuss optical shop techniques such as grinding and testing mirrors or lenses.

This book is organized so that, for understanding what different telescopes can do, it is not necessary to read beyond chapter 19. For the amateur who simply wishes to learn a bit about the performance of the different telescope types, we recommend studying chapters 4 through 19. These cover the properties of Newtonians, refractors, Schmidt and Maksutov cameras, Cassegrains—and, in fact, just about every type of telescope available today.

For the reader who is interested in investigating other systems or in designing new systems with his own computer, we have provided the more technical chapters 20, 21, and 22, which, when used with the computer programs, provide the tools to explore the field of telescope optics. The literature references have been specifically selected with the amateur in mind.

Chapter 2

Development of the Amateur Telescope

2.1 Early Developments

More than 300 years have passed since the invention of the telescope. In the following chapters, we shall consider many telescope designs of long standing plus a number of promising new designs. It is worthwhile, before delving into the optical properties, to consider the telescope's historic past as well as its flowering in the twentieth century.

From historical investigations, it appears that until 1600 the telescope was unknown, with the exception of some individuals who were probably unaware of its potential. History tells us that, in the year 1608, Hans Lippershey, a Dutch spectacle-maker of Middelburg, applied to the state of Holland for a patent for an instrument he had devised for looking at distant objects more closely. The patent, however, was not granted because Zacharias Janssen, also a Dutchman, claimed *he* had built such an instrument earlier than Lippershey. He claimed, in fact, that Lippershey's instrument was a copy of his own design.

The original Dutch telescopes were constructed with a positive objective lens and a negative eyelens—a design which provides upright images. "Dutch" telescopes were soon being made and sold in large numbers, and they soon spread throughout the whole of Europe.

The Italian Galileo Galilei also manufactured refracting telescopes, but only after he had heard about them from others. It would be interesting to know whether he re-invented the instrument on hearing that they existed, or whether he also heard how to construct them. In any event, in 1610 he announced the existence of the Jovian satellites, the phases of Venus, and lunar craters—and thereby revolutionized astronomy.

In the beginning, the color aberration that results from using a simple (i.e., single element) objective was suppressed by building refractors of extremely long focal lengths. Telescopes with apertures of 60 mm and lengths of 5 meters were common. In an extreme case, Christiaan Huygens made a simple refractor 60 meters long, with an aperture less than 200 mm.

Meanwhile, in 1663, a Scot named James Gregory designed the first mirror telescope. It consisted of a parabolodial concave primary mirror and an elliposidal concave secondary mirror to relay the image through a hole in the center of the primary. The atempts to carry out this design were unsuccessful however, since no one then knew how to test the complex mirrors.

Credit for the first *useful* reflector belongs to the Englishman Isaac Newton, who constructed a prototype of modern large telescopes in 1668. Newton's design is the same one used by many amateurs today. Newton had built lens telescopes, but after experimenting with the refraction of light and its dispersion into a spectrum, Newton became convinced that an achromatic (color-free) telescope based on lenses was impossible. His conclusion, though faulty, spurred his experiments with reflectors. Nine years after Newton announced his telescope, the Frenchman Guillaume Cassegrain proposed placing a convex mirror in front of the primary. Such a system is called a Cassegrain telescope.

Newton's pronouncement on color aberration in lens telescopes delayed further development of refracting telescopes. It was not until 1729 that the Englishman Chester M. Hall succeeded in making an achromatic objective by combining two lenses with different dispersions. Though he found the lens curvatures empirically, Hall eliminated roughly 95% of the color aberration present in simple refractors.

In 1758, the Englishman John Dollond was the first to construct an achromat based on calculations; thus, he is quite possibly the first optical designer.

Around 1774, William Herschel, a German who had moved to England, began constructing telescopes, first as an amateur, later, after his discovery of the planet Uranus with a home-built 150 mm reflector, as a professional. Herschel's reflecting telescopes—his largest had an aperture of 1.2 meters—became very famous, and the 1.2 meter instrument was the largest telescope in the world for many years.

During the nineteenth century, major progress took place in making large glass blanks sufficiently uniform for use in a telescope. By the end of the nineteenth century, it had become possible to make objectives up to one meter in diameter—and indeed, large refractors were the crowning glory of all major observatories at the beginning of the twentieth century.

But one meter remains the practical upper limit of objective lenses even today. During the World Exhibition in Paris in 1900 an objective with an aperture of 1.25 meters and a focal length of 40 meters was shown mounted horizontally. Optically, however, this telescope appears to have been a failure because of the sag of the glass under its own weight.

Until the middle of nineteenth century, all telescope mirrors had been made from metal. Even when Lord Rosse built a Newtonian telescope with an aperture of 1.8 meters and a focal length of 17 meters, he used metal "specula." The Frenchman Léon Foucault achieved a major step forward by making glass mirrors using the silvering process invented by Justus von Liebig to attain high reflectivity. This also meant that heavy metal mirrors could be replaced by much lighter glass mirrors. Furthermore, since glass mirrors can be resilvered quickly and easily while metal mirrors must be repolished at great time and expense, reflectors became far more practical than they had been. Within 30 years, reflectors had ousted refractors as the instrument preferred for serious research. Foucault also developed an important method for testing mirrors which amateur telescope makers still use today.

Looking over the early history of telescopes, it is striking that many developments in telescope making are the result of the efforts of amateurs and other individuals not occupied professionally with astronomy.

2.2 Twentieth Century Developments

Until the end of the nineteenth century, telescopes had either lenses or mirrors as objectives, but not both. In the twentieth century, telescopes combining both, or catadioptric systems, have come into use, driven in large part by the development of astronomical photography. Soon after photography became possible, astronomers realized that it was not possible to build a purely refracting or reflecting system with adequate aperture and speed with a large useful field. Reflectors, though large, suffered from off-axis aberrations which limited their angular coverage, and lens systems, though capable of covering relatively wide angles, suffered from severe color aberrations if the aperture was large or the focal ratio short.

The revolution began in 1930, when Bernhard Schmidt built the first Schmidt camera. He used a large spherical mirror plus a correcting lens placed at the center of curvature of the mirror. In this way, he combined the focusing power inherent in the reflector with the aberration-correcting ability of a specially made weak lens. Schmidt's invention offered astronomers the unheard-of combination of large aperture, wide field, and sharp images. The Schmidt camera is discussed extensively in chapter 8.

Despite their unprecedented performance, Schmidt cameras have some disadvantages. The most obvious is that the focal plane lies between the corrector and the mirror, inside the system. During the '30s and '40s, James Baker, Edward Linfoot, and Karl Slevogt investigated ways to improve access to the image plane. When a convex secondary mirror is placed between the Schmidt corrector and the primary mirror, the image is formed near the mirror—and in some designs the focal surface can be flat. These systems, while capable of excellent photography, are generally unsuitable for visual use on low-contrast objects (e.g., planets) because of the very large central obstruction of the secondary.

In the meantime, designers sought more compact telescopes. One of the most successful compact designs is the Schmidt-Cassegrain telescope. The corrector is closer to the primary mirror than in the Schmidt camera, and a small Cassegrain secondary mirror is used to form an image behind the primary. In 1962, Ronald R. Willey, Jr., showed that a compact Schmidt-Cassegrain could give excellent image sharpness both axially and off-axis. Encouraged by this analysis, Tom Johnson began to manufacture a 200 mm *f*/10 Schmidt-Cassegrain called the Celestron.

Some ten years later, Robert Sigler investigated a whole family of Schmidt-Cassegrain systems. He concluded that certain combinations of Schmidt correctors and mirrors have minimal image aberrations on and off-axis. We discuss a variety of Schmidt-Cassegrain forms in chapter 9, and in chapter 21 give the procedures for designing them.

In the 1930s, opticians found it exceedingly difficult to make the compound curves of Schmidt corrector lenses, so the designers investigated alternative correctors. Dimitri Maksutov in Russia, Albert Bouwers in the Netherlands, Dennis Gabor in England, and Karl Penning in Germany independently discovered that a deep meniscus lens has much the same effect as a Schmidt lens: it can correct the spherical aberration of a spherical mirror. Maksutov systems—including Maksutov cameras and Maksutov-Cassegrain telescopes—are popular today among amateur telescope makers. This is true especially for a Maksutov-Cassegrain designed in 1957 by John Gregory. Maksutov systems are discussed in chapters 10 and 11.

Most reflecting and catadioptric systems have a central obstruction caused by the secondary mirror. Horace Dall, William Pickering, and other experienced observers have found that the obstruction produces lowered contrast and image sharpness. These effects are especially pronounced when objects with low inherent contrast, such as planetary detail, are observed. Refractors, though unobstructed, suffer from residual chromatic aberration. What was needed was a system both unobstructed and color-free.

In the 1950s, Anton Kutter developed a class of unobstructed reflectors called schiefspieglers. They are the best known configuration of the larger family of tilted component telescopes, or TCTs. Chapter 12 deals with the schiefspieglers designed by Kutter, and chapter 18 covers the loss in contrast and resolution caused by the secondary obstruction.

The twentieth century has brought not only catadioptric instruments, but also further development of the classic two-mirror Cassegrain system. Developed by George Ritchey, an American, and Henri Chrétien, a Frenchman, the Ritchey-Chrétien configuration offers improved photographic performance. However, because this design has deeper aspherics than the Classical Cassegrain and suffers from spherical aberration at its prime focus, the instrument is not frequently built by amateurs.

Classical achromatic refractors have long focal ratios, typically *f*/15, in order to suppress chromatic aberration adequately. Using more than two lens elements, or certain special improved glasses, apochromatic refractors can have greatly reduced color aberration and be constructed in shorter focal ratios. Some of these developments are discussed in chapter 6.

An important part of every telescope, sadly neglected in the past, is the eyepiece. Since the 1970s, new designs have brought increased apparent field and improved edge sharpness. Chapter 16 surveys the historical development of telescope eyepieces and the best of the modern designs.

Before 1960, optical design was carried out by hand or with electromechanical calculators. In either case, it was a time-consuming operation. In the 1960s, large computers became available to professional designers and revolutionized optical designing. By the 1980s, computer technology had advanced to the point where an amateur can write and run his own computer programs. Such programs can ray-trace existing systems and design new systems. These subjects are covered in chapters 20, 21, and 22.

Chapter 3

First Order Optics: Lenses and Mirrors

3.1 Refraction and Reflection

The basic functions of telescopes are two: to enlarge the apparent angle subtended by a distant object, and to increase the amount of light reaching the observer's eye. As a result of the first function, the image of a planet appears larger when viewed through a telescope; as a result of the second, a star observed with a telescope appears brighter than with the unaided eye. Both functions are provided by an optical system, a combination of lenses, mirrors, or lenses and mirrors, called a telescope.

All optical systems are based on the refraction and/or reflection of light. Light is a wave phenomenon which propagates through space—and optical systems. Although light does not actually consist of rays, the calculation and design of optical systems is based on geometric optics, in which we treat light as a fan of rays, each ray following a straight line as long as it encounters no obstacles.

However, when a light ray traveling in one medium enters another medium, for instance, a ray in air strikes a glass surface, its direction of travel changes. This is because the velocity of light in glass is lower than the velocity of light in air. The measure of this change, called the index of refraction, is inversely proportion al to the velocity of light in the medium relative to the velocity of light in a vacuum. The index of refraction is, therefore, exactly unity for vacuum. For denser materials, the index of refraction is greater than unity: 1.00029 for air, 1.33 for water, and 1.51 for common soda-lime window glass. Every optical medium has its own index of refraction; moreover, the index depends on the wavelength of the light in a complex way.

When Willebrord Snell investigated the refraction of light in the early 17th century, he discovered a simple relationship between the angle of the incident ray with the surface normal, ε, and the angle of the refracted ray ε' (see fig. 3.1). According to Snell's law:

$$\frac{\sin\varepsilon}{\sin\varepsilon'} = \frac{n'}{n} \tag{3.1.1}$$

where n and n' are the indices of refraction of the two media. For instance, in a situation such as is shown in fig. 3.1, when the angle of incidence $\varepsilon = 30°$, and the index of refraction for the wavelength under consideration is 1.5, then

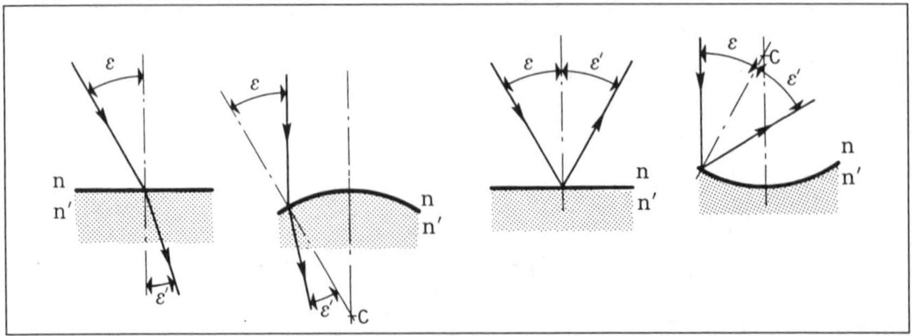

Fig. 3.1 *Refraction and Reflection.*

$$\frac{\sin 30}{\sin \varepsilon'} = \frac{1.5}{1}. \tag{3.1.2}$$

Solving for sin ε′,

$$\sin \varepsilon' = \frac{\sin 30}{1.5} = \frac{0.5}{1.5} = 0.3333\ldots \tag{3.1.3}$$

and therefore

$$\varepsilon' = 19.4712° . \tag{3.1.4}$$

Note that although the index for air is actually 1.00029, it is normally taken as 1 in the glass catalogs.

When a light ray travels from a dense medium (such as glass) to a less dense medium (such as air), according to Snell's law, the sine of the angle of refraction, ε′, for large angles of incidence ε, may be larger than 1. In this particular case, the ray does not exit the medium, but is, instead, reflected back into the medium (fig. 3.2). Since sin ε′ cannot exceed 1, the "critical angle" is arcsin($1/n$). For ordinary glass ($n = 1.5$), the critical angle is 41.8°; for an ultradense glass ($n = 1.9$), the critical angle is 31.8°. Total internal reflection is useful; it is used in some types of prisms (see section 3.4).

Reflection follows a simpler law than refraction: the angle of incidence equals the angle of reflection. Mathematically, this is:

$$\varepsilon = -\varepsilon' \tag{3.1.5}$$

It is important to note that the sign of the angle changes. We may also treat reflection as a special case of Snell's law in which:

$$n' = -n . \tag{3.1.6}$$

This is shown in fig. 3.1.

3.2 Image Formation

Telescopes, and indeed most optical systems, are made with more than one optical

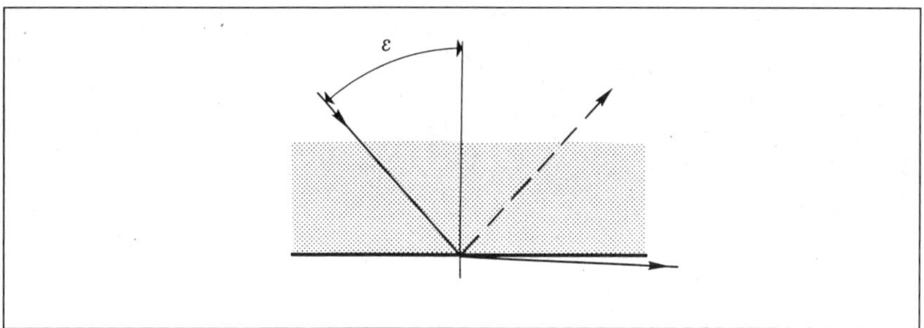

Fig. 3.2 *Critical Angle for Total Reflection.*

element. These are called compound systems. The result of such a combination may be, for example, a short tube length or a more favorable image position. However, the most important reason for compound systems is that they offer greater opportunity to control image aberrations.

In this section, we introduce first-order optics, a simplified and idealized approach that is very helpful in roughing out compound systems. First-order optics deals with rays and images close to the optical axis, in the so-called paraxial region. In first-order optics, the elements of the optical system have rotationally symmetric surfaces and share a common axis of rotation called the optical axis. Furthermore the optical elements (lenses and mirrors) are assumed to be infinitely thin, thus avoiding further complexities found in real optical systems.

Of course, real systems do not conform to the assumptions of first-order optical theory. Instead, they have relatively large apertures and relatively large ray-angles. In real systems, the focal surface is usually not a plane but a curved surface, and image aberrations can and do occur. We will examine aberrations in chapter 4. First order optics also ignores the diffraction of light. We will discuss diffraction and examine its consequences in chapter 18.

Although the formulae derived from first-order optics are insufficiently accurate for an exact design or analysis of an optical system, they are useful and important for making initial calculations or obtaining an overview of the paths of the rays in a telescope.

Telescopes are generally used on very distant objects; starlight is an excellent approximation of an infinitely distant point source. A fan of rays from a star can be considered perfectly parallel. When the star is on the extended optical axis (i.e. when the telescope is pointed at the star), the star will be imaged as a small disk of light on the optical axis. When the star is positioned away from the optical axis, the entering parallel bundle of rays makes an angle with the optical axis. The resulting image appears off the optical axis also.

Now consider fig. 3.3. Four optical elements each intercept a beam of rays parallel to the optical axis, possibly coming from a star, entering from the left side. The beam is imaged on the axis as the focal point F. The distance from F to the lens or the mirror is called f, the focal length, or focal distance. The focal ratio of a lens or mirror is defined as f/D, where D is the diameter of the beam, and is also

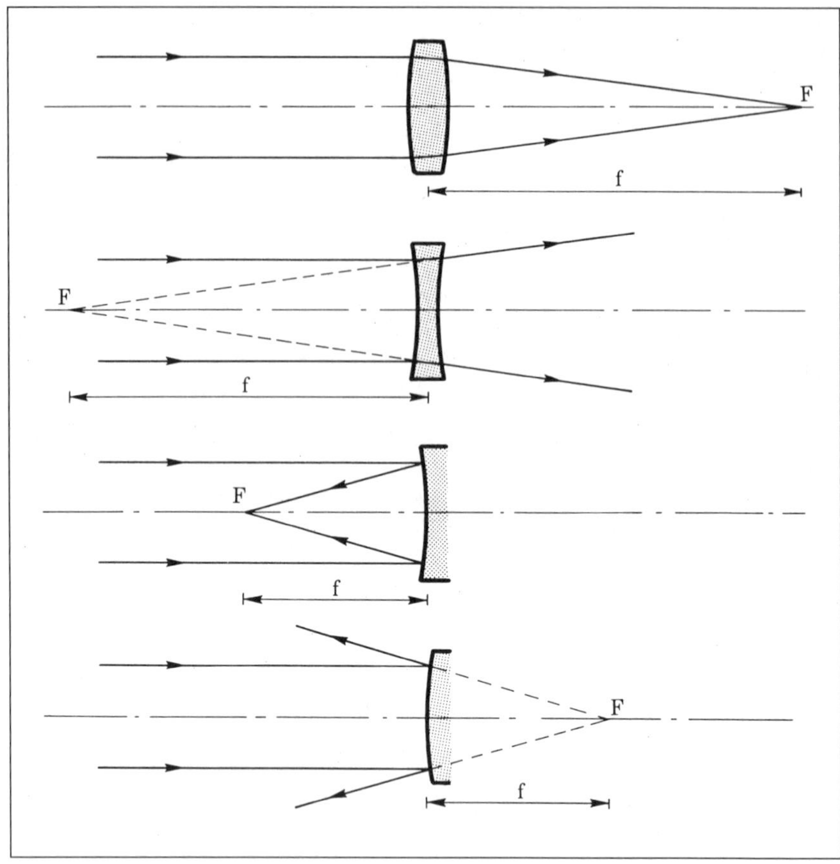

Fig. 3.3 *Focal Lengths of Lenses and Mirrors for a Constant Radius of Curvature. Note that refracting systems have longer focal lengths. The assumed refractive index is 1.500.*

the aperture of the system. When the focal length of a system is 1000 mm and the aperture D is 100 mm, then the focal ratio is 10. This system is therefore called an *f*/10 system. When comparing the focal ratios of different systems, the words "larger" and "smaller" should be avoided because they are ambiguous. To avoid confusion, we will use the terms "faster" and "slower" when comparing focal ratios. An *f*/5 system is, for example, faster than an *f*/10 system; an *f*/15 system is slower than an *f*/10 system.

For a thin lens having radii of curvature R_1 and R_2, and made of glass with refractive index n, the approximate focal length may be calculated as follows:

$$\frac{1}{f} = (n-1) \cdot \left(\frac{1}{R_1} - \frac{1}{R_2}\right). \tag{3.2.1}$$

For a biconvex lens with $R_1 = 50$ mm and $R_2 = -100$ mm (note that the sign of the second surface means it is concave to the incoming light) and refractive in-

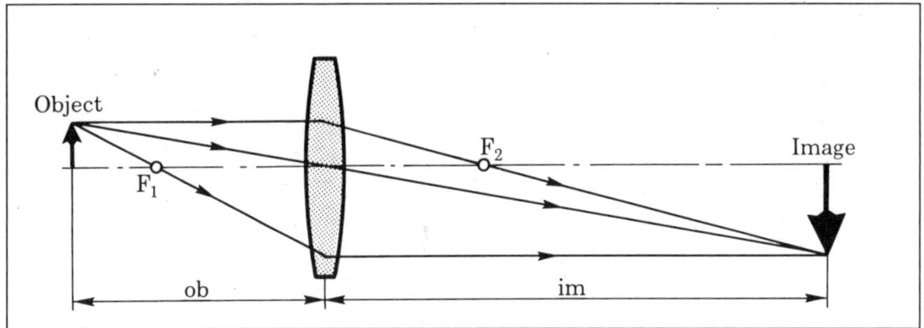

Fig. 3.4 *Image Formation for a Thin Lens (Finite object distance).*

dex $n = 1.5$, the focal length is calculated as:

$$\frac{1}{f} = (1.5 - 1) \cdot \left(\frac{1}{50} - \frac{1}{-100}\right) \qquad (3.2.2)$$

$$f = 66.667\text{mm}.$$

For a concave mirror with radius of curvature $-R$, the focal length is:

$$f = \frac{-R}{2}. \qquad (3.2.3)$$

Positive lenses and concave mirrors both act to converge a collimated light beam; either will focus a beam of parallel rays to a real focal point. A negative lens or a convex mirror will cause a parallel bundle of rays to diverge. Their focal points are not real but virtual, as shown in fig. 3.3. Their focal lengths can also be calculated with the formula above; the focal length will be a negative number.

In most cases a telescope is used for observing distant objects. We can generally treat the object distance as infinite; hence, the bundle of rays originating from every object point can be considered parallel. When the object distance is finite, the image distance lies farther from the system than the focal distance (see fig. 3.4). For this case we find the object and image distances from:

$$= \frac{1}{ob_{\text{dist.}}} + \frac{1}{im_{\text{dist.}}} \qquad (3.2.4)$$

where f is the focal length, $ob_{\text{dist.}}$ is object distance, and $im_{\text{dist.}}$ the image distance. If the object is 100 mm in front of the biconvex lens with focal length of 66.667 mm we mentioned above, then the image distance becomes:

$$\frac{1}{66.667} = \frac{1}{100} + \frac{1}{im}. \qquad (3.2.5)$$

Solving for the unknown quantity, we find $im_{\text{dist.}} = 200$ mm.

The magnification of the system is:

$$M = \frac{im}{ob} = \frac{\text{image height}}{\text{object height}}. \quad \textbf{(3.2.6)}$$

For this example, $M = -200/100$; therefore the magnification is –2.

In the case of a finite object distance, the construction of the image distance for a thin lens is easy if the focal length is already known (fig. 3.4). We draw three rays:

1. the first through the center of the lens called the principal ray;
2. the second parallel to the optical axis, which passes through the focal point, F_2, behind the lens;
3. the third through the focal point, F_1, in front of the lens, which becomes parallel to the axis after refraction.

The object and image planes lie at the intersections of the three rays.

Optical systems usually consist of more than one optical element. When two lenses with focal lengths f_1 and f_2 are spaced distance d apart, the effective focal length, f_{eff}, of the system becomes:

$$f_{\text{eff}} = \frac{f_1 \cdot f_2}{f_1 + f_2 - d}. \quad \textbf{(3.2.7)}$$

When the distance between the lenses is zero, the formula reduces to:

$$f_{\text{eff}} = \frac{f_1 \cdot f_2}{f_1 + f_2}. \quad \textbf{(3.2.8)}$$

If the position of the converging beam from the system is known, the effective focal length of the combined system is constructed by extending the converging rays back to the entering bundle, as shown in fig. 3.5. Note that the effective focal length may be greater or less than the physical length of the optical system, depending on the type and spacing of the optical elements.

This has a useful application to telescopes. Two mirrors may be combined as a Cassegrain telescope, shown in fig. 3.6. The effective focal length of the system becomes:

$$f_{\text{eff}} = \frac{f_1 \cdot f_2}{f_1 + f_2 - d}. \quad \textbf{(3.2.9)}$$

where d is the distance between the two mirrors. Note that the focal length of the convex mirror, f_2, is negative. The figure shows where a lens of the same focal length as the compound optical system would be: the Cassegrain telescope is considerably shorter than the equivalent simple telescope.

First-order optics can handle the passage of light rays through a thick lens, shown in fig. 3.7. A thick lens has four cardinal points, namely two focal points,

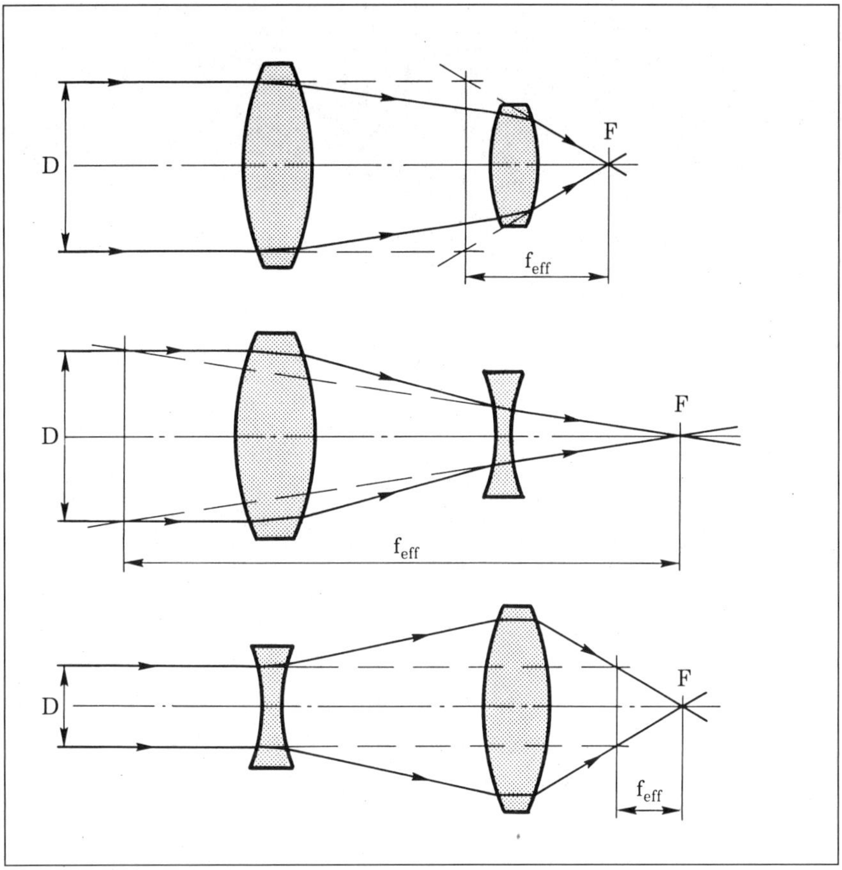

Fig. 3.5 *Effective Focal Lengths for Lens Combinations.*

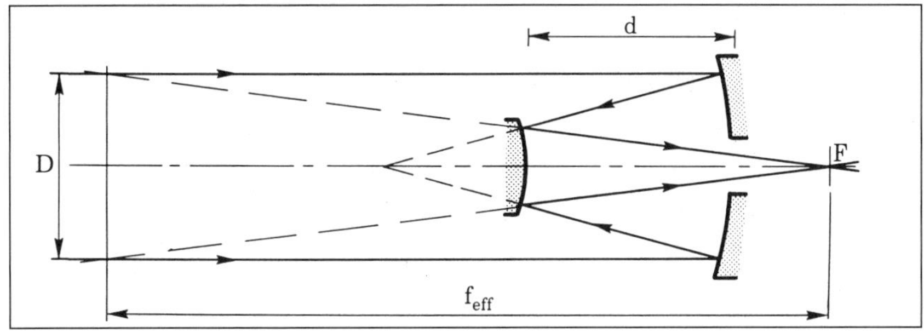

Fig. 3.6 *Effective Focal Length of a Cassegrain Telescope.*

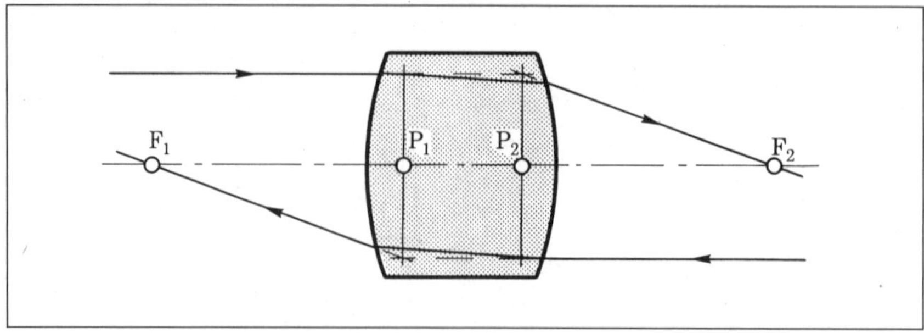

Fig. 3.7 *Principal and Focal Points for a Thick Lens.*

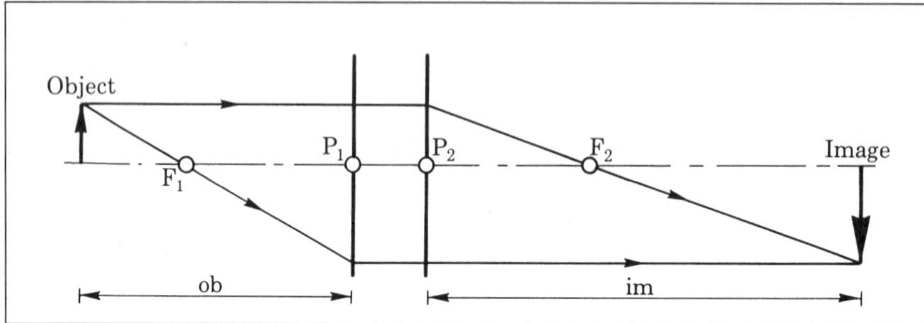

Fig. 3.8 *Image Formation for a Thick Lens.*

F_1 and F_2, and two principal points, P_1 and P_2. The principal points represent the intersection of two principal planes with the optical axis. The principal plane is the plane where the converging or diverging beam has the same diameter as the entering bundle. (We will show how P_1, P_2, F_1, and F_2 are calculated in chapter 20.) When the principal points and focal points are known, we can construct the image location as shown in fig. 3.8. The separated thin lens formula remains valid when the distance between the principal planes is taken into consideration. The position of the principal planes is quite different for the various lens types in fig. 3.9.

3.3 The Optical System of the Telescope

Optical systems which focus a parallel entering bundle of light into a point are called focal systems, or objectives. When an objective is used with photographic film, for example, it is called a camera.

A telescope consisting of an objective and an eyepiece is afocal; that is, it does not form an external image. Both the entering and the exit bundles of rays are parallel. Fig. 3.10 shows the principle of a telescopic optical system: two lenses separated by the sum of their focal lengths. A parallel bundle enters the objective and a parallel exit bundle leaves the eyepiece. Both the objective and the eyepiece are drawn as single elements for the sake of clarity.

In a simple telescope, the diameter of the entering bundle, D_e, is the entrance pupil. (In more complex systems, the entrance pupil is the front image of an aper-

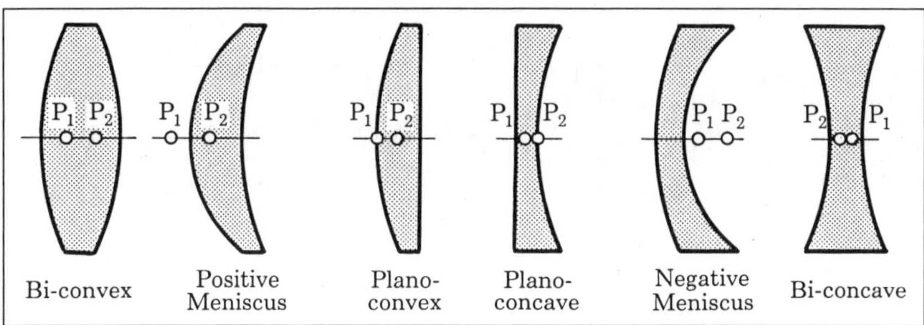

Fig. 3.9 *Position of the Principal Points for Various Lens Bendings.*

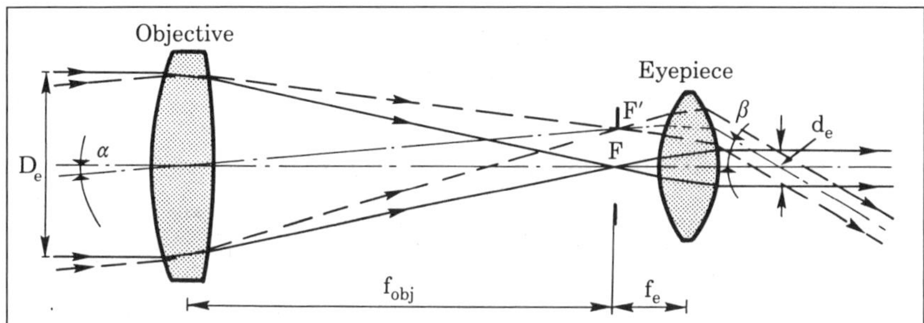

Fig. 3.10 *Image Formation in a Telescope.*

ture stop.) Let us now follow the edge rays of a bundle of starlight which enters the objective parallel to the optical axis. After passing through the objective, the rays converge and finally intersect at the focal point, *F*. After passing through *F*, they diverge again until they reach the eyepiece. From the eyepiece, the rays emerge parallel again. The lens of the observer's eye, which is located behind the eyepiece, can also be considered as an objective. It will focus the parallel bundle to a point on the retina.

The eyepiece diaphragm, or field stop, is located in front of the eyepiece. Point *F′*, located at the edge of this diaphragm, is the focal point of the most oblique beam visible to the eye. This oblique cone of rays, indicated in the figure by dotted lines, exits as a parallel bundle inclined to the optical axis by angle β after passing through the eyepiece.

An important function of the telescope is increasing the angle subtended by the object viewed. Since a star must be displaced by the angle α from the optical axis to form an image at point *F′*, the magnification is:

$$M = \frac{\tan\beta}{\tan\alpha}. \tag{3.3.1}$$

When these angles are small, the formula simplifies to:

Table 3.1

Observational Task	Exit Pupil (mm)
Wide Fields under dark skies: (large galaxies, Milky Way, diffuse nebulae, open clusters)	5 to 7
General Viewing: (nebulae, clusters)	3 to 4
Best match to eye's resolution: (moon, globular clusters, planetary nebulae, double stars)	2
Maximum planetary detail (according to Texereau)	0.8
Close double stars under best skies	0.5

$$M = \frac{\beta}{\alpha}. \quad (3.3.2)$$

The magnification of the system is also obtained from:

$$M = \frac{f_{obj}}{f_e}. \quad (3.3.3)$$

The diameter of the exit pupil is:

$$d_e = \frac{D_e}{M}. \quad (3.3.4)$$

The plane of the intersecting exit bundles is the exit pupil. This is also the image of the aperture stop formed by the eyepiece.

The greatest magnification possible is seldom the best for viewing a celestial object through a telescope. Experienced observers have established the guidelines for the diameter of the exit pupil for different types of observing (See table 3.1). In older observers, the pupil of the eye may not open to the full 7 millimeters we assume for youthful observers.

3.4 Flat Plates and Prisms

In addition to lenses and mirrors, plane parallel plates and prisms refract and reflect light. Although these elements do not have the ability to focus light, they can displace and deviate rays (fig. 3.11). For small-angle cones of light, the longitudinal beam displacement, or change of focal position, s, is approximated by:

$$s = t \cdot \frac{n-1}{n} \quad (3.4.1)$$

where t is the thickness of the plate and n is its refractive index. Although the thin-lens equation (eq. 3.4.1) does not indicate it, the displacement is dependent on the cone angle of the beam, and a thick glass plate will introduce detectable aberrations in a fast light beam. In addition to the longitudinal displacement, a tilted plate

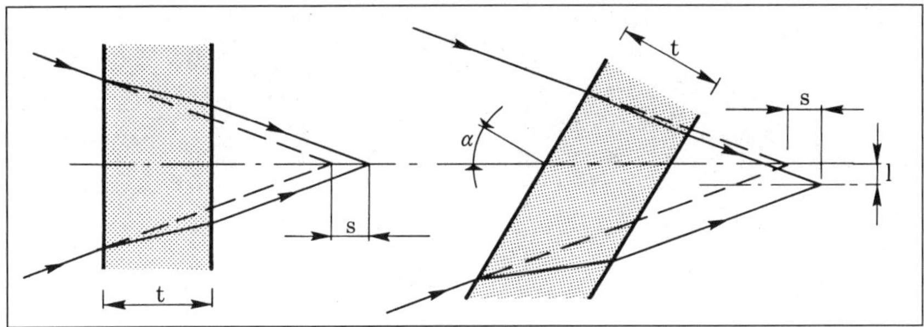

Fig. 3.11 *Image Displacement by a Plane-Parallel Plate.*

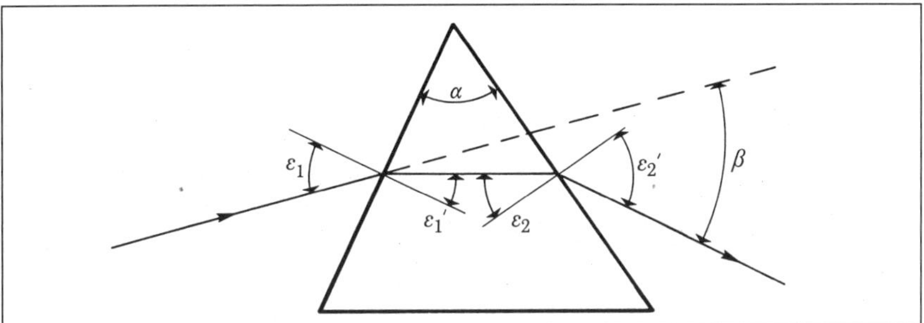

Fig. 3.12 *Refraction of a Ray in a Dispersing Prism.*

introduces a small lateral displacement of the beam.

Fig. 3.12 shows a dispersing prism. A ray entering the prism exits deviated from its original direction by angle β:

$$\beta = \varepsilon_1 + \varepsilon_2' - \alpha . \tag{3.4.2}$$

The exit angle, ε_2', is calculated from the angle of incidence, ε_1, and the top angle of the prism, α, with the aid of Snell's law. Since the index of refraction depends on the wavelength of the light passing through the prism, the angle of deviation is different for different wavelengths. This is discussed further in chapter 4.

For the modern amateur astronomer, the most useful prisms are the right angle prism and the pentaprism (fig. 3.13). These prisms are commonly used as "star diagonals" to provide the astronomer a comfortable and convenient viewing position when observing objects high in the sky. In the right angle prism, total internal reflection occurs at the surface, so that this surface need not be silvered, though it often is in wide-angle or fast optical systems. A pentaprism requires two silvered surfaces.

Both prisms deviate the optical axis by 90°. A right angle prism, however, inverts but does not reverse the image because there is a single reflection in it. A pentaprism, with two reflections, shows the image in its initial orientation. Note that a tilt error in positioning a right angle prism causes a ray deviation twice the tilt angle error; because there are two internal reflections, a tilted pentaprism in-

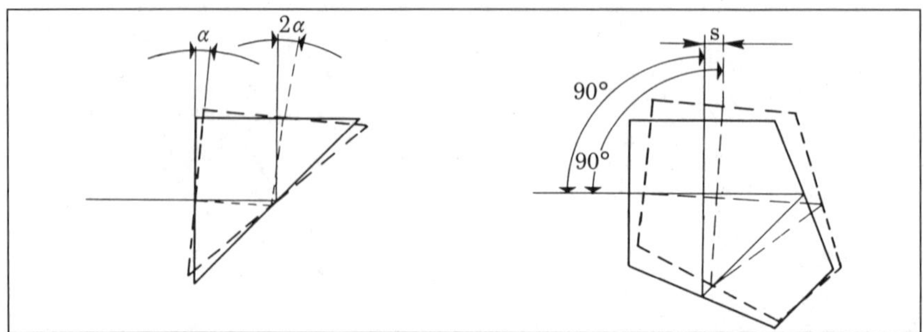

Fig. 3.13 *Reflection by a Right Angle Prism and by a Pentaprism.*

troduces an image shift but does not influence the direction of the exit rays.

Placing a prism in the light path of a telescope is equivalent to introducing a thick, plane-parallel glass plate. The displacement of the image and possible aberrations must be considered when designing a telescope for use with a prism.

Chapter 4

Image Aberrations and Their Presentation

"Experienced amateur astronomers tend to see image aberrations rather than stars through colleagues' telescopes."

Dr. W. J. P. van Enckevort

4.1 The Spot Diagram

Apart from errors in manufacture and assembly, the quality of an optical system is determined by the design's residual image aberrations. In the absence of diffraction, a perfect optical system would produce a point image in the focal surface if the source were a point. Ideally this would be the case both on the optical axis and away from it.

In practice, however, this goal is seldom attained over the entire focal surface even when it is achieved on the axis. In the following pages, we will see that various types of image aberrations exist. In the presence of aberration, the image of a point source becomes a blur called the scattering figure. The scattering figure, or "blur circle," almost always arises from a *combination* of aberrations rather than any single aberration. It is therefore often difficult to decide exactly which part of the scattering figure results from which aberration.

In order to judge the image quality of an optical system the designer must compute the magnitude of the aberrations and somehow evaluate the results. Until the mid–1950s, before the age of electronic computers, optical analysis was carried out with graphs or tables which showed or listed each aberration separately. Because the optical quality of a system is determined by the combination of its aberrations, graphs and tables seldom present the information in a form suitable for evaluating the overall optical quality. This difficulty is all the more true for those not involved daily in the design and evaluation of optical systems.

Since digital computers have become widely available, the situation has improved considerably. An ordinary microcomputer programmed in a slow, interpreted high-level language such as BASIC can handle the enormous number of calculations necessary to compute spot diagrams—and spot diagrams display image quality in a way that is intuitively understood by practical workers and optical professionals alike.

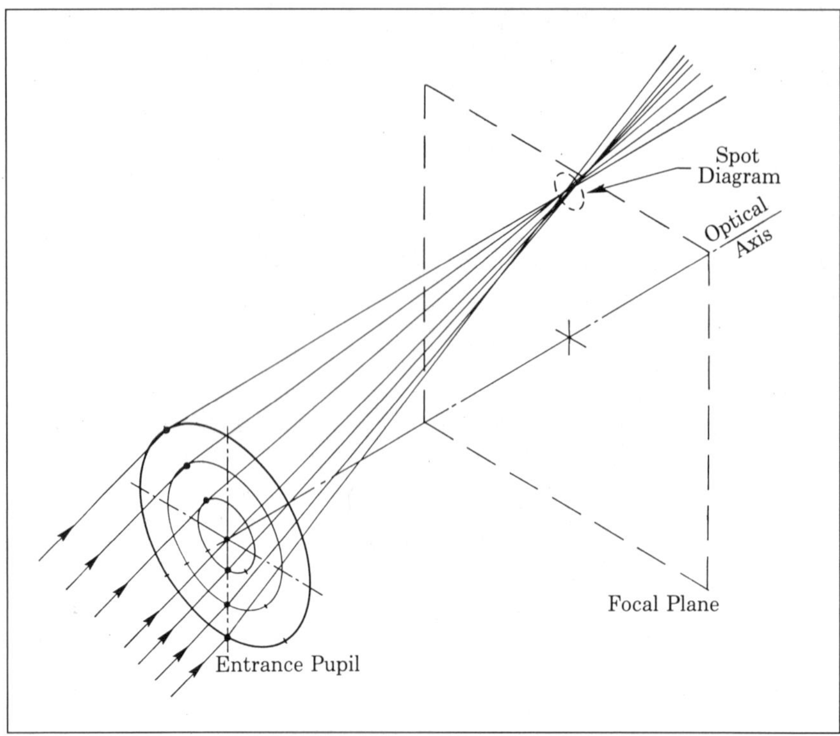

Fig. 4.1 *The Principle of the Spot Diagram.*

Spot diagrams simulate the size, shape, and light distribution in the image surface. The principle of the spot diagram is illustrated in fig. 4.1. Using the equations of geometric optics, we follow a bundle of rays through a system of lenses and mirrors to the focal plane. Since we typically trace several hundred rays distributed in a regular array on the entrance pupil through each of the surfaces of the elements in the optical system, the total number of ray and surface intersections may easily run into the thousands. The distribution, in the focal plane, of rays from a point source is called a spot diagram.

We need not restrict ourselves to calculations for a bundle of rays, or pencil, that is parallel to the optical axis, but can also investigate imagery at various angles off the axis. When the system contains lenses, we may repeat the calculations for various colors to assess the influence of chromatic aberrations.

By shifting the focal plane forward or backward, we can examine the effect of focusing on the size of the spot diagram, and also see how it influences the shape of the image blur. Since the focus may not have an exactly defined position, we define the "best focus" as the focal position yielding the smallest scattering figure, using several trials. Calculating spot diagrams is explained in chapter 20.

In our treatment of the optical systems investigated in chapters 5 through 16, we often use spot diagrams. In those cases where it is beneficial, we have also em-

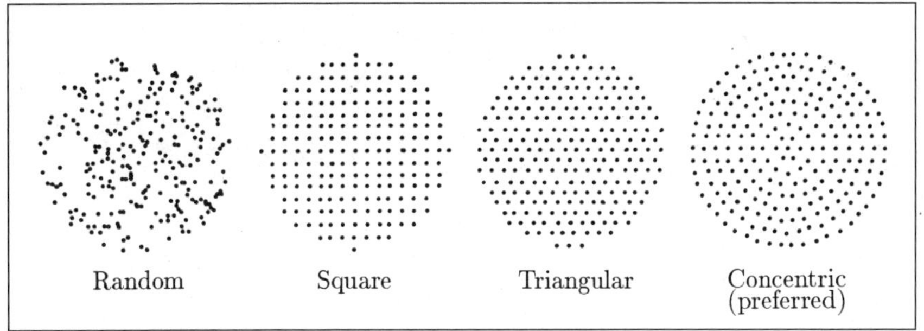

Fig. 4.2 *Common Distributions of Rays at the Entrance Pupil.*

ployed aberration graphs.

In calculating a spot diagram it is important to trace a sufficient number of rays, and to be sure that the rays are uniformly distributed over the entrance pupil. In most cases, 200 rays is sufficient to obtain a good idea of the size and shape of the scattering figure. Occasionally, when there is a wide range of spot density in the spot diagram, it may be necessary to trace considerably more rays to gain insight into the intensity distribution of light in the image.

In principle there are numerous possibilities for distributing the incoming rays in the entrance pupil, but in practice only a few are satisfactory. Those that we have used—the random, square, triangular, and concentric distributions—are shown in fig. 4.2.

In the random distribution, positions of the entering rays are determined by a random number function in the computer program. Unfortunately, a random distribution offers no guarantee that a uniform distribution will be obtained. Moreover the distribution pattern changes every time. Only when a very large number of random rays is used does the distribution become acceptable.

The triangular distribution is equally unsatisfactory because spot diagrams made with it appear jagged, quite unlike the appearance of a blurry image. Square and concentric distributions are almost equivalent in giving smooth images with a relatively small number of rays. We have used a concentric distribution because we feel it produces spot diagrams that are visually more attractive than those of the square distribution.

Fig. 4.3 shows, on a greatly exaggerated scale, spot diagrams for a star image 12.5 mm off-axis in a 200 mm *f*/8 Newtonian telescope. These show not only the size and shape of the light blur caused by image aberrations, but also the light distribution within the blur. The number of spots on a certain unit of area is a measure of the light intensity at that place. A relatively faint comatic tail can clearly be seen in both spot diagrams. As the total number of spots in the diagram increases from 250 to 1000, however, less obvious features of the comatic blur are more readily distinguished.

Interpreting spot diagrams is straightforward compared to graphic presentations. However, there is a pitfall the designer should always keep in mind: the rel-

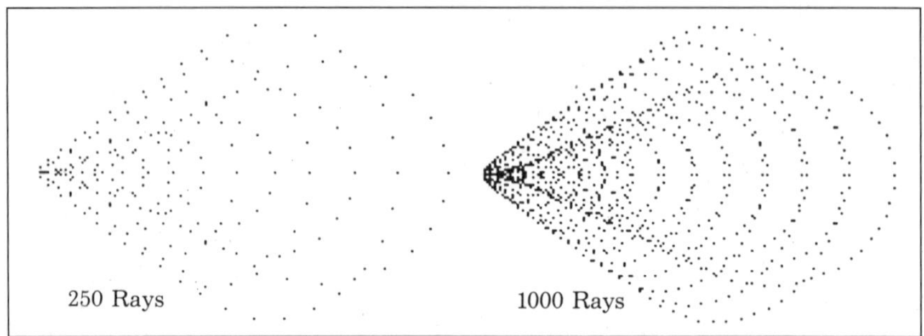

Fig. 4.3 *Spot Diagrams for a 200 mm f/8 Newtonian, 12.5 mm Off-axis.*

ative intensity of parts of an image. Fig. 4.3, and, indeed, all such representations, under-represent the intensity in the brightest parts of the image. Fig. 4.4 shows the intensity distribution within the same comatic blur as a "three-dimensional" histogram of 25,000 rays. Note the intensity of the tail compared to that of the peak. If the detector has a threshold for response, as a photographic emulsion does, then the tail may not record at all. Anyone using the longest dimension of the comatic blur to assess the effective image size would estimate it approximately three times larger than normally records.

Furthermore, since part of the light in an image is lost in a tail or blur below the threshold value, the light which should have been in the image is not, resulting in the loss of faint stars. If the star is bright, the whole blur will be recorded, producing unacceptably "soft" star images. Only spot diagrams showing a high concentration of light in the core image are truly satisfactory for astrophotography.

The designers should also remember that spot diagrams do not include the effects of diffraction. Sophisticated computer programs are necessary to calculate the combined effects of diffraction and image aberrations. Fortunately, the influence of diffraction may be neglected in most cases. In section 4.4, we discuss further the criteria determining image quality for spot diagrams.

4.2 Image Aberrations

We divide image errors, or aberrations, into the same two broad classes as most optical textbooks do:

1. Monochromatic aberrations, which may occur in both refracting and reflecting systems when light of one wavelength is involved; and
2. The chromatic aberrations, which occur because different wavelengths behave differently in the system. Chromatic aberrations appear only when a system contains refractive elements.

The monochromatic aberrations were analyzed in the 1850s by the German mathematician Ludwig von Seidel, and are named the Seidel aberrations after him. Von Seidel distinguished the following: (See also section 20.6.)

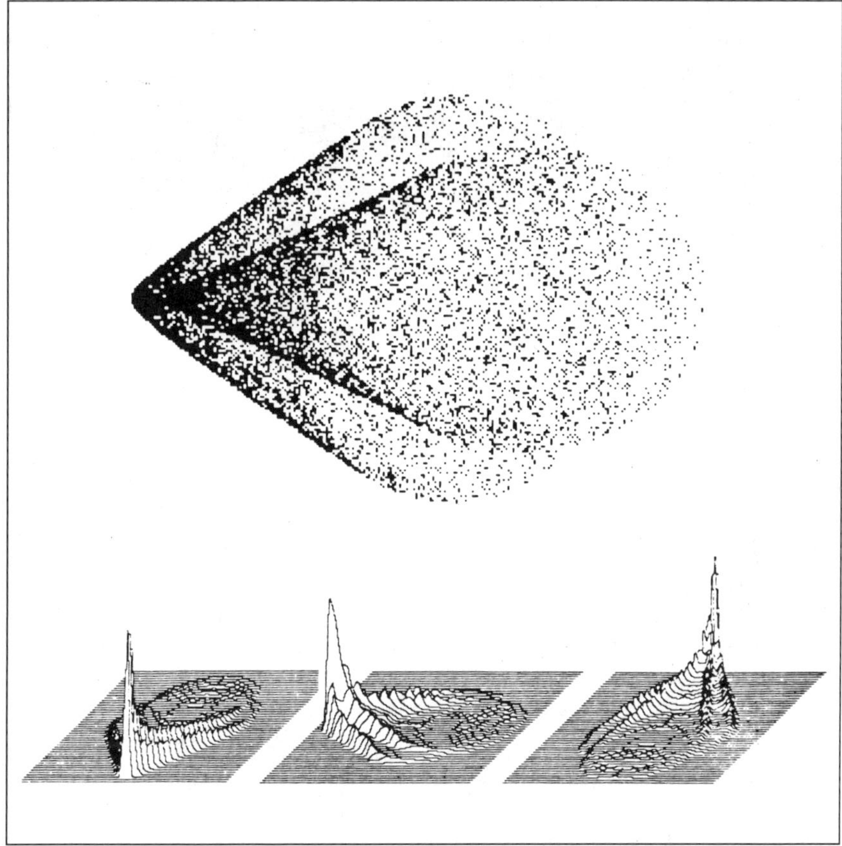

Fig. 4.4 *Comparison of a Spot Diagram and 3-D Distribution of 25,000 Randomly Traced Rays.*

1. spherical aberration,
2. coma,
3. astigmatism,
4. curvature of field, and
5. distortion.

Of the five Seidel aberrations, only spherical aberration is an axial one. Coma, astigmatism, curvature of field, and distortion are all off-axis aberrations.

4.2.1 Spherical Aberration

Spherical aberration occurs when light rays parallel to the optical axis entering a system at different heights come to a focus at different points along the axis. In a single lens or spherical mirror the outer rays intersect closer to the lens or the mirror than the inner rays. Rays entering very close to the optical axis intersect at F,

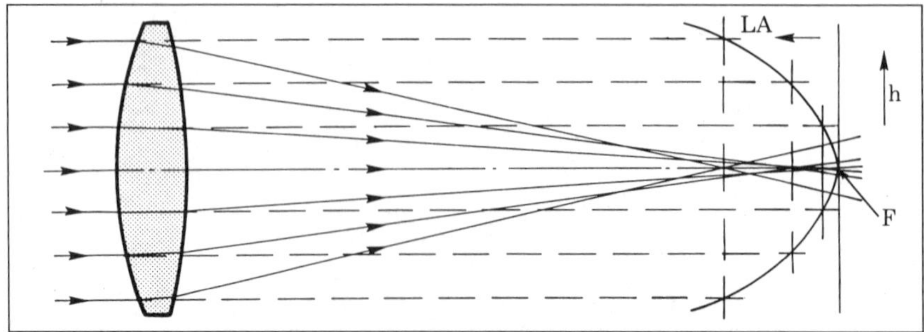

Fig. 4.5 *Longitudinal Spherical Aberration (LA) in a Positive Lens.*

the paraxial focus (fig. 4.5). (See section 20.5) We may graph this aberration as a plot of longitudinal aberration against the height of the zone at the entrance pupil. The longitudinal aberration, or *LA*, is the difference between the intersection of the rays at the optical axis and the paraxial focus.

In fig. 4.6, we present another type of graph of spherical aberration. Transverse aberration, *TA*, is shown for five intercepting planes. The graphs plot the distance from the ray to the axis for various zones of the entrance pupil in a specific plane. Because spherical aberration is symmetrical, graphs of longitudinal aberration are usually drawn for the upper part of the entrance pupil only. From both graph and spot diagrams, we see that the spread of light is smallest at plane *c*.

Spherical aberration can be reduced by diaphragming the aperture. The outer rays will then be intercepted, resulting in an improvement of the sharpness but a loss in light grasp. The plane of minimum spread, plane *c*, will shift away from a simple lens or mirror, toward plane *d*.

4.2.2 Coma

Coma occurs in an oblique bundle of light when the intersection of the rays is not symmetrical, but is shifted with respect to the axis of the bundle. Off-axis light rays passing through the lens near its edge (i.e., marginal rays) intersect the image surface at different heights than those that pass through the center of the aperture. This results in an image with a comet-like shape, a bright core of light accompanied by a spreading tail (see fig. 4.7). Graphs of *TA* against ray height for pure coma are shown for five focal planes. Transverse curves give the intercept distances from the oblique axis for every zone of the entrance pupil.

Coma is a very troublesome aberration. Not only is the off-axis image unsharp, but it is also asymmetric. As a result, it is impossible to measure accurate star positions from photographic plates taken with systems having coma.

In order to be free of coma, a system which is free of spherical aberration already must comply with the Abbe sine condition. Discovered by Ernst Abbe, the sine condition requires that in a system free of coma, every exit ray of an oncoming beam of rays parallel to the axis will satisfy the condition:

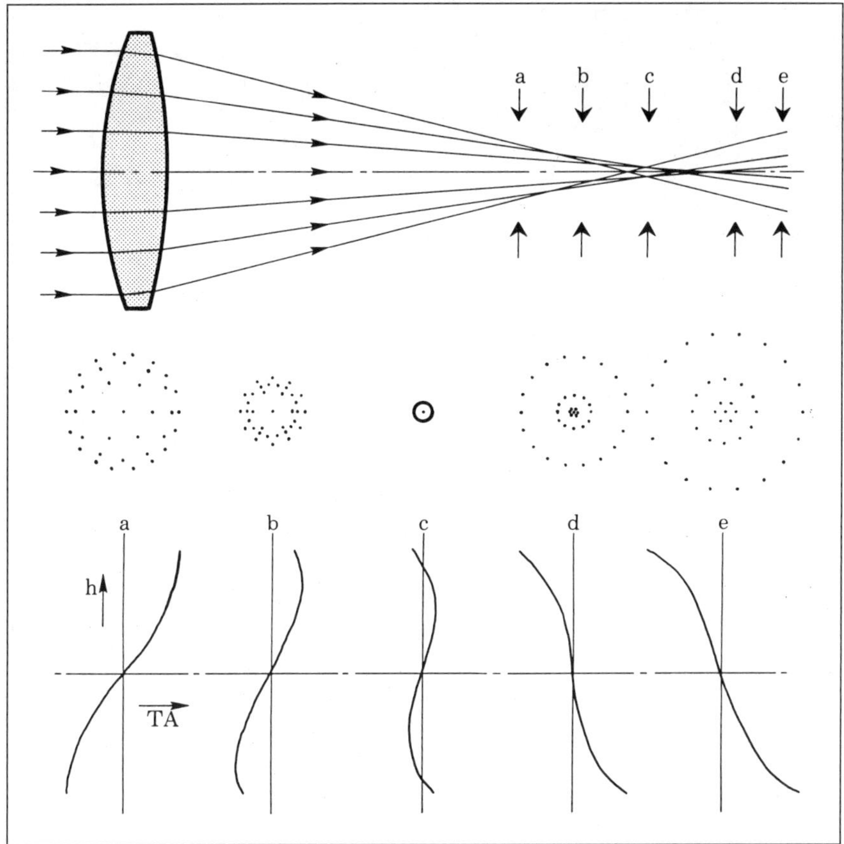

Fig. 4.6 *Transverse Spherical Aberration in a Positive Lens.*

$$\frac{h}{\sin U'} = C\,, \tag{4.2.1}$$

where h is the ray height of the ray before it enters the system, U' is the angle between the ray and the optical axis as it travels toward focus (fig. 4.8), and C is a constant which we will define below. This implies that somewhere between the entrance pupil and the focal plane there exists an imaginary spherical surface connecting the parallel entering rays and the exit rays with its center at F, the focal point. When such a surface exists and the system has no spherical aberration, then all distances between the imaginary spherical surface and the focal point F are the same, and there is no asymmetry present in the system. The constant distance is, in fact, C, the constant we saw above, and it is the effective focal length of each zone of the entrance pupil. Abbe called a system free of spherical aberration and coma "aplanatic," and so will we.

When a system has coma, then the "Offense against the Sine Condition," or OSC, is expressed as the fractional difference between the axial value of C and the

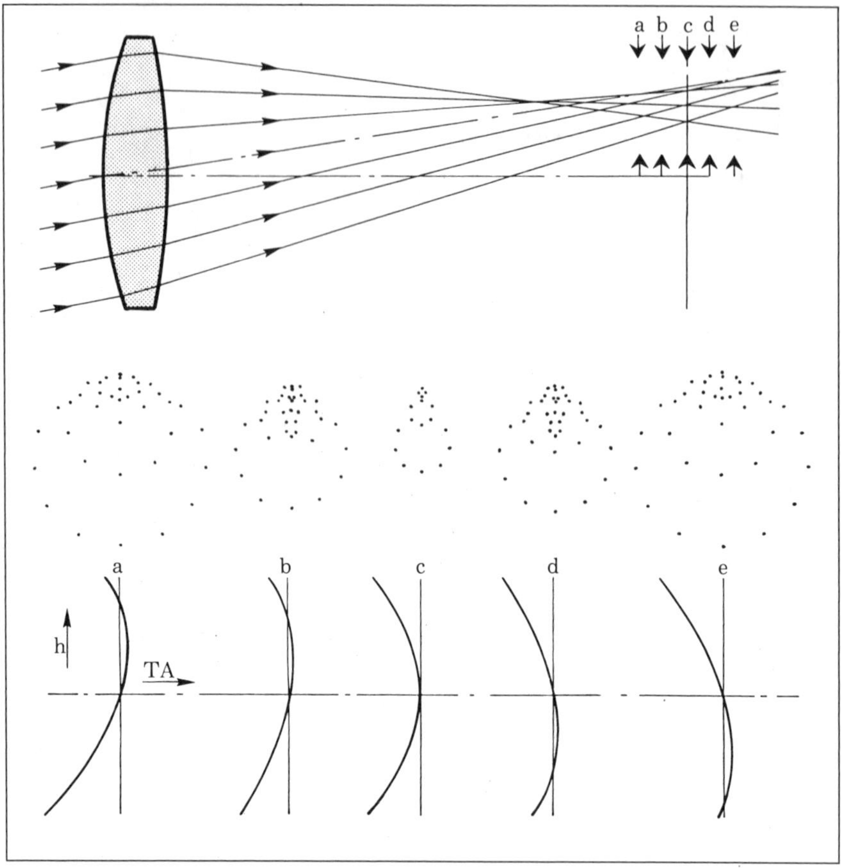

Fig. 4.7 *Coma in a Positive Lens.*

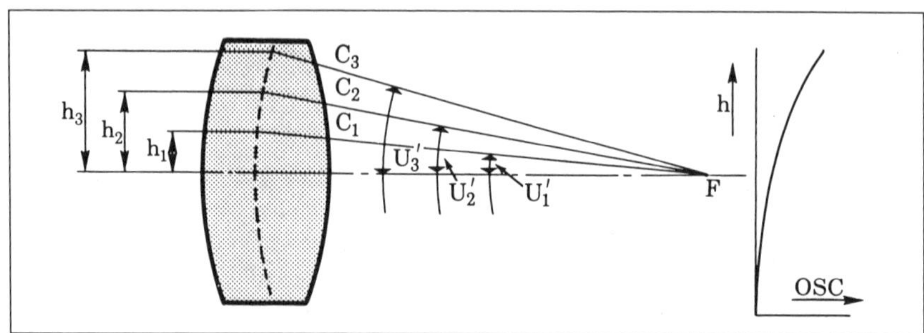

Fig. 4.8 *The Abbe Sine Condition and the OSC.*

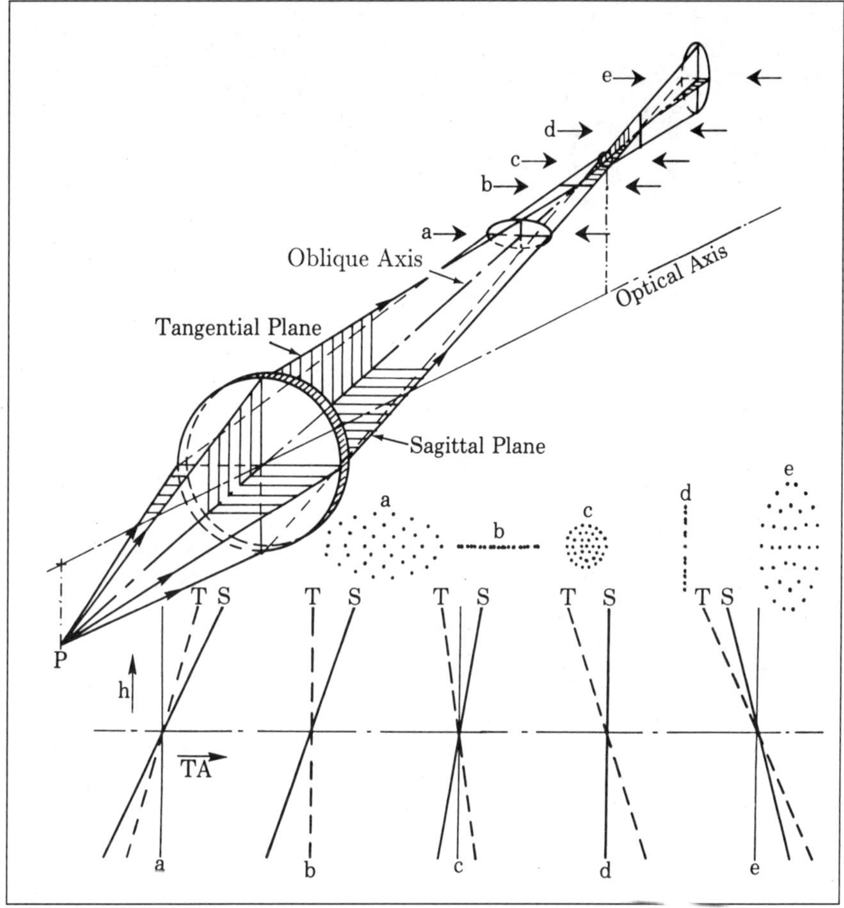

Fig. 4.9 *The Origin of Astigmatism.*

value of C at height h:

$$OSC = \frac{C(h)}{C(o)} - 1 . \quad (4.2.2)$$

These values may be plotted against the ray height as shown in fig. 4.8. Departure from a perpendicular line indicates the presence of coma.

For proper correction of coma in the presence of spherical aberration, the position of the limiting stop or entrance pupil is critical. Sometimes a non-aplanatic system can be made aplanatic only by repositioning the stop.

4.2.3 Astigmatism

Astigmatism is probably the most difficult aberration to understand. It exists whenever there is a difference between the optical power of the system in the tan-

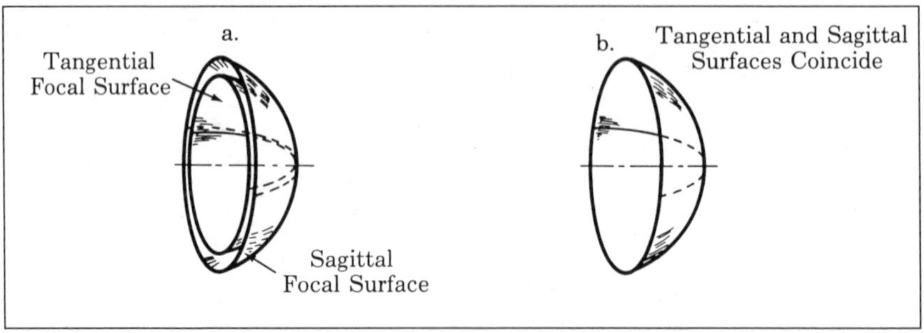

Fig. 4.10 *Tangential and Sagittal Focal Surfaces.*

gential plane and in the sagittal plane. This means that there are two focal surfaces, one formed by the focus of rays from the tangential plane and the other formed from rays in the sagittal plane of the lens.

In fig. 4.9 we see an oblique bundle of rays entering an optical system. Two planes are drawn:

1. A plane through the object point and the lens axis called the tangential or meridional plane, normally shown as the plane of the drawing; and
2. A plane at right angles to the tangential plane called the sagittal plane.

The tangential rays come to focus in *b*, while the sagittal rays focus in *d*.

Midway between the tangential and the sagittal focus lines, in plane *c*, the blur of light reaches its smallest dimensions. Here it is approximately circular, as shown in the spot diagrams, but this is true only for pure astigmatism. The diameter of the smallest blur circle depends on the focal ratio of the exit bundle: the faster the focal ratio, the larger the blur.

Fig. 4.9 also shows transverse aberration curves for pure astigmatism. The graph plots the interception distance for the tangential and sagittal planes against the ray height with respect to the oblique axis for every zone of the entrance pupil.

The tangential and sagittal focal surfaces touch each other on the optical axis. Fig. 4.10a shows the focal surfaces of a single positive lens. In the middle, between these two focal surfaces lies another focal surface where the unsharpness caused by astigmatism is at its smallest. When astigmatism is fully corrected, so the tangential and sagittal focal surfaces coincide as shown in fig. 4.10b, the system is called anastigmatic.

Depending on how the system has been corrected, the tangential and sagittal focal surfaces may be concave or convex. All four possible combinations of the tangential and sagittal focal surfaces have been drawn in fig. 4.11. Vertical lines are planes perpendicular to the optical axis. In cases 1 and 2, when the position of the tangential focal surface lies to the left of the sagittal focal surface, the system is said to have positive, or undercorrected, astigmatism. In cases 3 and 4, the astigmatism is negative, or overcorrected. Astigmatism is undercorrected for a single positive lens.

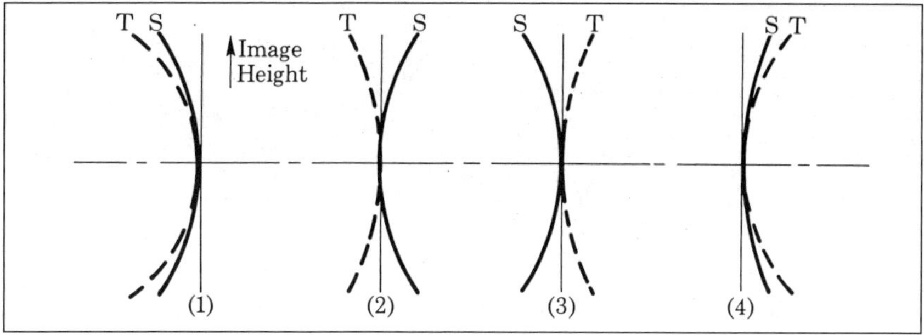

Fig. 4.11 *Possible Tangential and Sagittal Focal Surfaces.*

In the presence of astigmatism it is impossible to obtain a sharp off-axis image. If a star is imaged by an astigmatic system, it will appear as a horizontal line in the tangential focal surface at *b* and as a vertical line in the sagittal focal plane at *d* (fig. 4.9). At *c*, the star will be imaged unsharply as a round or somewhat squarish blur.

4.2.4 Curvature of Field

Pure curvature of field means that the sharpest image is formed on a curved focal surface rather than a flat focal plane. If a flat photographic plate is used, the off-axis image will be unsharp because it is out of focus. The transverse aberration curves in fig. 4.12 show the interception distances for every zone of the entrance pupil with respect to the oblique axis for five intercepting planes.

Many of the systems discussed in the following chapters suffer from curvature of field. Generally speaking, unless deliberate measures are taken to prevent it, the focal surface will be curved. As mentioned in section 4.1, image aberrations, to the extent they have not been corrected, always occur simultaneously. For this reason spot diagrams and aberration curves generally differ from the idealized patterns shown for a single aberration.

4.2.5 Distortion

Distortion is not an image aberration in the normal sense because it influences image scale rather than image sharpness. When the image scale is too large, it is referred to as pincushion, or positive distortion. If it is too small, we have negative, or barrel distortion. These are shown in fig. 4.13. The sides of a square are curved because the image scale varies with the distance to the axis. The graphs in fig. 4.14 show the image scale growing with axial distance of the image (positive distortion) and shrinking (negative distortion).

4.3 Chromatic Aberrations

Chromatic aberrations occur because the refractive index of glass is different at

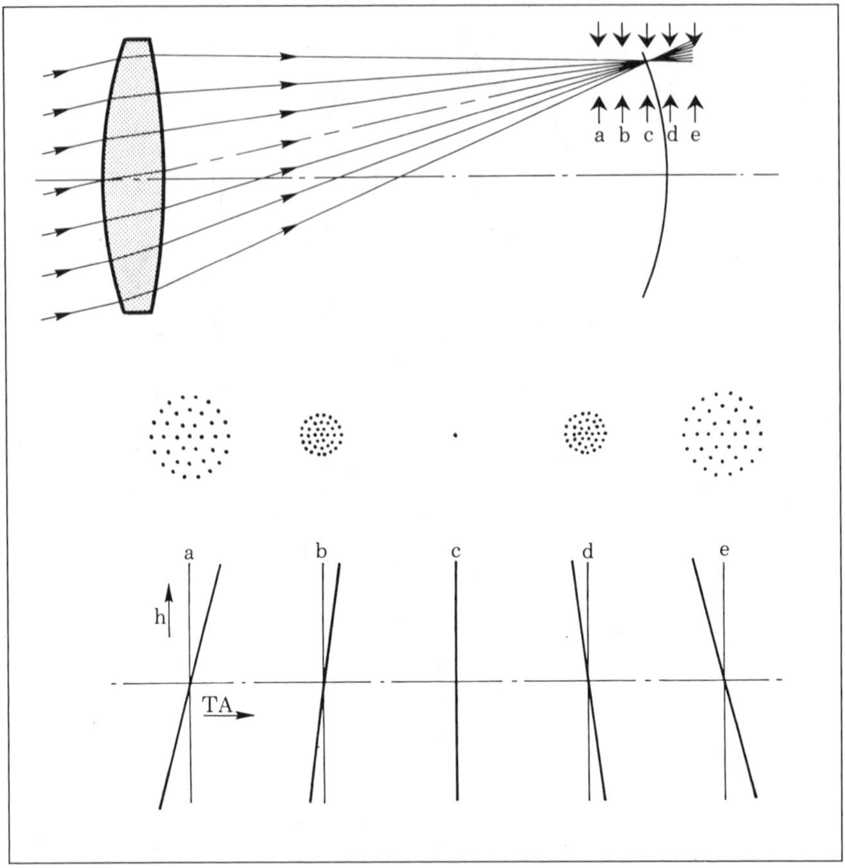

Fig. 4.12 *Curvature of Field.*

different wavelengths. Since wavelength determines the color seen by the eye, these aberrations render star images as colored blurs. The human eye responds to wavelengths as in tables 4.1a and b.

Table 4.1a Color Sensation		**Table 4.1b Color Sensitivity**		
			Relative Response	
Wavelength[a]	Color	Wavelength	Photopic[b]	Scotopic[c]
360 to 440 nm	violet	436 nm	<.02	.28
440 to 495 nm	blue	486 nm	.20	.86
495 to 580 nm	green	513 nm	.60	1.00
580 to 600 nm	yellow	555 nm	1.00	.40
600 to 630 nm	orange	587 nm	.80	.06
630 to 780 nm	red	656 nm	.08	<.01

a. 1nm = 10^{-6}mm = 10^{-9}meter
b. Photopic: bright enough to see color
c. Scotopic: too dim for color

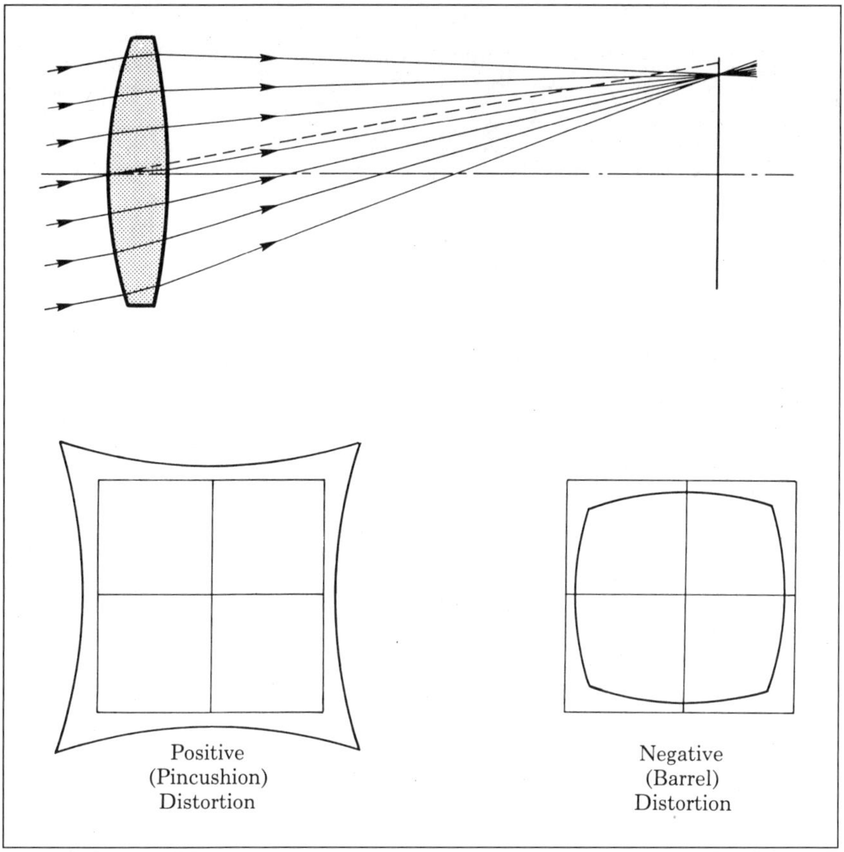

Positive (Pincushion) Distortion

Negative (Barrel) Distortion

Fig. 4.13 *Positive and Negative Distortions.*

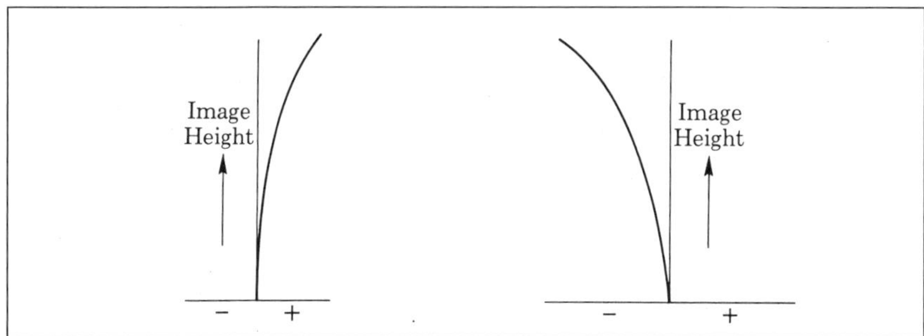

Fig 4.14 *Presenting Positive and Negative Distortion.*

Table 4.2
Standard Wavelengths and Their Sources

Line	Wavelength (nm)	Element	Color
h	404.66	Hg	violet
g	435.83	Hg	violet
F′	479.99	Cd	blue
F	486.13	H	blue
e	546.07	Hg	green
d	587.56	He	yellow
C′	643.85	Cd	red
C	656.27	H	red
r	706.52	He	deep red

In their catalogs, optical glass manufacturers list the refractive indices for many kinds of optical glass for certain standard wavelengths. Some of these wavelengths are listed along with the chemical elements that produce them in table 4.2.

White light is dispersed in a prism as a result of different refractive indices for various colors, shown in fig. 4.15. Refractive index increases as wavelength decreases; red is refracted less than violet. The rate of change of refractive index with wavelength is called the dispersion.

Optical glass manufacturers supply a large variety of glasses with a wide range of refractive indices and dispersions. The refractive index for each type of glass is given for the d–line (yellow, 587.56 nm) or the e–line (green, 546.07 nm). Dispersion numbers V_d and V_e are based on the difference between the refractive indices for the F and C lines and the F′ and C′ lines respectively. The dispersion measures—called the Abbe numbers—are calculated as follows:

$$V_d = \frac{n_d - 1}{n_F - n_C} \tag{4.3.1}$$

and:

$$V_e = \frac{n_e - 1}{n_{F'} - n_{C'}} . \tag{4.3.2}$$

We will apply these formulae in chapter 21. Note that if the refractive indices at F and C are only slightly different, indicating a small rate of change of refractive index with wavelength, the Abbe number is large.

The d–line refractive index for modern optical glasses varies from 1.44 to 1.96; the Abbe number varies from 20 to 90. In fig. 4.16, we see a plot of glass types by their refractive index and dispersion. Glasses with an Abbe number higher than 55 are generally called crowns; those lower than 50 are called flint glasses.

In Appendix A, we list important modern optical glasses. Optical glass now bears an international standardized six digit code, for example: 517642. This indi-

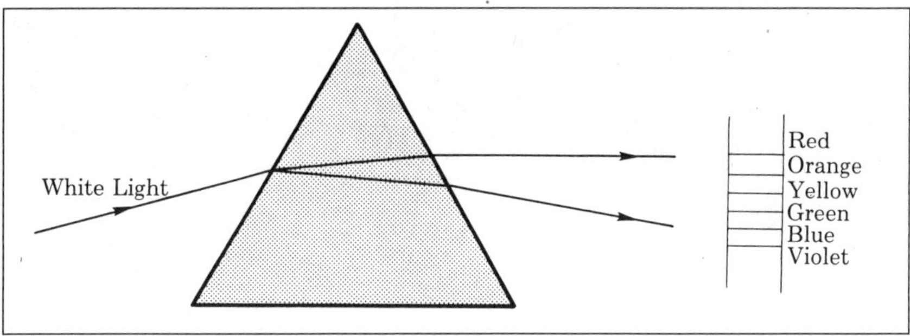

Fig. 4.15 *Refraction and Dispersion of Light by a Prism. (The dispersion is exaggerated for clarity.)*

cates the first three digits after the decimal point in the refractive index, n_d, and the first three digits of the Abbe number, V_d. Glass 517642 has, for example, a refractive index of 1.517 and an Abbe number of 64.2. The catalog shows that this glass is BK7, a borosilicate crown glass made by Schott Optical Glass, Inc.

4.3.1 Longitudinal Chromatic Aberration

This aberration is also called axial chromatic aberration. We have seen that the focus of a lens on the optical axis is different for various colors because the blue light is more strongly refracted than the red. Fig. 4.17 illustrates this. Longitudinal chromatic aberration expresses itself as color fringes on the optical axis. In chapter 21, we will show how this aberration may be corrected for a refracting objective. This aberration is usually shown by a graph of the focal position at various wavelengths relative to the focal position at a standard wavelength versus the wavelength.

4.3.2 Lateral Color

Lateral color occurs off the optical axis, and is shown in fig. 4.18; the foci for various colors outside the optical axis simply do not coincide. This is because the image scale varies with the wavelength. You may be most familiar with this aberration from inexpensive binoculars; it produces color fringes outside the optical axis. This aberration can be expressed graphically as the difference between the red and the blue image heights versus image height, or off-axis distance, in green light. Lateral color appears as off-axis color fringes.

4.4 Presentation of Image Aberrations with Spot Diagrams

In chapters 5 through 16, we use spot diagrams to show the performance of optical systems. How are these to be interpreted? Quite obviously, spots fall within some diameter. To a first approximation, the smaller the spread of the spots, the better the image quality will be. However, it is important to define criteria for acceptable image quality, and this, in turn, implies we must consider the use of the telescope. We define two criteria, one for visual observation (which we take to include ob-

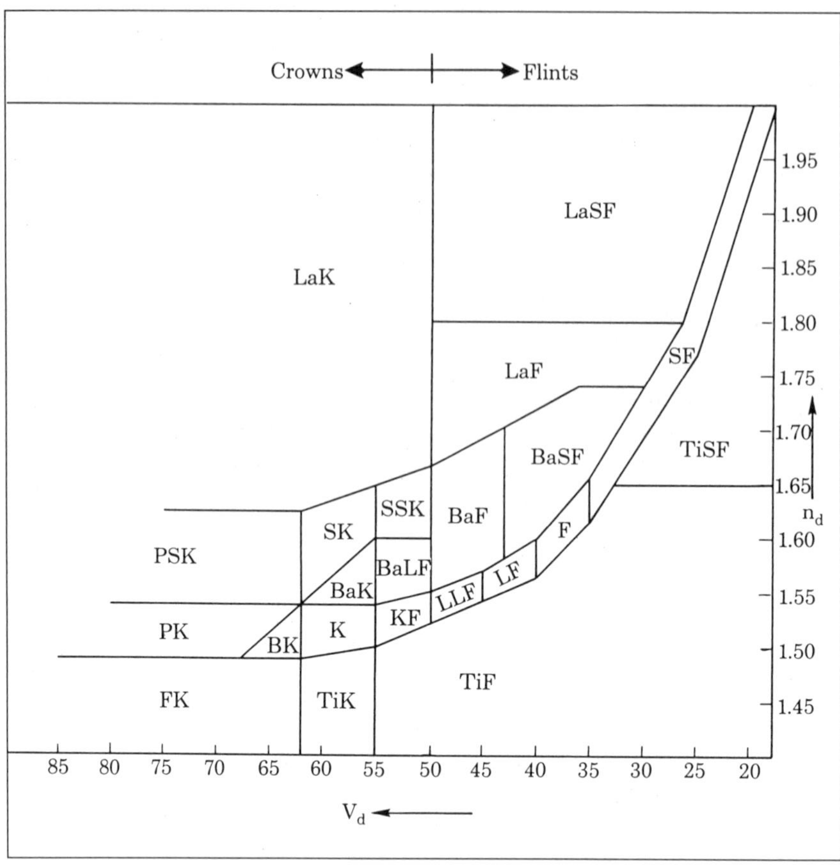

Fig. 4.16 *Chart of Schott Optical Glasses.*

servations at high magnification, lunar and planetary photography), and the other for use as an objective in astrophotography.

For low-power visual observation, the maximum effective semi-diameter of the field is 10 mm for 1¼–inch eyepieces and 20 mm for 2-inch eyepieces. Frequently, however, the eyepiece aberrations dominate the aberrations of the objective. This is discussed in chapter 16.

For visual use, especially when high powers may be used, the spread of the spots in the focal plane of the objective must be very small, but because the field is small at high magnification, only the image sharpness near the optical axis is important. Between 90 and 95% of the geometrically traced rays must be concentrated within a circle no larger than the Airy disk in e–light (see section 18.2). The linear diameter of the Airy disk depends entirely on the focal ratio of the objective and the wavelength of the light; it is independent of the aperture (although a central obstruction slightly affects the Airy disk size). This important point we discuss further in chapter 18.

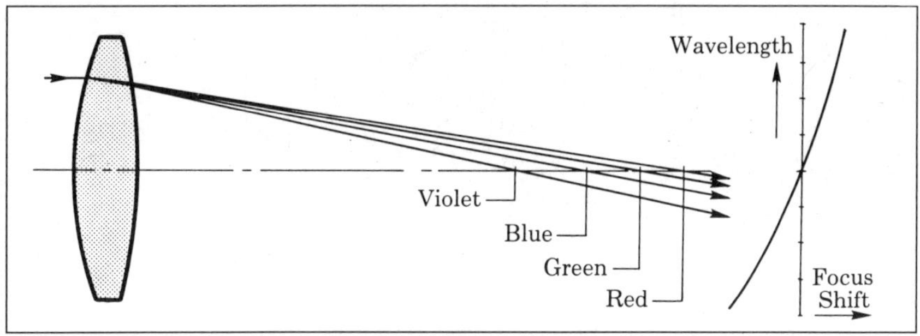

Fig. 4.17 *Longitudinal Chromatic Aberration.*

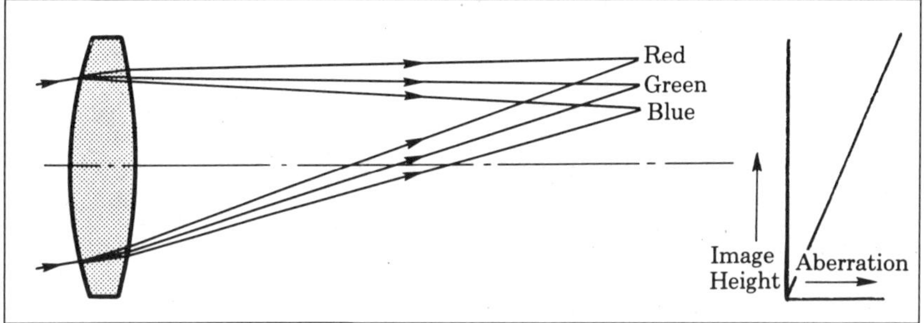

Fig. 4.18 *Lateral Color Error.*

Table 4.3
Airy Disk for Various Focal Ratios

Focal Ratio	Airy Disk (mm)
f/20	0.026
f/15	0.020
f/10	0.013
f/8	0.011
f/6	0.008
f/5	0.007

When a telescope is used as an astrocamera for stellar and nebular photography, the requirements change. For photography, the image quality necessary is entirely dependent on the resolving power of the photographic emulsion. We have taken this as 0.025 mm, or 0.001 inches. This criterion is based on extensive investigations (ref. 4.4), which showed that the smallest diameters of star images on professional astrophotographic plates were never smaller than 0.025 mm. Although certain modern emulsions do give smaller images, a concentration of 90 to 95% of the spots within this 0.025 mm circle is sufficient in almost all cases. (The designer may, of course, choose a more demanding criterion if he wishes.) The remaining 5 to 10% of the light must be well spread out or the star image will be

Table 4.4
Visual Analysis

Color	Wavelength	Line
red	656.27 nm	C
green	546.07 nm	e
blue	486.13 nm	F

Table 4.5
Photographic Analysis

Color	Wavelength	Line
red	656.27 nm	C
blue	486.13 nm	F
violet	404.66 nm	h

asymmetric. For photographing large areas of the sky, these requirements are a goal to be met over the entire field of interest.

When an optical system contains lenses, the calculation of the spot diagrams must be carried out for various colors. Of the systems used visually, we compute a spot diagram for the following colors (see table 4.4).

The peak sensitivity of the eye falls in the green; the red and blue wavelengths are limits where the sensitivity of the eye has fallen to about 10% of its peak sensitivity. For the calculation of the spot diagrams, we pick the focal plane so the size in the spot diagram is at its smallest in green light. For a system to give good performance, however, the red and blue images must be reasonably small.

When the optical system is to be used as an astrocamera, the range of important wavelengths is larger than it is for the eye (see table 4.5).

Red light is particularly important for photographing emission nebulae since they give off much of their energy in the H–alpha line. However, galaxies and young stars emit most of their light in the blue and violet regions. The position of the focal plane is usually chosen so that blue is in focus.

Because evaluating a system for astrophotography is far from simple, we will give an example. Consider a Maksutov-Cassegrain telescope, the Simak, using optical data published in ref. 4.5. This system is intended primarily for photographic use. It has an aperture of 200 mm and a focal ratio of 5.6. The image is behind the mirror. We consider several common film formats that might be used with this instrument:

1. 24 by 36 mm semi-diagonal 21.6 mm
2. 45 by 60 mm (in practice 45 by 57 mm with a semi-diagonal of 36.3 mm)

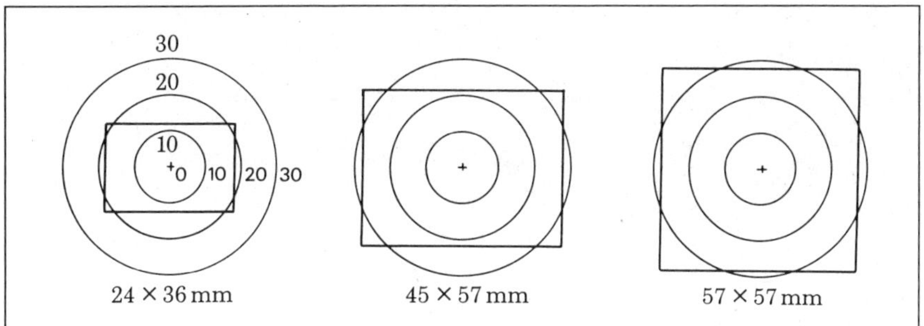

Fig. 4.19 *Various Film Formats and Off-Axis Distances.*

3. 60 by 60 mm (in practice 57 by 57 mm with a semi-diagonal of 40.3 mm):

We first investigate the image sharpness for a flat focal plane both on the optical axis and for distances of 10, 20, and 30 mm off-axis. In fig. 4.19 we show these off-axis distances superimposed on the film formats. The 20 mm off-axis distance corresponds rather closely to the extreme corners of a 24 by 36 mm format.

Fig. 4.20 shows the spot diagrams for a flat focal plane compared to the 0.025 mm image size criterion; spot patterns larger than 0.15 mm are not shown because of their size. From this illustration, we see that at off-axis distances greater than 10 mm the image sharpness on the flat focal surface does not meet the 0.025 mm criterion.

Before concluding that the Simak is not good enough, however, we must take a closer look. The enlargement of the spot diagrams is a result of field curvature; when we calculate spot diagrams for an optimally curved field (fig. 4.21), the size of the spot patterns easily meets the 0.025 mm criterion over the whole field.

Note, though, that the spot diagrams for red, green, blue and violet are shifted with respect to each other. These shifts were drawn on the same scale as the spot diagram itself; they show that the Simak system has lateral color toward the edge of the field. This aberration means off-axis star images will appear as color blurs. Because image sharpness is determined by the *combined* size of the blur for all four colors, the image of a star 30 mm off-axis would be insufficiently sharp to meet the criterion of 0.025 mm.

We evaluated the image quality for a curved focal surface, and films are usually flat. To transfer the full sharpness of this system to a photographic emulsion, the designer must resort to one of two methods:

1. Some provision for curving the film, thus forcing it to conform to a radius of curvature of R = –510 mm; or

2. Using a field-flattening lens in front of the focal surface.

These methods are discussed in chapter 14.

We can also look into using the system with flat film by refocusing it some-

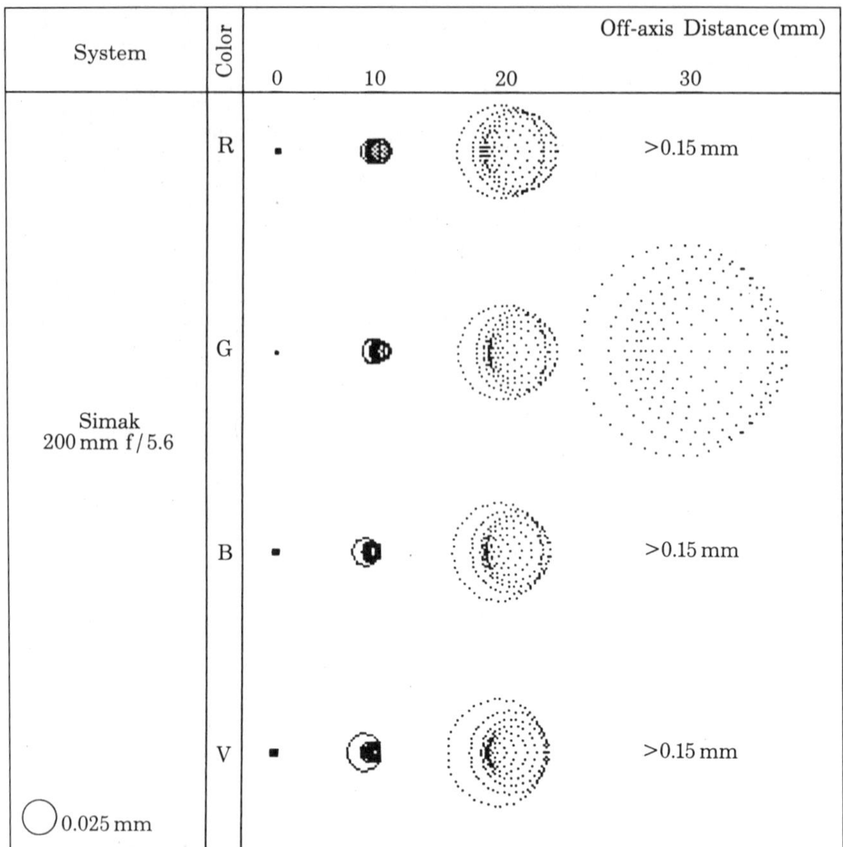

Fig. 4.20 *Spot Diagrams for the Flat Focal Surface of a Simak Focused for Blue.*

System	Color	Off-axis Distance (mm) 0	10	20	30
Simak 200 mm f/5.6	R				
	G				
R_{field} = −510 mm	B				
0.025 mm	V				

Fig. 4.21 *Spot Diagrams for the Curved Focal Surface of the Simak Focused for Green.*

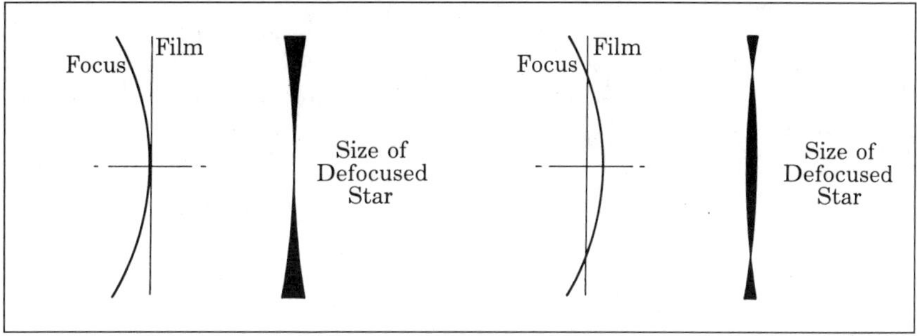

Fig. 4.22 *Enlarging the Useful Field of an Optical System with a Curved Focal Surface.*

what (see fig. 4.22). If we focus for the axial image, the extrafocal disk expands off-axis. However, if the flat film is shifted so that the diameter of the spot diagram on the optical axis becomes as large as we care to tolerate, 0.025 mm, then at some point toward the edge of the frame, the film will intersect the surface of best focus. This technique can enlarge a mildly curved field considerably, though the limiting magnitude will vary nonlinearly with field angle.

4.5 Scaling Optical Systems

The objective systems discussed in chapters 5 through 13 all have, with a single exception, an aperture of 200 mm. Suppose the designer wants a system with an aperture of 100 mm or 400 mm: how will it perform? Fortunately, this can be derived rather easily from the spot diagrams calculated for the 200 mm instrument. This basic rule applies:

> For a system of the same focal ratio, all dimensions of the instrument must be increased or decreased proportionally by the same factor. This is true not only for the diameters, thicknesses, radii of curvature, and distances between the optical components, but also for the off-axis distances and diameters of the spot diagrams. However, the criteria for visual use (i.e., the diameter of the Airy disk) and for photographic use (0.025 mm) remain the same.

The linear size of the Airy disk remains the same because this diameter depends solely upon the focal ratio and wavelength, and not on the aperture. The photographic criterion, of course, depends only on the photographic emulsion and does not depend on the aperture of the instrument.

Suppose that a designer wishes to double the aperture of a system. All diameters, thicknesses, radii, and spacings must be doubled. So, too, must the computed off-axis distances and diameters of the spot diagrams. If the spot diagram is 0.020 mm diameter 20 mm off-axis in the 200 mm system, satisfying the 0.025 mm criterion for photography, in a 400 mm aperture version of the same system, it will be 0.040 mm diameter 40 mm off-axis, and will not satisfy the criterion,

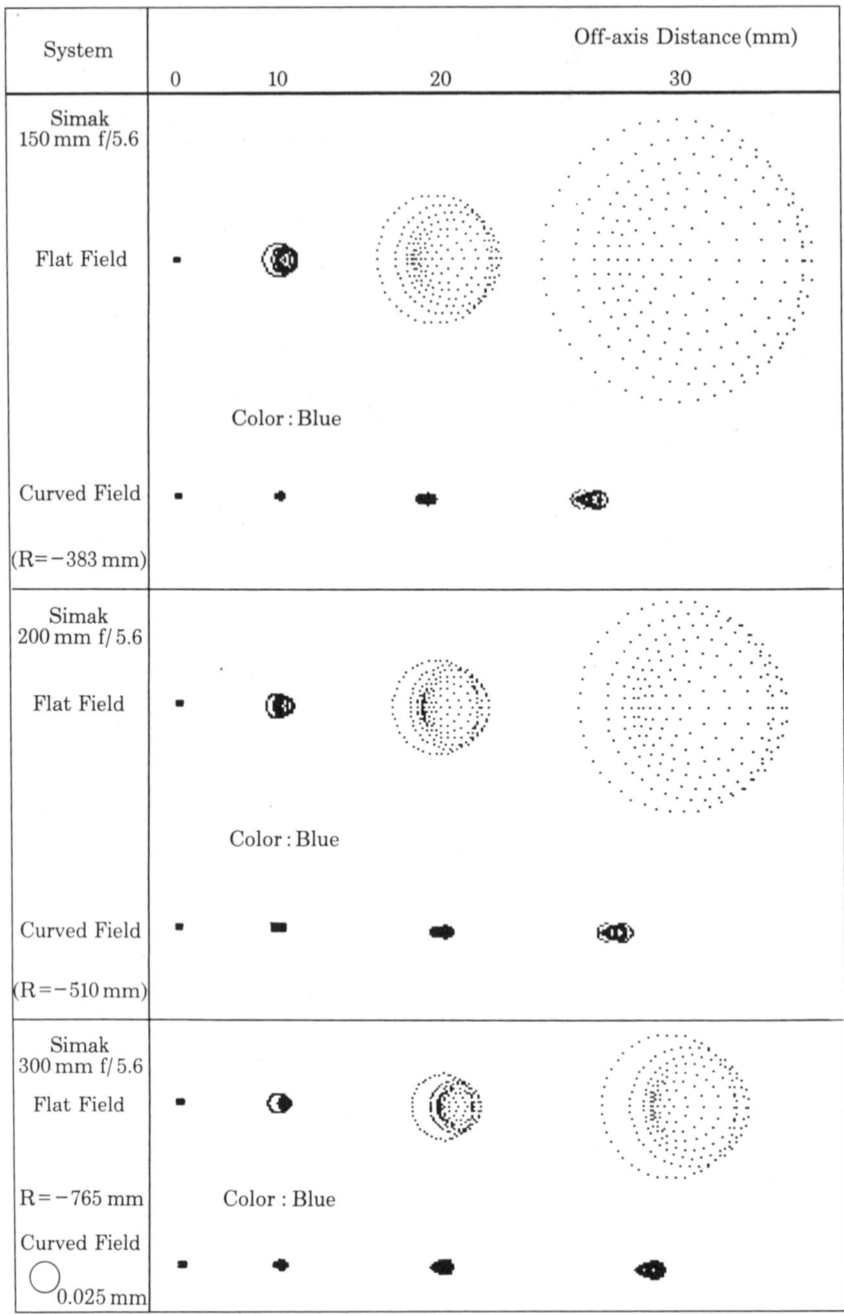

Fig. 4.23 *Spot Diagrams for the Simak with Three Apertures.*

which remains 0.025 mm. To find the diameter of the spot diagram 20 mm off-axis, the designer can compute the size of the spot diagram 10 mm off-axis in the 200 mm system, then double it for the 400 mm system.

We show spot diagrams in fig. 4.23 for the Simak instrument scaled to 150 mm, 200 mm, and 300 mm aperture, for both the flat and curved focal surfaces, in blue light, at off-axis distances of 10, 20, and 30 mm. With this design, we see that the linear size of the useful photographic field increases somewhat in the larger instruments, but that the increase is by no means proportional to the increase in aperture. Such a result is frequently the case.

Note that it is not always possible to scale an optical system to a larger size. Consider a typical 50 mm focus lens for a 35 mm camera: its aberrations are low and tolerable. However, if such a lens is scaled up to an aperture of 200 mm, the residual aberrations will be enlarged by the same scaling factor applied to the lens, and its performance would, in all likelihood, be intolerable. Moreover, such a system would be extremely heavy because of its thick lenses.

Downscaling is usually permissible, or at least not prohibited, by image aberrations. However, downscaled systems may not be practical if the axial thickness of lens elements or spacing between lenses becomes too small. Another problem often encountered in downscaling is that too little space for a focusing device may remain.

4.6 Concluding Remarks

In perusing chapters 5 through 13, the reader will find many comparisons of objective systems with the same aperture, but often with different focal ratios. The question of how to compare the off-axis performance of these systems soon arises. Should the comparison be based on linear off-axis distance (expressed in mm), or on an equal angular off-axis distance (expressed in degrees)?

While both methods of comparison have merit, in this book we have deliberately chosen comparison based on linear off-axis distance. The reason is that most amateurs choose a certain film format irrespective of the objective's focal ratio or angular coverage. They want the system to produce a sharp image over that whole film frame. This does not mean that angular coverage may be ignored: the designer must give full and careful consideration to the desired angular coverage. For those who wish to, it is easy to convert the linear scale into an angular scale for a given focal length. In order to facilitate comparison based on a fixed angular field, every listing of optical data gives the linear size of a one-degree field.

Another important consideration for the designer is curvature of field. As the reader will see in the following chapters, many objective systems suffer from it—Newtonians, refractors, Cassegrains, and Schmidt- and Maksutov-Cassegrains all do. Only systems for which deliberate and special design steps have been taken have truly flat fields.

In order to demonstrate the effects of curvature of field for the Simak, we have shown spot diagrams for both the flat and the curved field. The reader must understand, however, that we could not show all possible spot diagrams for all these systems. We wish to emphasize that spot diagrams for the optimally curved focal surface give a much better idea about the optical performance of an objective system than the spot diagrams on the flat field can. In most of the systems treated in the following chapters, we have given spot diagrams for the curved focal surface, and when we have done so, we have provided the radius of curvature of the focal surface to the left of the spot diagrams.

Keep in mind the following:

1. For visual use of the telescope with an eyepiece, field curvature is, in most cases, easily compensated by the accommodation power of the eye. For visual use, it is usually not a problem.

2. For photographic use, the focal surface must be flat or flattened by means of a field flattener, or the film must be bent to conform to the focal surface.

Chapter 5

The Newtonian Telescope

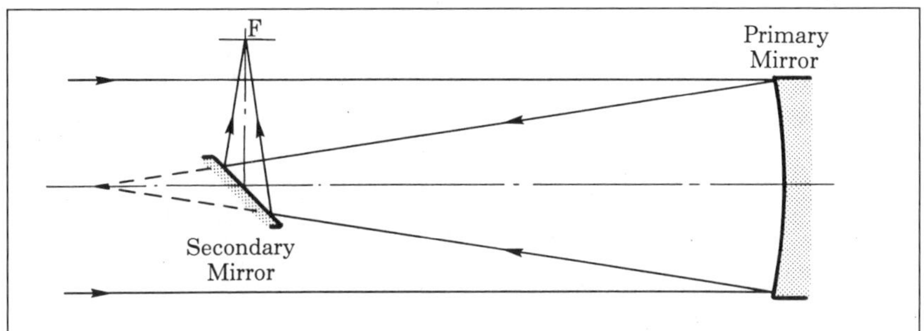

Fig. 5.1 *Optical Layout of the Newtonian Telescope.*

5.1 Introduction

The Newtonian is the simplest and probably the most popular type of telescope for amateur astronomers. It is, without a doubt, the most commonly home-built amateur instrument. As such, the Newtonian is the *de facto* standard against which the performance of other types of telescope is measured.

The Newtonian optical system consists of a paraboloidal (often mistakenly called "parabolic") primary mirror and a flat secondary mirror which directs the converging cone of light out of the tube for visual or photographic applications. The usual arrangement is shown in fig. 5.1.

What are the characteristics of this instrument? Focal ratios generally lie between *f*/4 and *f*/12. *F*/4 is a practical lower limit because of the rapid increase of coma away from the optical axis; *f*/12 is the upper limit because the length of the instrument becomes excessive. Those with slow primaries (between *f*/7 and *f*/12) are generally considered best for lunar and planetary observing, while those with fast primaries (*f*/6 to *f*/4) are better suited for observing deep-sky objects and for prime-focus astrophotography.

The primary is centrally obstructed by the diagonal mirror. Although this element must be large enough to illuminate the required region of the focal surface, it should be no larger than necessary because any obstruction degrades the system's optical performance. In Newtonians made for visual use, a fully illuminated region 10 mm in diameter is more than adequate, while a Newtonian telescope used as an astrocamera usually requires a significantly larger diagonal.

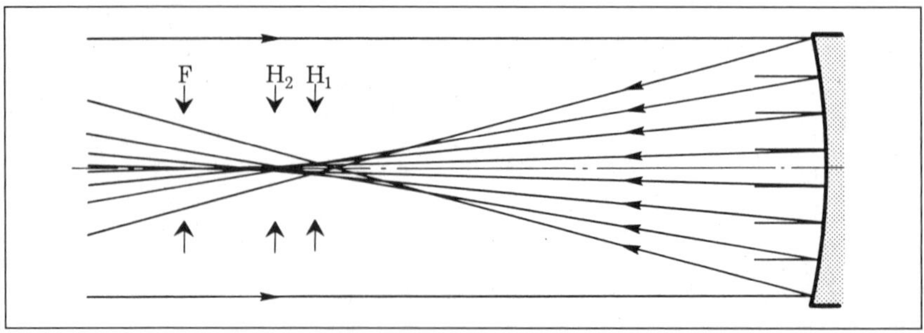

Fig. 5.2 *Spherical Aberration in a Spherical Mirror.*

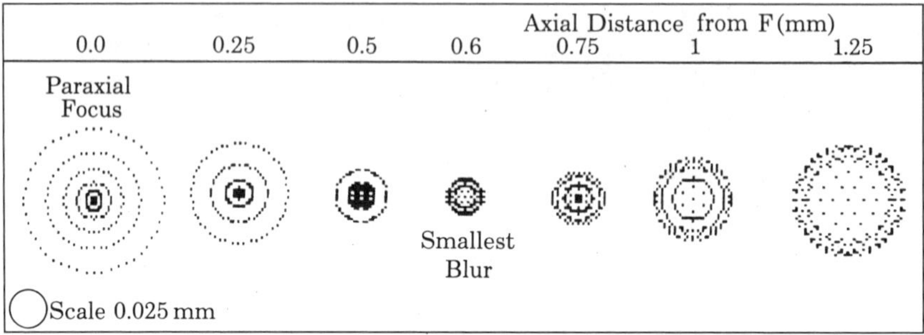

Fig. 5.3 *Spot Diagrams for a 200 mm f/8 Spherical Mirror.*

For relatively slow and small Newtonians, the primary mirror need not be paraboloidal. Although most Newtonians today do have a paraboloidal mirror, we will begin by examining the characteristics of a Newtonian with a spherical mirror. Because the sphere is quite easy to make, this simplest of all telescopes is often the amateur's first.

We wish to remind the reader that the diagrams which follow are based on a standard aperture of 200 mm. The performance of larger and smaller instruments may be obtained from the scaling rule given in section 4.5.

5.2 The Spherical Mirror

In a spherical mirror, paraxial rays (i.e., those close to the optical axis) intersect at point F, in fig. 5.2. Rays at greater distances from the axis focus closer to the mirror; the image suffers from spherical aberration. The bundle of converging rays reaches its smallest cross-section at plane H_1. However, somewhere between H_1 and F, at H_2, there is a considerably higher concentration of rays in the bundle; this bright "core" produces sharper images than any other focus.

Fig. 5.3 shows the sequence of axial spot diagrams at different focal planes for a 200 mm spherical *f/8* mirror. The sequence begins with the paraxial focus, at 0.0 mm. The blur attains its smallest diameter 0.6 mm inside paraxial focus. The pattern of changes in the diameter of the blur and the concentration of light in the

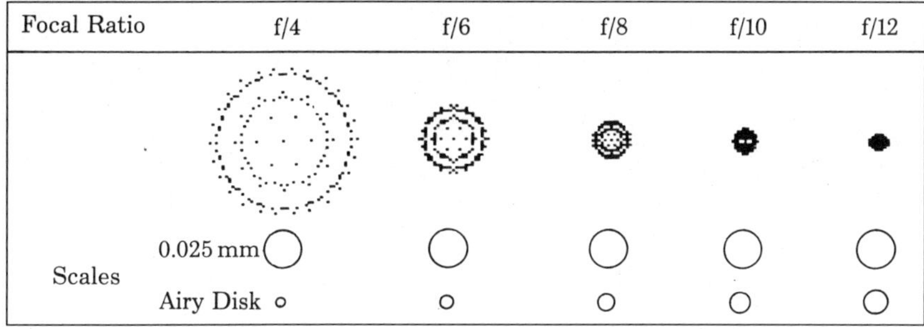

Fig. 5.4 *Smallest Blur Spot Diagrams for 200 mm Spherical Mirrors.*

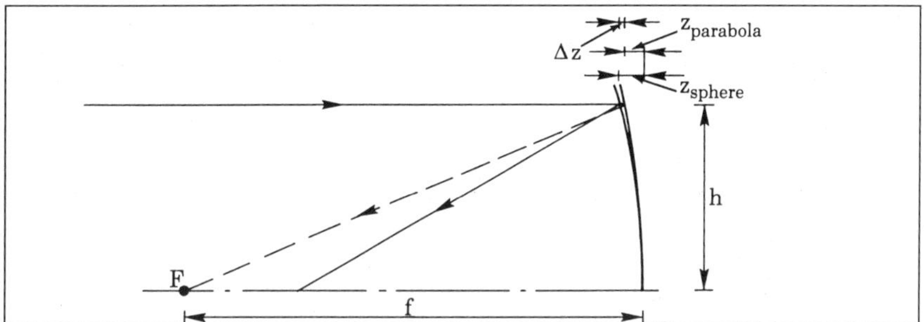

Fig. 5.5 *Spherical and Paraboloidal Mirrors Compared.*

core of the image is characteristic of pure spherical aberration.

Slow Newtonians need not be parabolized if the spherical aberration blur is smaller than the Airy disk. This is clearly shown in fig. 5.4. The diameters of spot diagrams in plane H_1 for various focal ratios are compared with the corresponding Airy disks and also with the photographic criterion of 0.025 mm. For the 200 mm Newtonian, a spherical mirror is acceptable for visual use in *f/12* and slower systems; parabolizing is not needed for photographic instruments *f/8* and slower.

5.3 The Paraboloidal Mirror

We compare a paraboloidal and a spherical mirror in fig. 5.5. Close to the optical axis both mirrors have the same radius of curvature. The paraxial focus for both surfaces is F; the paraxial sphere "touches"the paraboloid. At greater off-axis distances, however, it is necessary to deform the sphere backward if the rays are to intersect at F. The new curve is a paraboloid.

The equation of the sphere is:

$$z_{\text{sphere}} = r - \sqrt{r^2 - h^2} \tag{5.3.1}$$

and that of the paraboloid:

Table 5.1
Δz for 200 mm Mirrors

Focal Ratio	Δz (mm)
f/4	0.00306
f/6	0.00091
f/8	0.00038
f/10	0.00020
f/12	0.00011

$$z_{\text{paraboloid}} = \frac{h^2}{2r} = \frac{h^2}{4f} \tag{5.3.2}$$

where r is the paraxial radius of curvature; z, the deviation from vertical plane (also called the "sag" or sagittal); and h, the off-axis distance of an entering ray, or ray height.

Based on these formulae, we can find the difference, Δz, between the edges of a spherical and a paraboloidal mirror, measured parallel to the optical axis, from:

$$\Delta z = z_{\text{paraboloid}} - z_{\text{Sphere}} \,. \tag{5.3.3}$$

For mirrors of 200 mm aperture, we have tabulated this difference for focal ratios between *f*/4 and *f*/12 (see table 5.1).

For comparison, the wavelength of green light is 0.00055 mm. The table shows that for *f*/8 and slower mirrors, the difference between the edges of a spherical and a paraboloidal mirror is less than one wavelength of green light.

The dominant image aberration of the Newtonian is coma. Astigmatism plays a role at relatively large image angles, and curvature of field is present. When the entrance pupil is at the mirror, which is usually the case, the best focal surface lies midway between the tangential and sagittal focal surfaces, and has a radius of curvature equal to the focal length. It is inward curving. Because coma so strongly dominates the other aberrations in Newtonians, there is little to be gained from the use of curved film or a field flattener.

Fig. 5.6 shows spot diagrams calculated for focal ratios between *f*/4 and *f*/12 for distances of 0, 10, 20, and 30 mm off-axis for the optimally curved focal surface. Strong coma and astigmatism show at large off-axis distances, especially for fast primaries. Coma is the primary aberration near the axis; farther away astigmatism is visible as a second tail in the comatic blur figure.

Evaluation of the photographically useful field of a Newtonian is complicated since the emulsion may not record the faint outer parts of the comatic tail because it is faint, as discussed in section 4.1. Photographic images are, therefore, considerably smaller than the spot diagrams indicate, though the loss of light from the core of the image results in a degraded limiting magnitude. The 0.025 mm criterion should be applied when the highest demands are made on photographic image quality. For less critical work, a spread of 0.100 mm in the corner of the

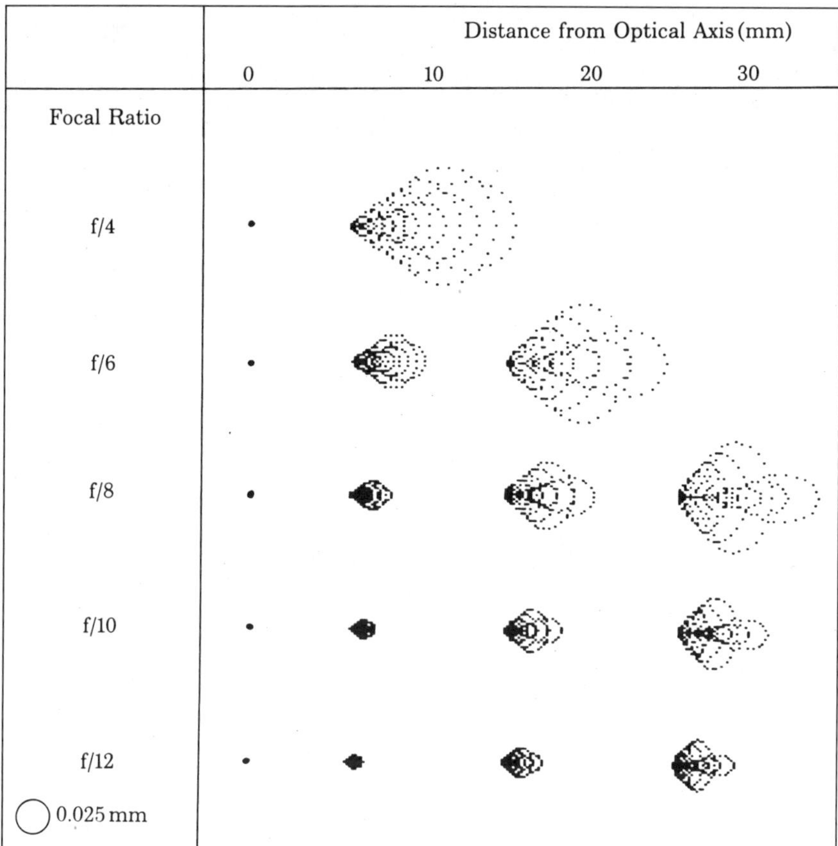

Fig. 5.6 *Spot Diagrams for 200 mm Paraboloidal Mirrors.*

film frame will give images that most astrophotographers find acceptable.

5.4 The Size of the Secondary Mirror

In principle, a telescope maker wishes the smallest secondary mirror possible, not only to avoid the loss of light, but also to prevent loss of image contrast through diffraction. (This point is further discussed in chapter 18.) Critical observers see a detectable loss with an obstruction 20% of the aperture, and all agree that the obstruction should not exceed 30%. From fig. 5.7 we show that the minimum diameter for a secondary mirror (i.e., the minor axis of an elliptical mirror) is:

$$d_{\min} = \frac{a}{N} \tag{5.4.1}$$

where N is the focal ratio, f/D

A secondary of the minimum size is impractical. Aside from the danger of figure errors at the edge of the diagonal mirror, the whole light cone cannot reach

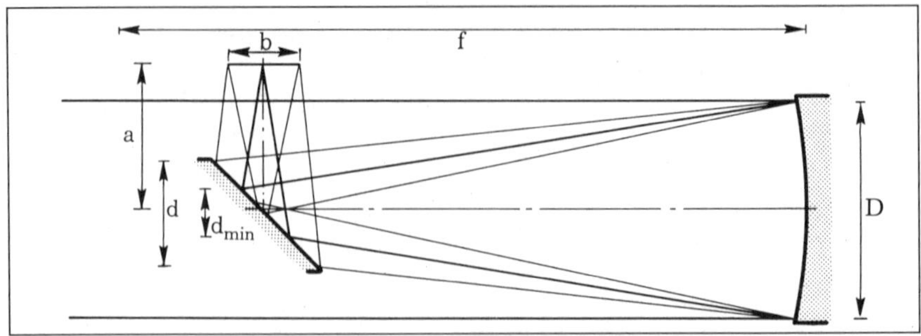

Fig. 5.7 *Size of the Secondary Mirror in the Newtonian.*

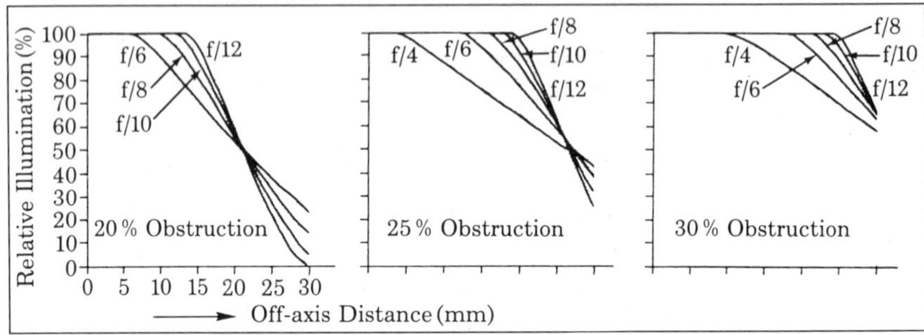

Fig. 5.8 *Illumination in the Focal Plane of a 200 mm Newtonian.*

the focal plane except on the optical axis, and thus the image is not fully illuminated. To illuminate a focal surface of diameter *b*, the minor axis of the diagonal should be:

$$d = \frac{a}{N} + b - \frac{a \cdot b}{f} \qquad \textbf{(5.4.2)}$$

For fast Newtonians, the secondary mirror must also be shifted slightly toward the primary and away from the focus in order to illuminate the focal plane symmetrically about the axis. (See refs. 5.1 and 5.4. For greater precision, refer to ref. 5.5.)

In fig. 5.8, we have plotted the relative illumination at various off-axis distances at a variety of focal ratios for secondaries of 40 mm, 50 mm, and 60 mm—20%, 25%, and 30% obstructions of a 200 mm aperture, respectively. Diagonal to focus distance, *a*, is taken as 180 mm, a figure which depends on the minimum height of the focusing device and the diameter of the telescope tube.

Note that for an *f*/4 mirror, a 40 mm secondary is too small to reflect the entire light cone of the axial beam; it is not plotted. For an *f*/12 mirror, a 40 mm secondary is too small to catch any part of the light cone 30 mm off-axis, and the illumination drops to zero there.

As a rule of thumb, the drop in illumination at the corner of the photographic negative should not exceed 30 or 40%. When a telescope is used visually, a fully illuminated field 10 mm diameter is usually sufficient. Vignetting and light falloff are treated more fully in chapter 19.

Chapter 6

The Refractor

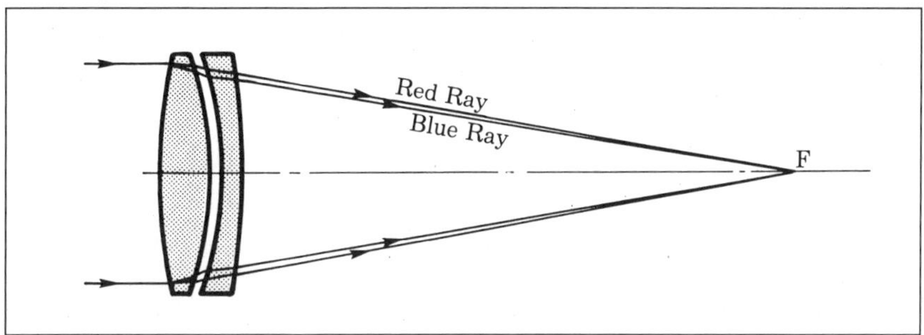

Fig. 6.1 *Achromatic Doublet Lens Corrected for Two Colors.*

6.1 Correction of Aberrations

In chapter 4 we saw that a single lens suffers from a number of aberrations. A refractor objective must have more elements to reduce them. In particular, two especially disturbing aberrations—longitudinal color and spherical aberration—must be corrected. Furthermore, coma should be corrected over a large field if possible.

Chromatic aberrations result from the different refractive indices of glasses at different wavelengths; blue light focuses closer to the lens than green, and red light focuses still farther from the lens. In a negative lens the sequence is the same, but the focal points (which are virtual) lie in front of the lens.

To obtain an idea how achromatizing is accomplished, imagine two thin lenses, one positive and the other negative, very close together. Choose glass for the lenses such that both elements have the same refractive index but different dispersions. What must be done to make the focal points of red and blue coincide?

Suppose we make the positive lens half the focal length of the negative lens—but pick the glass so the dispersion of the negative lens is twice that of the positive lens. The combination will be achromatic. Why? The negative lens (because of its higher dispersion) compensates for the short focus of the blue and long focus of the red rays with errors of the opposite sign, bringing the rays to focus together. This is shown in fig. 6.1.

In the real world, we deal with lenses of finite thickness and finite curvature. Such lenses suffer from spherical aberration. When the refractive indices for the

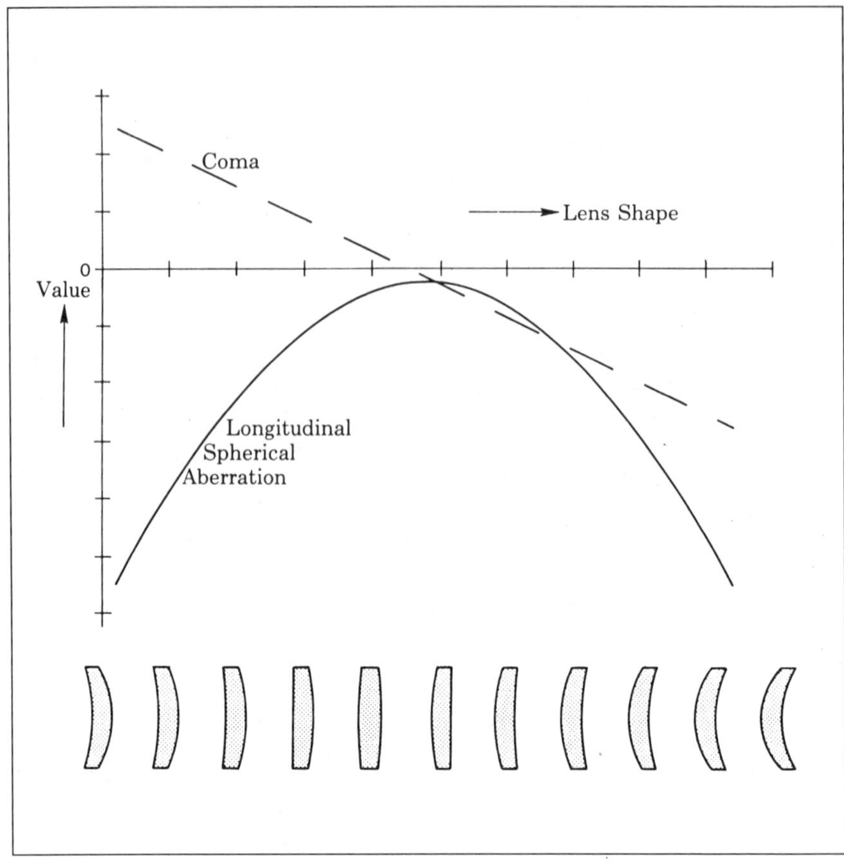

Fig. 6.2 *Spherical Aberration and Coma of a Simple Lens.*

design color are the same for both the positive and negative lens elements, it becomes difficult to correct spherical aberration, especially if the doublet is cemented. Therefore, the glass for the negative lens will usually have somewhat higher refractive index than the positive lens.

For a classical two-lens achromat, we choose crown glass for the positive lens, with a refractive index near 1.5 and relatively low dispersion (an Abbe number of 60). The negative lens is flint glass, with a refractive index near 1.6 and a relatively high dispersion (an Abbe number of 35).

Before looking into the correction of spherical aberration and coma in an achromatic doublet, let's examine first the spherical aberration and coma of a simple positive lens. These aberrations can be influenced considerably by the choice of the radii of curvature of that lens. In fig. 6.2 we graph spherical aberration and coma for a progression of positive lenses having the same power, but different shapes. For normal glass (i.e., $n \approx 1.5$), both aberrations are minimized when the lens has front and back radii in the ratio 1:6.

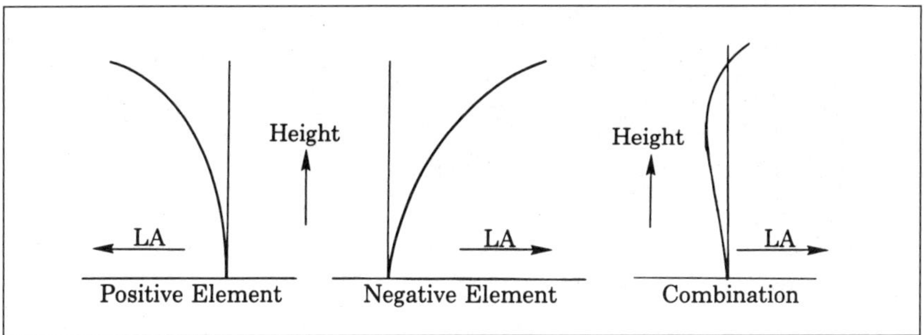

Fig. 6.3 *Correction of Longitudinal Spherical Aberration (LA) in Doublet Objective.*

For an optical designer, "bending" a lens without changing the power of its elements is a very important tool in controlling aberrations. As fig. 6.2 shows, spherical aberration cannot be eliminated completely in a single lens. However, by combining a positive lens with a negative lens having the same amount of spherical aberration, but of the opposite sign, we can greatly reduce it. This principle is shown schematically in fig. 6.3.

The third aberration which must be corrected is coma. This aberration is absent if the lens meets the sine condition. When the elements are separated by an air space, the radii of curvature of the inner surfaces can be different or the air space itself can be used to correct coma. This gives the designer enough freedom to correct both spherical aberration and coma. Small objectives are cemented, but those larger than 60 mm are generally airspaced to avoid temperature stresses as well as to correct aberrations.

From a designer's viewpoint, the focal length of a doublet depends on the outer radii of curvature; the inner radii are used mainly to correct spherical aberration and coma.

Summarizing the foregoing:

1. In achromatizing a two-lens objective, a positive lens is combined with a negative lens of lower power, while the positive lens has a lower dispersion (i.e., higher Abbe number) than the negative lens.

2. For a better correction of spherical aberration glasses are usually chosen so the negative lens has a higher refractive index than the positive lens.

3. Spherical aberration is minimized by bending both lenses.

4. Coma is also corrected by bending the lenses.

5. When the objective is air-spaced, the inner radii need not be the same, so the designer can correct both aberrations more easily.

Astigmatism and curvature of field cannot be corrected in a normal achromatic doublet. Doublets are not, therefore, suitable as objectives where the desired end is sharp imagery over a wide angle on flat film. For visual use field curvature is

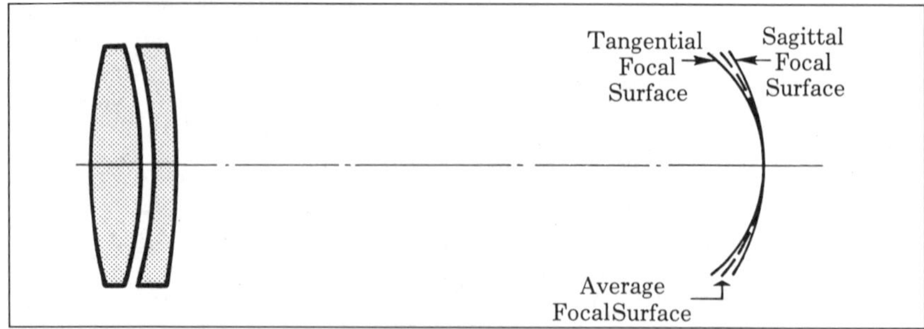

Fig. 6.4 *Tangential and Sagittal Focal Surfaces for a Doublet Objective.*

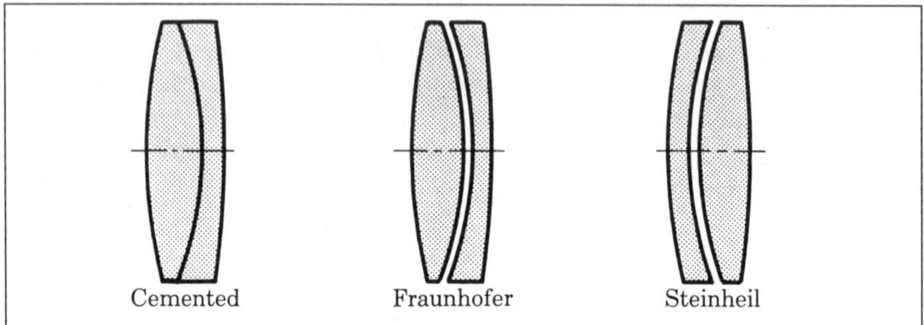

Fig. 6.5 *Achromatic Doublet Designs.*

not generally bothersome over small fields because the eye compensates by refocusing. When the image angle exceeds a few degrees, as it does in a binocular objective, astigmatism and field curvature may become sources of unsharpness at the edge of the field.

The tangential and sagittal focal surfaces of a doublet curve inward as shown in fig. 6.4. The radius of curvature lies between 0.30 and 0.35 times the focal length. Lateral color can be safely neglected so long as the entrance pupil coincides with the lens. Distortion is very low because of the small image angles and because the aperture stop is in contact with a relatively thin lens.

Over the course of time, many kinds of achromatic doublets have been designed; some bear the name of their designers. The best known, at least in astronomy, is the Fraunhofer doublet. The positive crown element of this air-spaced doublet is in front, the negative flint at the rear; the interior radii are unequal, and the air space is small. In a Steinheil doublet, the negative element is the front lens (fig. 6.5), and the positive element, the back lens. The Steinheil design is used when it offers better correction of aberrations or when the positive element has poor resistance to weathering (as fluorite does). However, because it has stronger curves, it is seldom used unless necessary.

In specifying a lens, a designer takes advantage of a number of variables to influence aberrations. Given an aperture and focal length, ten parameters remain free:

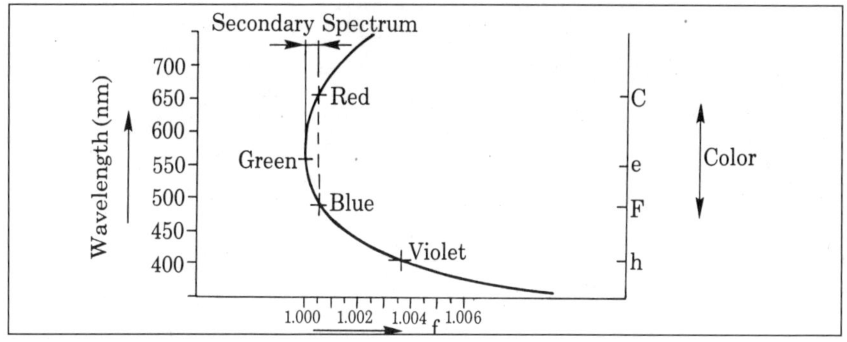

Fig. 6.6 *Color Curve for a Normal Visually Corrected Doublet.*

- types of glass (2)
- radii of curvature (4)
- airspace (1)
- sign of the front element (1)
- axial thicknesses (2).

Because of the numerous combinations possible, different manufacturers' designs are seldom the same. The detailed procedure for designing an achromatic doublet and apochromatic triplet are discussed in chapter 21.

6.2 Residual Aberrations in Objective Lenses

6.2.1 Chromatic Aberration

When we speak of correcting chromatic aberration, the term should not be taken literally, as meaning "eliminate," but should be understood more in the sense of "suppress." Complete elimination of color aberration in lens objectives is not possible. This is true, although to a much lesser extent, of spherical aberration when the lens surfaces remain spherical.

The degree of correction of the color aberration in a lens objective is often indicated in a graph of focus shift against wavelength. The color aberration curve of a typical two-lens refractor objective, corrected for visual use, is given in fig. 6.6. For visual use, the system is designed so the focal length is shortest in yellow-green light (555 nm), with the focal lengths of blue (the F–line at 486.13 nm) and red (the C–line at 656.27 nm) longer and coincident. It is then said that the objective has been corrected for two colors, or that it has been C–F corrected.

This color correction is closely connected with the color sensitivity curve of the eye, shown in fig. 6.7. Its peak sensitivity (for bright light) lies at 555 nm, corresponding roughly to the wavelength of e–light. Although the eye's sensitivity to C– and F–light is roughly equal, it is much lower for them than it is for e–light.

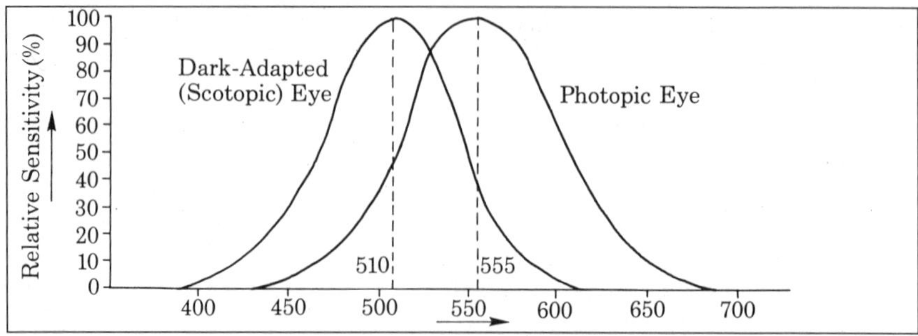

Fig. 6.7 *Relative Spectral Sensitivity of the Human Eye.*

The focus for violet light (fig. 6.6) is much longer than either C or F, but this is not bothersome for visual use because of the low sensitivity of the eye to the violet. However, when a visually corrected achromatic objective is used for photography, it is advisable to use a filter to absorb violet light which would otherwise form an out-of-focus image. Photographic emulsions are much more sensitive to blue and violet light than the eye is. Systems made for photography must be corrected so the shortest focus lies in blue or violet light.

The difference between the green and the red/blue focus is called secondary spectrum. For normal Fraunhofer-type astronomical doublets, secondary spectrum amounts to $0.0005f$, or 1/2000 the focal length of the lens. The very best visually-corrected doublets possible with slightly abnormal glasses are "semi-apochromats" or "half-apochromats." In these designs, secondary spectrum is reduced to approximately $0.00025f$, or 1/4000 the focal length.

Still better color correction can be obtained with certain new but expensive glasses. With them, secondary spectrum is suppressed to 1/8000 the focal length. We give an example of such an objective, the Apoklaas, in section 6.3. However, the best color correction possible in two-element objectives comes only with the use of an extremely expensive optical material called fluorite. Fluorite is man-made monocrystalline calcium fluorite. It has a low refractive index ($n = 1.43$) and very low dispersion ($V = 95.6$). With this material, the secondary spectrum can be reduced to 1/16,000 the focal length.

Although fluorite is ideal as an optical material, it has its drawbacks. Not only is it expensive and the size of blanks severely limited, but fluorite is highly subject to weathering. Despite its drawbacks, a number of telescope manufacturers now offer fluorite objectives.

When the optical designer wishes to bring three colors—such as red, blue, and violet—to the same focus, the objective requires three lenses made of different materials. An objective that does this is termed an apochromat. The secondary spectrum, i.e., the difference between the green and red/blue focal points, is typically $0.0001f$, or 1/10,000 the focal length. It must be emphasized, however, that not all three-lens objectives, nor all lenses labelled "apochromatic" have this excellence of color correction. Secondary spectrum in the three-lens Zeiss Schwer-

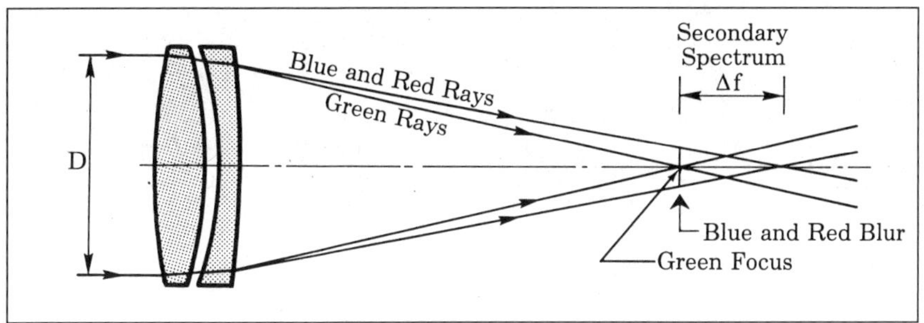

Fig. 6.8 *Secondary Spectrum in a Normal Doublet Lens.*

flint Apochromat, for example, amounts to 1/3500 the focal length—not as good as the best normal doublets. Since an apochromat usually focuses a wider range of wavelengths than an achromat does, apochromatic objectives are more suitable for photography than doublets.

How much secondary spectrum is objectionable? In conventional doublets, secondary spectrum is acceptable when the red and blue blurs do not exceed three times the diameter of the Airy disk in green light. This much blur is tolerable because the eye is relatively insensitive to blue and red. Fig. 6.8 shows how green, blue, and red light focus in a normal doublet. Since the difference in focal length, Δf, is typically $0.0005f$ when the objective is focused for green, the blue and red blurs have a diameter of $0.0005D$. In chapter 18, we show that the diameter of the Airy disk for green light is $280/D$ seconds of arc, where D is the aperture in millimeters. Since one second of arc is $f/206265$ mm diameter in the focal plane, the diameter of the Airy disk, d_A, becomes:

$$d_A = \frac{280}{D} \cdot \frac{f}{206265} = \frac{1}{735} \cdot \frac{f}{D} = \frac{N}{735} \text{mm} \qquad \textbf{(6.2.1)}$$

where N is the focal ratio, f/D. From this equation, then, it is easy to see that the condition for freedom from chromatic aberration is:

$$\frac{3N}{735} = 0.0005D \qquad \textbf{(6.2.2)}$$

hence:

$$N_{min} = 0.122D \qquad \textbf{(6.2.3)}$$

where D is expressed in millimeters. Thus we see that the minimum focal ratio, N_{min}, increases with increasing aperture. For an aperture of 100 mm, the fastest focal ratio giving full achromatism is $f/12.2$; for an aperture of 200 mm, $f/24.4$. Because of the need for stability and ease of movement, however, telescope objectives seldom exceed $f/20$. This means the greatest aperture for a practical fully achromatic doublet refractor made with normal glass is approximately 160 mm.

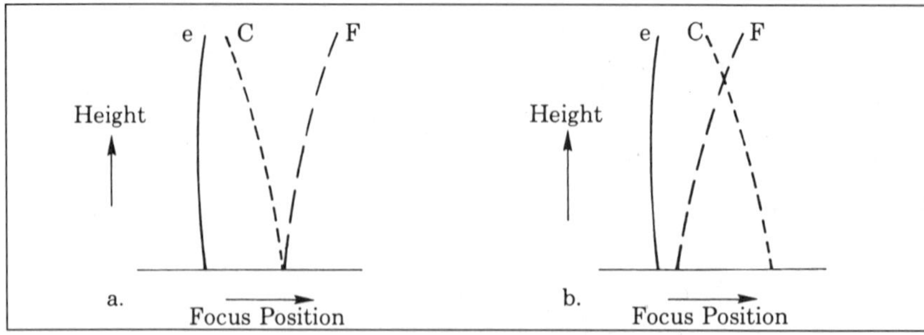

Fig. 6.9 *Spherochromatism in the Achromatic Doublet.*

We also see that all large astronomical refractors suffer from serious secondary spectrum.

6.2.2 Spherical Aberration and Spherochromatism

When a doublet has been corrected for spherical aberration in green light, the lens will be undercorrected in red light (C–line) and overcorrected in blue (F–line). This means that in red light, outer rays focus closer to the lens than the paraxial rays, while the opposite is true in the blue. This occurs for all cemented doublets and doublets with narrow air gaps—indeed, a large air gap is sometimes used to control the problem. This change of spherical aberration with wavelength is called spherochromatism.

In a doublet designed so that the paraxial focal length is the same for red and blue, spherochromatism introduces considerable difference in the focal lengths of the outer rays in these two colors, resulting in poor color correction, as shown in fig. 6.9a. To minimize it, a designer may choose to make the paraxial focus slightly longer in red than it is in blue, shifting the spherical aberration curves for the two wavelengths so they intersect at the 70.7% zone (fig. 6.9b). This zone is chosen because the areas inside and outside it are equal.

It is important to note that color correction curves for lens objectives are often referenced to the 70.7% zone rather than the paraxial zone. The curve shown in fig. 6.6 is an example of this practice. The designer should be aware that it may not be the best zone for determining the state of color correction since the overall color correction also depends on the shapes of the C- and F-curves.

6.3 Evaluation of Lens Objectives

We now compare three refractor objectives. In table 6.1 we give the optical characteristics—axial thickness of the lenses, radii of curvature, distances, and glass types—for systems of 200 mm aperture. The *f*/15 Fraunhofer doublet employs normal crown and flint glasses while the *f*/10 Apoklaas utilizes two special glasses, reducing its secondary spectrum to 0.00012*f*. The *f*/8 fluorite objective offers half the secondary spectrum of the Apoklaas. Because of its poor weathering resistance, the fluorite element has been placed behind the negative lens, so this lens

Table 6.1
Three 200mm Doublet Refractor Objectives
(all dimensions in millimeters)

	Fraunhofer *f*/15	Apoklaas *f*/10	Fluorite *f*/8
R_1 RC	2009.753	1175.900	546.292
T_1 Axial dis.	31.336	24	12
M_1 Medium	517642	487845	720504
R_2	–976.245	–513.920	356.146
T_2	3.315	1	1
M_2	Air	Air	Air
R_3	–985.291	–522.606	348.299
T_3	25.109	12	24
M_3	613370	558542	Fluorite
R_4	–3636.839	–2756.828	–6670.445
T_4 Back Focal Length	2968.12	2000	1599.996
M_4	Air	Air	Air
EFL	3000	2020	1600
1° Field	52.4	35.3	27.9

is a Steinheil doublet. The Apoklaas and the fluorite objectives were designed by Klaas Compaan.

Apart from reducing secondary spectrum, a primary advantage of the newer glasses, and particularly fluorite, is that faster refractor objectives are possible, resulting in considerably shorter telescopes. Conventional 200 mm refractors have focal ratios between *f*/15 and *f*/20; the new glasses permit shortening this to *f*/10. Spherochromatism limits further shortening. With fluorite, lenses may be *f*/8 or even faster.

Three-lens apochromatic objectives are seldom amateur-made because the glass is expensive. In addition, making such an objective is beyond the skills of most ATMs because of the close tolerances in grinding, polishing and centering such a system. Many of these problems are avoided with a special type of triplet: the oiled objective. The spaces between the lenses are filled with a special oil having a refractive index near that of the glass, so only the front and rear surfaces need to be fully polished and figured. Because of the unusual construction, this objective is discussed in chapter 13.

In the spot diagrams in fig. 6.10, we show green light (e-line) in focus. The spot diagrams for the field have been calculated for the optimally curved focal surface shown in the left-hand column. Because of their focal ratios, the diameters of the Airy disks differ for the different lenses. When judging the performances for visual use, therefore, be sure to compare the size of the blue and red spot diagrams with the appropriate green-light Airy disk. For photographic use, apply the 0.025 mm criterion throughout. The off-axis swelling of the spot diagrams results mainly from astigmatism which cannot be corrected for these systems.

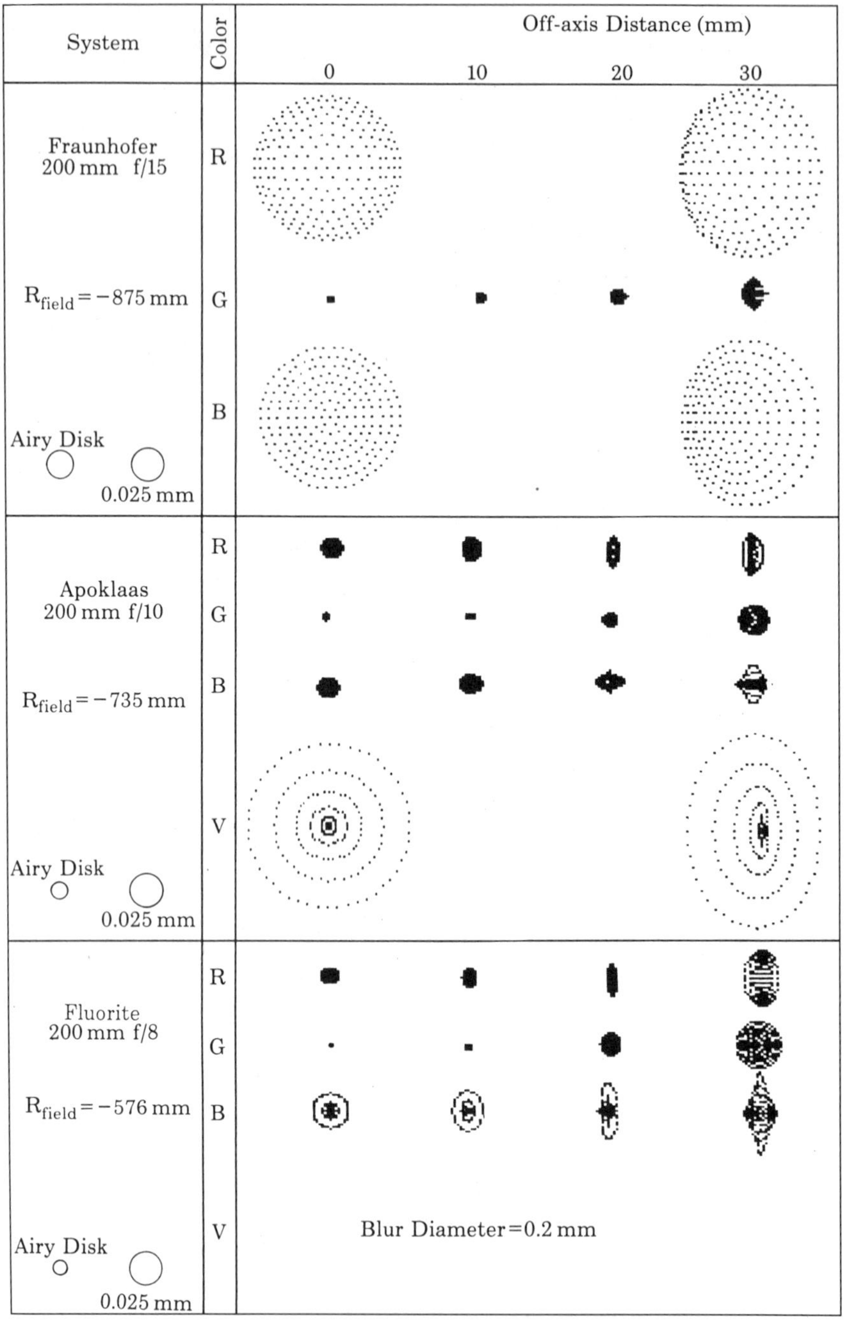

System	Color	Off-axis Distance (mm) 0	10	20	30
Fraunhofer 200 mm f/15	R				
$R_{field} = -875$ mm	G				
Airy Disk 0.025 mm	B				
Apoklaas 200 mm f/10	R				
	G				
$R_{field} = -735$ mm	B				
Airy Disk 0.025 mm	V				
Fluorite 200 mm f/8	R				
	G				
$R_{field} = -576$ mm	B				
Airy Disk 0.025 mm	V	Blur Diameter = 0.2 mm			

Fig. 6.10 *Spot Diagrams for Various 200 mm Refractor Objectives.*

In the case of the *f*/15 Fraunhofer achromat, note that the blue and red spot diagrams are roughly five times the diameter of the Airy disk in green. Since the limit of good color correction is three times the Airy disk, this lens suffers from considerable color aberration. With a smaller aperture or a longer focal length, this lens would yield better images, but it is clear that observers are willing to tolerate this much color error.

The *f*/8 fluorite and *f*/10 Apoklaas objectives display considerably better color correction. Their blue and red spot diagrams are only 1.5 times the Airy disk diameters. The fluorite has the lowest secondary spectrum, but greater spherochromatism, resulting in enlargement of the blue and red spot diagrams. Note that the fluorite objective delivers the same performance at *f*/8 as the Apoklaas delivers at *f*/10. For optimum correction of aberrations, the designer combined the fluorite lens with an expensive high-index glass (720504).

From fig. 6.10, we see dramatic evidence that visually-corrected two-element refractor objectives suffer a serious disadvantage in astrophotography. Because these systems are corrected for red and blue, but not for violet, the violet image is very large. Even the fluorite objective, which is exceptionally well-corrected for visual use, shows a violet blur (h–light) of 0.20 mm, eight times the photographic criterion. For this reason, photography with these objectives at this aperture will result in considerable unsharpness unless they are used with a filter that absorbs violet light. For the fluorite and Apoklaas, the filter should have a cutoff wavelength of 450 nm.

The reader may wonder how improved color correction is possible with special glasses and fluorite. In particular, what do these materials offer that is not available in the commonly used crown and flint combinations? The answer is that they have more favorable dispersion characteristics. We explore this matter more fully in section 21.13.3.

Chapter 7

The Cassegrain Telescope

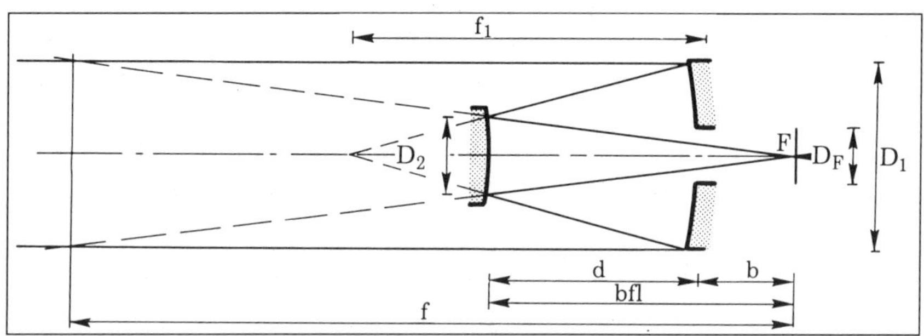

Fig. 7.1 *Typical Layout of a Cassegrain Telescope.*

7.1 Introduction

All Cassegrain telescopes consist of a concave primary mirror with a small secondary mirror inside the focus of the primary; the secondary redirects starlight toward the primary. The image, in most cases, lies behind the primary where it is easily accessible for visual observing and photography. The convex secondary multiplies the focal length by a factor M. This factor M is termed secondary magnification, and is defined:

$$M = \frac{\text{focal length of system}}{\text{focal length of primary}}.$$

Fig. 7.1 shows the layout of the Cassegrain telescope with the main dimensions. The focal length of the system, f, is equivalent to the distance found by extending the rays in the exit cone to their height in the incoming beam, as shown in the figure by dotted lines. The instrument is quite short with respect to its focal length. The major parameters of a Cassegrain telescope are related by the following formulae:

$$f = M \cdot f_1 = \frac{f_1 f_2}{f_1 + f_2 - d} \tag{7.1.1}$$

$$bfl = (d+b) = \frac{(f_1 - d) \cdot f_2}{f_1 + f_2 - d} \qquad (7.1.2)$$

$$f = d + b + M \cdot d \qquad (7.1.3)$$

$$f_1 = d + \frac{d+b}{M} \qquad (7.1.4)$$

$$f_2 = \frac{-(d+b)}{M-1} \qquad (7.1.5)$$

where f_1 is the focal length of the primary, f_2 is the focal length of the secondary, negative in Cassegrains, and d is the separation. The quantity $(d + b)$ is the back focal length (b.f.l.), and should not be mistaken for the effective focal length, f or e.f.l., of the system.

To use an existing mirror as the primary in a Cassegrain system, substitute the value f_1 and desired value of M and b to find the parameters for a secondary mirror:

$$d = \frac{M \cdot f_1 - b}{M+1} \qquad (7.1.6)$$

$$f_2 = -\frac{M}{M^2 - 1} \cdot (f_1 + b) . \qquad (7.1.7)$$

The smallest diameter for the secondary mirror, D_2, for which no axial loss of light occurs, is:

$$D_2 = D_1 \cdot \frac{d+b}{f} . \qquad (7.1.8)$$

When the designer cannot accept any light loss at the edge of the field, then given a field diameter at the focal plane of D_F, the diameter of the secondary mirror must be:

$$D_2 = \frac{D_1 \cdot (d+b)}{f} + D_F \cdot \frac{d}{f} . \qquad (7.1.9)$$

This formula is valid only when the entrance pupil of the system (i.e., the aperture limiting stop) coincides with the primary mirror, which is the case in most Cassegrain systems. When the entrance pupil lies in front of the primary (as is the case with most catadioptric systems), the diameter of the secondary must be larger than indicated by eq. 7.1.9, and the primary must also be enlarged.

Because the secondary blocks light which would otherwise reach the primary, all Cassegrain instruments suffer some light loss relative to unobstructed instruments. Furthermore the obstruction causes a loss of contrast and image sharpness due to diffraction. The designer usually attempts to make the secondary as small as possible—generally smaller than indicated by eq. 7.1.9 but larger than the minimum derived in eq. 7.1.8—in some cases accepting considerable light loss at the edge of the field.

7.2 Curvature of Field

Cassegrain telescopes should be designed in such a way that the resulting instrument has the following properties:

1. short tube length
2. small secondary mirror
3. flat focal surface
4. accessible focal surface.

Unfortunately it is impossible to satisfy all four constraints simultaneously. In particular, control of field curvature, associated with the strong curve of the secondary mirror, causes difficulty. Designers and users of Cassegrain telescopes should never forget the system's inherent curvature of field. Field curvature is stronger for many Cassegrains than it would be for a Newtonian or refractor telescope of the same aperture and focal length. The field is inward curving, that is, concave toward the sky. The radius of curvature of the focal surface of a Cassegrain, when no astigmatism is present in the system, is:

$$\frac{1}{R_F} = \frac{2}{r_1} - \frac{2}{r_2}. \tag{7.2.1}$$

Astigmatism will be discussed later. It is evident that in order to have a flat focal surface, the radii of curvature of the primary and secondary mirror must be equal. As we shall demonstrate, a short tube length, small secondary mirror, and flat focal surface are incompatible with placing the focal plane behind the primary.

Consider two *f*/10 Cassegrain telescopes of equal aperture, as shown in fig. 7.2. In one, make the primary mirror *f*/2, and set the secondary magnification, M, equal to 5.0. For the other, make the primary *f*/5 and set the secondary magnification equal to 2.0. Place the focus 150 mm behind the primary in both. With the help of the formulae above, we calculate the instrumental parameters listed in table 7.1.

It will be clear from the example that a configuration with a high secondary magnification has a short tube, small secondary, and strong field curvature. The short tube and small secondary are, of course, desirable features in an instrument intended for visual use, but strong field curvature is unfavorable for photographic use. We note again that these figures are approximate because eq. 7.2.1 is valid only when astigmatism is absent.

Since no Cassegrain can be entirely free of astigmatism, eq. 7.2.1 should not be used for exact results. When astigmatism is present, the curvature of field will be stronger than indicated by eq. 7.2.1. However, the designer would do well to keep in mind this general rule:

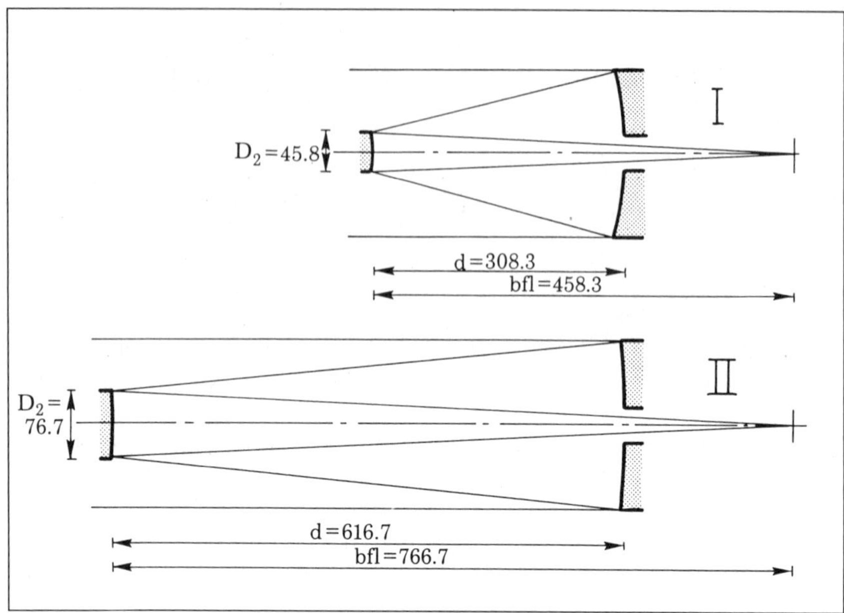

Fig. 7.2 *Two Cassegrains.*

Table 7.1
Comparison of Cassegrains
Primary Focal Length and Secondary Magnification

	I	II
Aperture D_1 (mm)	200	200
Focal ratio N	10	10
Focal length (mm)	2000	2000
Focal ratio primary	2	5
Secondary magnification M	5	2
Distance between mirrors (mm)	308.3	616.7
Back focal length (mm)	458.3	766.7
Minimum diameter secondary mirror (mm)	45.8	76.7
Radius of primary	–800	–2000
Radius of secondary	–229.16	–1533.333
Radius of curvature of focal field* (mm)	–160	–3287

*Theoretical values, not corrected for astigmatism

Table 7.2
Cassegrain Optical Combinations

System	Primary	Secondary
Classical Cassegrain	paraboloid	hyperboloid
Dall-Kirkham	prolate ellipsoid	sphere
Ritchey-Chrétien	hyperboloid	hyperboloid
Pressman-Camichel	sphere	oblate ellipsoid

Table 7.3
Schwarzchild Constants

Aspheric Surface	Schwarzschild Constant
sphere:	$SC = 0$ (i.e., no deformation)
ellipsoid (prolate):	$-1 < SC < 0$
paraboloid:	$SC = -1$
hyperboloid:	$SC < -1$
ellipsoid (oblate):	$SC > 0$

> For a given focal length, field curvature *increases* as the diameter of the secondary mirror *decreases* and the separation of the mirrors *decreases*.

Taking this rule into account, the designer of a Cassegrain makes a choice between a system intended for visual use having a small secondary mirror (diameter 25–30% of the entrance pupil) in which a relatively strong field curvature is accepted, or one intended only for photographic use with a weakly curved or flat focal surface and a large secondary mirror.

Field curvature may be compensated by using a curved film or a field flattener, as described in chapter 14. Note that field curvature is of practical significance mainly in amateur instruments. For large telescopes, Cassegrain field curvature is relatively unimportant because professional astronomers either use field flattening lenses or bend their photographic plates to conform to the focal surface.

7.3 Optical Performance

It must be emphasized that eqs. 7.1.1 through 7.1.7 are valid only in the paraxial region. Although they are extremely useful for making quick calculations of the main dimensions, they fail to give the designer any information about the off-axis aberrations of the system.

Any Cassegrain can be free of spherical aberration, permitting a sharp image on the axis, provided the proper combination of shapes of both optical surfaces is chosen, but off-axis image sharpness depends strongly on the shapes of the surfaces. The four most important combinations are shown in table 7.4.

We treat aspheric optical surfaces in chapter 20. Let it suffice to mention

Table 7.4
Optical Characteristics of Four 200mm *f*/8 Cassegrain Systems
(Dimensions in mm)

Type	Ritchey-Chrétien	Classical	Dall-Kirkham	Pressmann-Camichel
Focal Ratio *f*	8	8	8	8
Effective fl	1600	1600	1600	1600
Secondary mag. M	8/3	8/3	8/3	8/3
Primary RC	–1200	–1200	–1200	–1200
Primary SC	–1.13682	–1	–0.61328	0
Secondary RC	–628.36	–628.36	–628.36	–628.36
Secondary SC	–6.55243	–4.84	0	7.6755
Mirror sep.	–403.64	–403.64	–403.64	–403.64
Back focal length	523.64	523.64	523.64	523.64
1° field (mm)	27.9	27.9	27.9	27.9

here that the deviation with respect to a sphere can be characterized by a deformation factor, SC, the Schwarzschild Constant. For the conic aspheric surfaces, this factor is shown in table 7.4.

The Schwarzschild Constant, *SC*, and the paraxial radius of curvature, *R*, fully describe the "conic" aspheric mirror surfaces of the instruments evaluated in this chapter. See tables 7.4, 7.5a and 7.5b for examples of the Schwarzschild Constant.

The Classical Cassegrain is sometimes made as a modification of an existing Newtonian. The flat diagonal mirror is removed and replaced by a convex secondary, or the two mirrors are used interchangeably.

Some optical workers find it difficult to figure the convex hyperbolic secondary of the Classical Cassegrain, preferring the easy-to-make spherical secondary of the Dall-Kirkham system. The elliptical primary of the Dall-Kirkham is also relatively easy to make. Unfortunately, the Dall-Kirkham has far stronger coma than the Classical Cassegrain and is therefore used for smaller fields.

The Pressmann-Camichel offers the attractive possibility of a spherical primary mirror—but the secondary must be strongly deformed to remove spherical aberration. In practice, the design suffers from such severe coma that it is limited to a very narrow field.

In the Ritchey-Chrétien, the mirror shapes are chosen to eliminate coma. This system suffers only from astigmatism and, of course, curvature of field (as do most Cassegrain systems), permitting a relatively wide field. A Ritchey-Chrétien is more difficult to make than the other systems because the shapes of the two mirrors are more aspheric than the Classical Cassegrain. This system is not easily made by amateurs, but is highly regarded by professional astronomers.

In order to provide direct comparison between various systems, we calculated spot diagrams for four systems having the same primary focal ratio and effective focal ratio but different surfaces. The results are shown in fig. 7.3; system

Table 7.5a
Optical Characteristics of Six 200mm Classical Cassegrain Systems
(Dimensions in mm)

Focal Ratio *f*	10		15		20	
Effective fl	2000		3000		4000	
Secondary mag. M	2	4	2	4	2	4
Primary RC	–2000	–1000	–3000	–1500	–4000	–2000
SC	–1	–1	–1	–1	–1	–1
Secondary RC	–1600	–373.3333	–2266.6667	–506.6667	–2933.3333	–640
SC	–9	–2.7778	–9	–2.7778	–9	–2.7778
Mirror Separation	–600	–360	–933.3333	–560	–1266.6667	–760
Back Focal Length	800	560	1133.3333	760	1466.6667	960
1° field (mm)	34.9	34.9	52.4	52.4	69.8	69.8

Table 7.5b
Optical Characteristics of Six 200mm Dall-Kirkham Cassegrain Systems
(Dimensions in mm)

Focal Ratio	10		15		20	
Effective fl	2000		3000		4000	
Secondary mag. M	2	4	2	4	2	4
Primary RC	–2000	–1000	–3000	–1500	–4000	–2000
SC	–0.55000	–0.671875	–0.575	–0.703125	–0.5875	–0.71875
Secondary RC	–1600	–373.3333	–2266.6667	–506.6667	–2933.3333	–640
SC	0	0	0	0	0	0
Mirror Separation	–600	–360	–933.3333	–560	–1266.6667	–760
Back focal length	800	560	1133.3333	760	1466.6667	960
1° field (mm)	34.9	34.9	52.4	52.4	69.8	69.8

parameters are listed in table 7.4. In each case the aperture is 200 mm and the primary focal ratio *f*/2.67, a value that is often used for large professional telescopes. The focal surface lies 120 mm behind the primary mirror.

In evaluating these photographic systems, we calculated spot diagrams at the optimally-curved focal surface, that is, we show the smallest blur figures each system can produce. Because the Ritchey-Chrétien has no coma, the blurs of the star images are circular when the focal surface chosen is midway between the tangential and sagittal focal surfaces. Round star image positions are easier to measure than asymmetric images, the reason professional astronomers prefer the Ritchey-Chrétien over the Classical Cassegrain.

We also evaluated Classical and Dall-Kirkham Cassegrains at *f*/10, *f*/15 and *f*/20, for systems with secondary magnifications of 2.0 and 4.0 (see fig. 7.4). In these instruments, the focal surface lies 200 mm behind the primary; they are mainly intended for visual use. At these focal ratios, the Ritchey-Chrétien offers

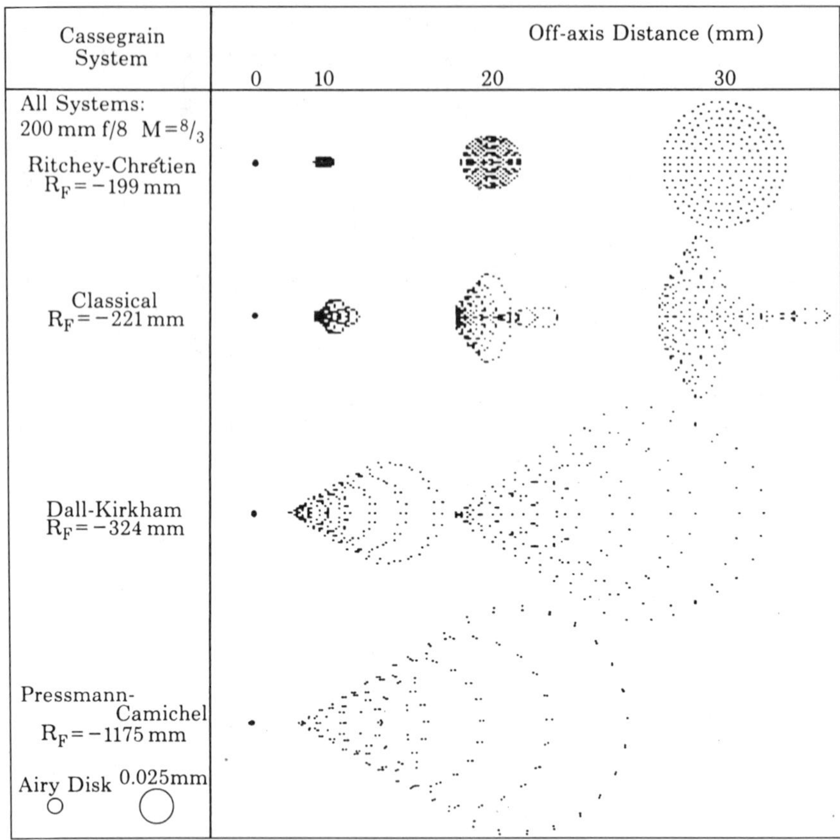

Fig. 7.3 *Spot Diagrams of Four 200 mm f/8 Cassegrain Telescopes.*

no advantage over the Classical Cassegrain and since it is more difficult to make, we omitted it.

Comparison of figs. 7.3 and 7.4 supports the following assertions:

1. The Dall-Kirkham and Pressmann-Camichel are unsuitable for wide-field photographic applications because of strong coma.

2. The Ritchey-Chrétien is the best for photographic use because of its round star images. Nevertheless this instrument shows considerably enlarged spot diagrams for off-axis distance greater than 20 mm (0.7°) as a result of astigmatism. This design has the strongest field curvature.

3. When Classical Cassegrains with the same focal ratio but different secondary magnification are compared at the same off-axis distance, it appears that greater secondary magnification leads to a slightly decreased image sharpness. In the Dall-Kirkham, image sharpness drops

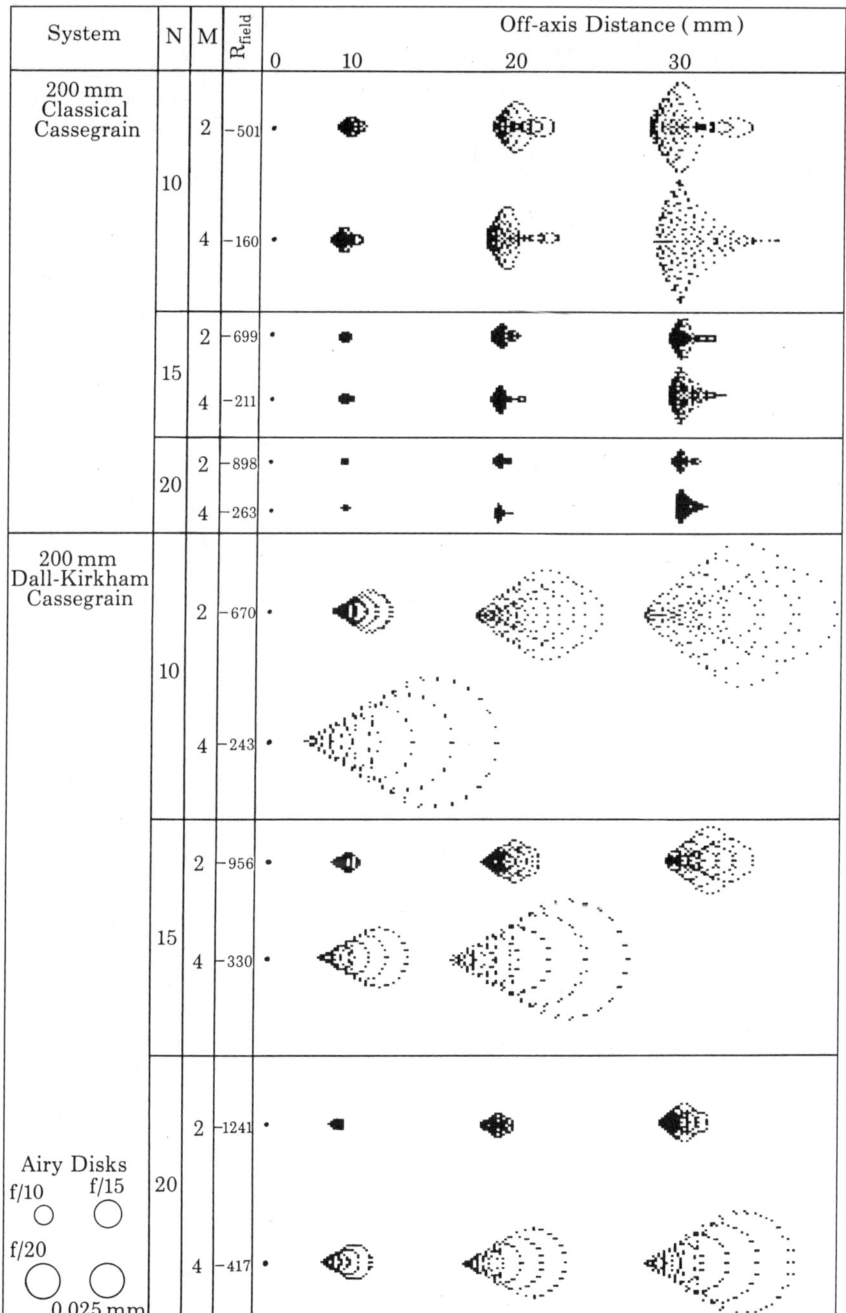

Fig. 7.4 *Spot Diagrams of Various 200 mm Cassegrains.*

much more. For a given focal ratio, low secondary magnification produces better off-axis image quality.

4. For a given focal ratio, greater secondary magnification increases the field curvature. This confirms the conclusion reached in section 7.2.

5. The Classical Cassegrain produces spot diagrams equal to those of a Newtonian having the same focal ratio (compare figs. 7.4 and 5.6), but the field curvature of the Cassegrain is much stronger.

7.4 Baffling

A troublesome feature of Cassegrain telescopes and their derivatives is sky flooding: light entering the tube can reach the focal surface without being reflected by the mirrors. Under certain circumstances this can lead to considerable loss of contrast. Even for observation or photography of the dark night sky, it is important to install baffling to block stray light. Baffling increases the central obstruction, and may also introduce or increase loss of light at the edge of the field, especially for the large fields desired in photography. We discuss baffle systems for telescopes in chapter 19.

Chapter 8

The Schmidt Camera

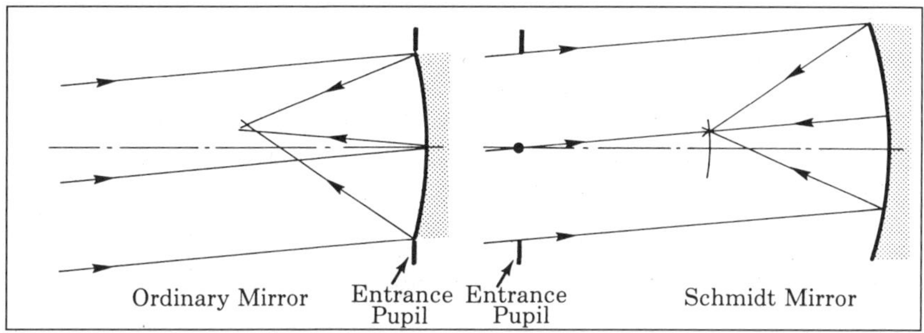

Fig. 8.1 *The Position of the Entrance Pupil in Ordinary and Schmidt Systems.*

8.1 Introduction

Schmidt cameras, though eminently suited for astrophotography, are used surprisingly little by amateurs. The fact that these instruments cannot be used visually may play a role in this, but another contributing cause is certainly the difficulty of making the Schmidt corrector plate. This effectively blocks all but a few optically skillful amateurs from constructing this instrument.

The Schmidt camera offers the unparalleled combination of a fast focal ratio and a large image angle. Compared to other astrocameras, the Schmidt camera offers an unparalleled image sharpness over the entire field. Its principal drawback is that the focal surface is curved and lies inside the instrument.

At the end of this chapter we discuss the lensless Schmidt camera, or "poor man's Schmidt." Although this system is suitable for amateur construction, it does not offer the speed of a genuine Schmidt camera.

8.2 Optical Principles

The Schmidt camera is based on the principle of symmetry and the ability of a corrector lens to suppress the spherical aberration of a spherical mirror. As we saw in chapter 5, the Newtonian's field is limited by coma. Coma occurs in both spherical and paraboloidal Newtonians because an oblique beam of parallel rays has no axis of symmetry with respect to the mirror. The image of this beam is not formed on the axis of the beam—but is instead formed to one side (see fig. 8.1). The image, therefore, will not be symmetrical.

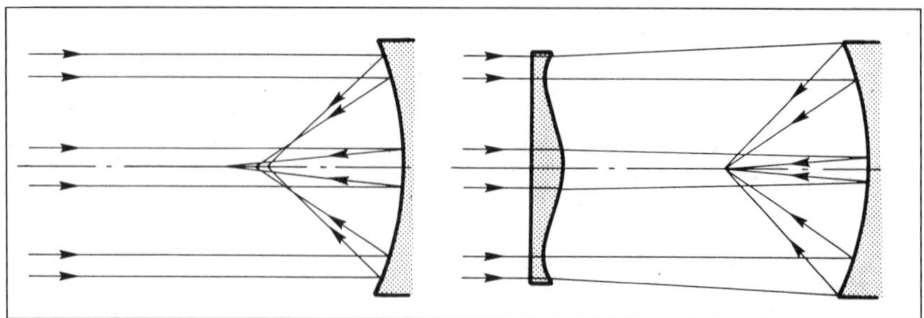

Fig. 8.2 *The Role of the Corrector Lens in a Schmidt System.*

The underlying cause of this asymmetry is that the entrance pupil, which forms the limiting stop for all entering bundles of light, coincides with the mirror itself. When the stop is moved from the mirror to its center of curvature, symmetry is obtained. It is clear that every beam of light passing through the stop, whether it is parallel to the optical axis or not, has its own axis of symmetry. This is not true when the stop lies at the mirror. The symmetry principle holds, of course, only when the mirror is spherical.

But the image formed by the spherical mirror still suffers from spherical aberration. Bernhard Schmidt solved this problem by placing a thin aberration-correcting lens at the center of curvature of the spherical mirror. Although similar schemes had been described by the Finnish astronomer Väisälä in 1924, Schmidt succeeded in designing and fabricating a working camera at the end of the 1920s.

8.3 The Schmidt Corrector

In a spherical mirror, the focal distance of the central rays is greater than that of the rays farther away from the mirror's center—in short, it suffers from spherical aberration. This can be eliminated by placing a specially shaped lens in front of the mirror, as shown in fig. 8.2.

The zone where the corrector is thinnest is called the neutral zone because rays pass through it without deviation. In order to bring light rays entering at other zones to the same focus, corrector zones outside the neutral zone have negative power while zones inside the neutral zone have positive power.

Since a Schmidt corrector is a refracting element, it causes chromatic aberration, but the aberration is not severe and can be minimized by the location of the neutral zone. Higher order residual monochromatic aberrations, coma and astigmatism, also depend on the place of the neutral zone.

The profile of a Schmidt corrector satisfies the following equation:

$$Z = Ah^2 + Bh^4 + Ch^6 + \dots \tag{8.3.1}$$

in which Z is the depth of the curve, h is the off-axis distance, and A, B and C are constants depending on the refractive index of the glass, the focal ratio, the aperture, and the position of the neutral zone.

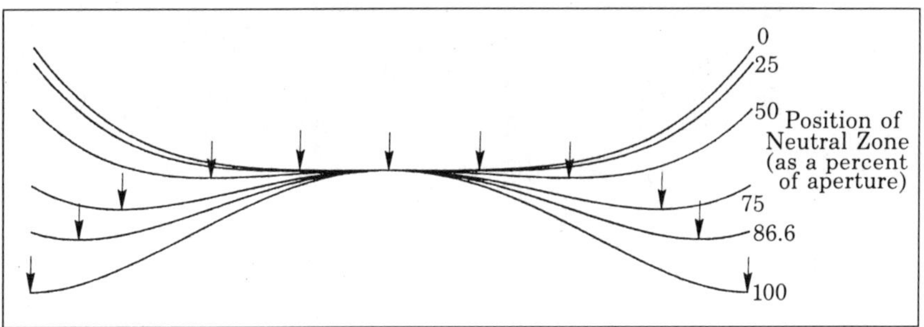

Fig. 8.3 *Profile Shapes for Schmidt Correctors.*

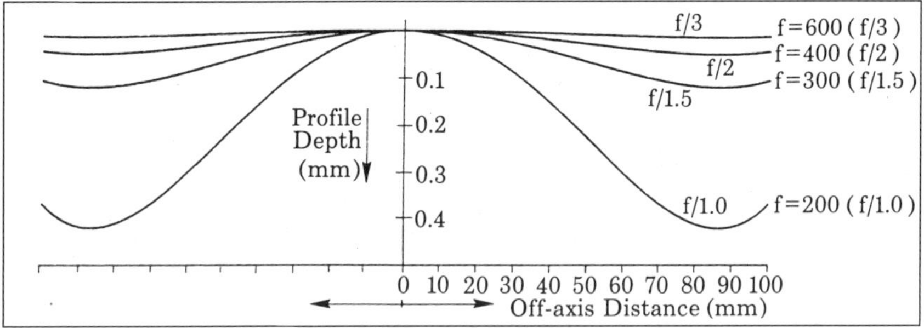

Fig. 8.4 *Corrector Profiles for 200 mm Schmidt Cameras.*

For focal ratios slower than *f*/3, the shape of the Schmidt corrector may be described with sufficient accuracy by:

$$Z = Ah^2 + Bh^4 . \tag{8.3.2}$$

For the neutral zone at 86.6% of the aperture of the corrector plate, the values of A and B are:

$$A = \frac{3D^2}{32 \cdot (n-1)R^3} \text{ and } B = -\frac{1}{4 \cdot (n-1)R^3} \tag{8.3.3}$$

where n is the design refractive index of the glass, R is the radius of curvature of the mirror, and D is the aperture of the corrector. These equations are those found most often in the literature; the designer should remember that eq. 8.3.2 is an approximation. For systems faster than *f*/3, three terms from eq. 8.3.1 must be incorporated to obtain adequate image quality. A more detailed method for determining the necessary lens shape is described in chapter 20.

Fig. 8.3 shows cross sections of six Schmidt correctors with neutral zones between 0% to 100% of the radius. The curve depths are strongly exaggerated; in reality they amount to only a few hundredths of a millimeter. Every one of these lenses enables a beam of parallel rays striking a spherical mirror to converge to a focal point. The most interesting corrector is the one with its neutral zone at 86.6%

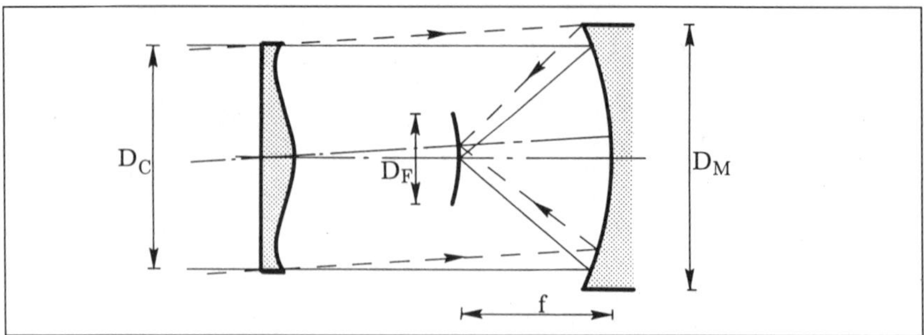

Fig. 8.5 *Main Characteristics of a Schmidt Camera.*

of the radius because its color aberration is smallest. For optimum handling of monochromatic off-axis aberrations, the neutral zone should be slightly different, but the difference is negligible for our purposes.

Fig. 8.4 shows the corrector profiles for four 200 mm aperture Schmidt cameras having focal ratios of *f*/3, *f*/2, *f*/1.5, and *f*/1. In the case of the *f*/1 camera, the maximum curve depth is roughly 0.4 mm; for the *f*/3 instrument, it is only 0.014 mm.

8.4 Characteristics of the Schmidt Camera

Fig. 8.5 shows the main characteristics of the Schmidt camera. The focus is midway between the corrector and mirror, and lies on a convex surface. Clearly, conventional cameras cannot be used with the Schmidt camera. Instead, individual pieces of film must be bent to the curve of the focal surface and placed at the focus in a special cassette. The radius of curvature of the film should be equal to half the radius of the mirror.

To prevent vignetting and light loss at the edge of the field the diameter of the mirror, D_M, should be:

$$D_M = D_C + 2 \cdot D_F \,. \tag{8.4.1}$$

The notation is that of fig. 8.5. In practice, though, some edge vignetting may be tolerated. The diameter of the mirror is often:

$$D_M = D_C + D_F \,. \tag{8.4.2}$$

Some suppliers of amateur instruments make the mirror diameter the same as the diameter of the corrector. While this may cut the cost, it causes considerable vignetting. In fig. 8.6, we show the light dropoff for a 200 mm aperture Schmidt camera for the case in which the mirror diameter is the sum of the corrector and the film diameters, and the case where the mirror diameter equals the diameter of the corrector. In fig. 8.6, the film diameter is 90 mm and the total central obstruction is 100 mm.

Because of the fast focal ratio, focusing a Schmidt camera is very demand-

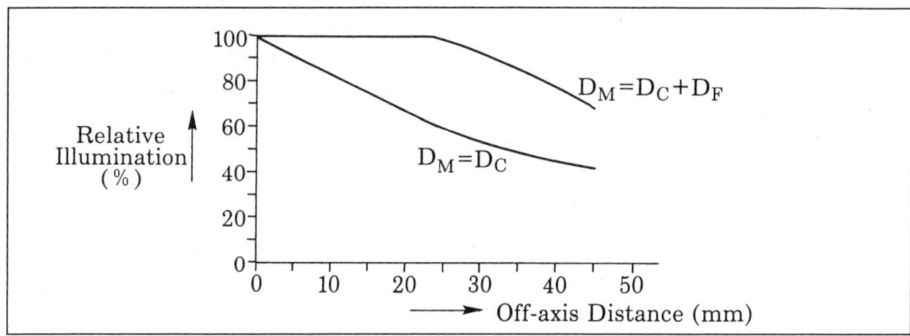

Fig. 8.6 *Vignetting in a 200 mm Schmidt Camera.*

ing. The focus tolerance seldom exceeds a few hundredths of a millimeter. In order to avoid defocusing as a result of temperature change, the filmholder should be supported longitudinally by a material with low thermal expansion such as Invar. The distance between the corrector and the mirror is not critical.

8.5 Results of Optical Ray Tracing

Spot diagrams in figs. 8.7 and 8.8 show the optical performance of four 200 mm aperture Schmidt cameras having focal ratios of *f*/3, *f*/2, *f*/1.5, and *f*/1, for three colors, red (656.27 nm), blue (486.13 nm), and violet (404.66 nm). The Schmidt corrector was designed for best correction in the blue. UBK7 glass was chosen for its relatively high transmission in the ultraviolet.

From the image blur sizes shown by the spot diagrams, the *f*/3 and *f*/2 systems give excellent image-sharpness over the whole focal surface—indeed, for the colors under consideration, the image blur is smaller than our standard photographic criterion. This is a result of the relatively small deformation of the corrector and the placement of the neutral zone at 86.6%.

In the *f*/1.5 system, color aberration at the edge of the field has become noticeable, and at *f*/1, monochromatic errors play an appreciable role in expanding the image. These focal ratios represent the effective limit of the speed of a Schmidt camera, a conclusion confirmed in the literature and in practical application. At focal ratios faster than *f*/1.5 to *f*/1, more complicated systems are used. These are usually combinations of a mirror with several spherical and aspheric lenses. It is also clear that making the corrector for an *f*/1 camera is difficult because of its large deformation. Lateral color does not occur in a Schmidt camera, since this camera is a symmetrical system with respect to all entering beams regardless of color. For that reason the spot diagrams for various colors overlap. Table 8.1 gives the construction parameters for the four Schmidt cameras we have analyzed.

8.6 The Field-Flattened Schmidt Camera

Because of the excellent image sharpness of the Schmidt camera, the inconvenience of a curved focal surface located inside the camera is generally tolerated.

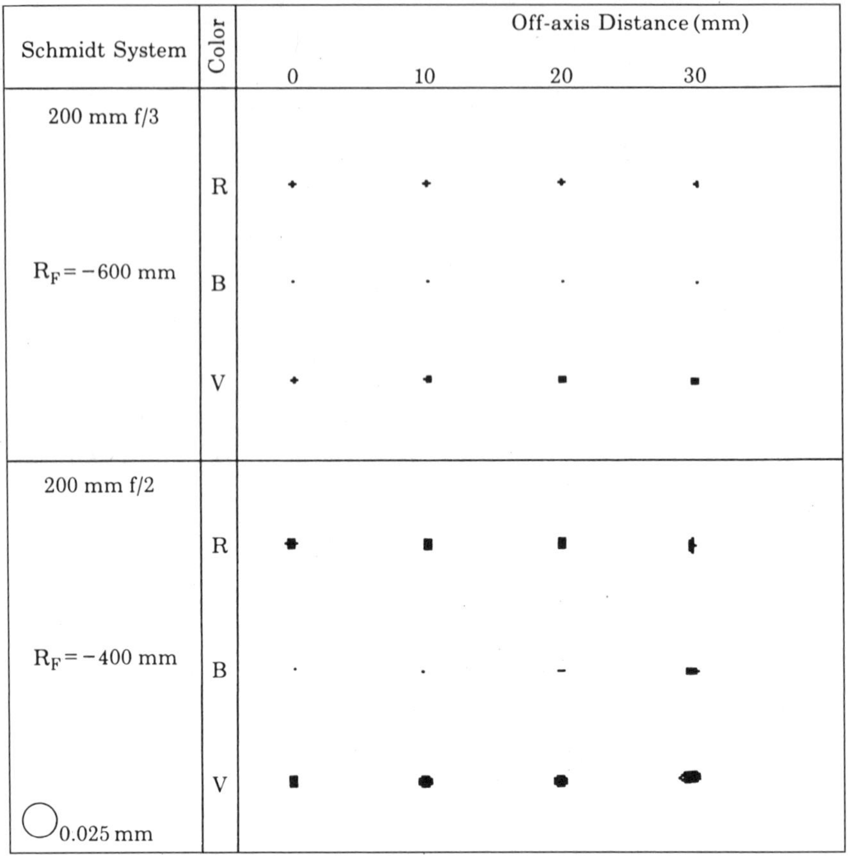

Schmidt System	Color	Off-axis Distance (mm) 0	10	20	30
200 mm f/3					
	R				
$R_F = -600$ mm	B				
	V				
200 mm f/2					
	R				
$R_F = -400$ mm	B				
○ 0.025 mm	V				

Fig. 8.7 *Spot Diagrams for Two 200 mm Schmidt Cameras.*

Table 8.1
Four Schmidt Cameras
(All dimensions in millimeters)

Geometric FR	*f*/1.0	*f*/1.5	*f*/2.0	*f*/3.0
Corr. Coeff.				
A:	$-1.13508538 \cdot 10^{-4}$	$-3.3408233 \cdot 10^{-5}$	$-1.40616246 \cdot 10^{-5}$	$-4.15958921 \cdot 10^{-6}$
B:	$6.73706618 \cdot 10^{-9}$	$2.12109462 \cdot 10^{-9}$	$9.1256521 \cdot 10^{-10}$	$2.74057192 \cdot 10^{-10}$
C:	$7.3836654 \cdot 10^{-14}$	$9.44513913 \cdot 10^{-15}$	$2.21625683 \cdot 10^{-15}$	$2.90241421 \cdot 10^{-16}$
Corr. Glass	517643	517643	517643	517643
Dist. mir./film	195.141	296.825	397.666	598.431
EFL	204.859	303.175	402.364	601.569
Mir. RC	400	600	800	1200
Dist. mir./corr.	400	600	800	1200
1° Field	3.6	5.3	7.0	10.5

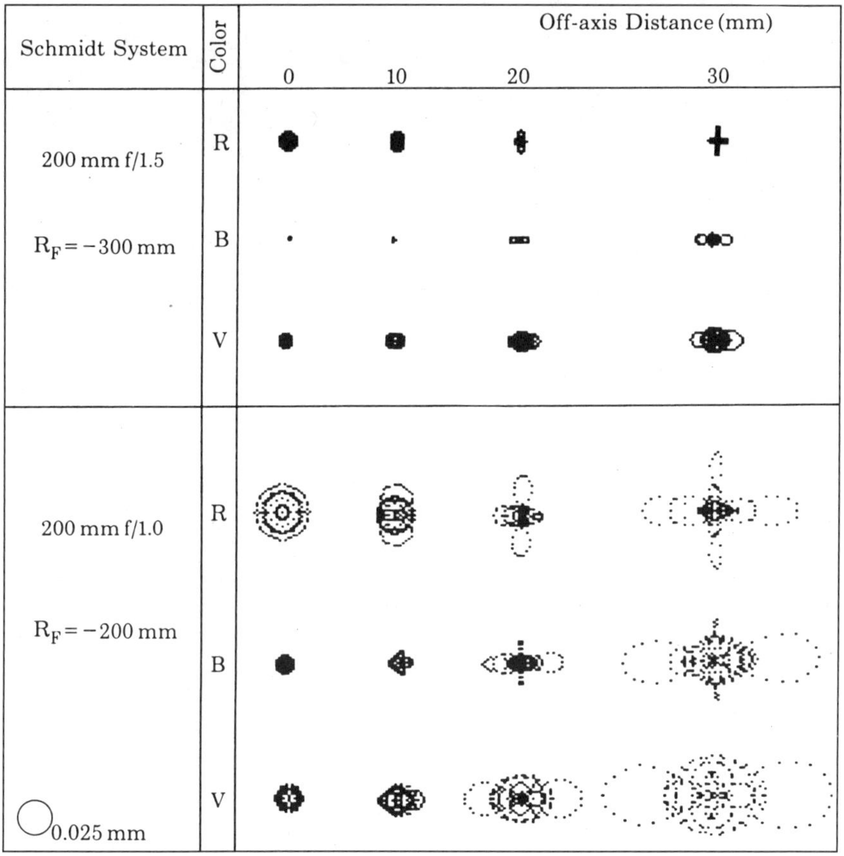

Schmidt System	Color	Off-axis Distance (mm)			
		0	10	20	30
200 mm f/1.5	R				
$R_F = -300$ mm	B				
	V				
200 mm f/1.0	R				
$R_F = -200$ mm	B				
0.025 mm	V				

Fig. 8.8 *Spot Diagrams for Two 200 mm Schmidt Cameras.*

However, it is possible to flatten the field using a single plano-convex lens placed with its plano side near or in contact with the flat film. The radius of curvature, *R*, of the convex surface of such a lens is:

$$R = \frac{f(n-1)}{n} \tag{8.6.1}$$

in which *f* is the focal length of the Schmidt camera, and *n* is the refractive index of the lens at the color for which it is designed. Although the lens introduces some coma, this may be reduced by shifting the Schmidt corrector slightly toward the mirror. The residual coma lies within tolerable limits when the speed of the system does not exceed *f/*3. Color aberrations are also somewhat increased by the field flattener; when the system is faster than *f/*3, it is necessary to design the Schmidt corrector and field flattening lens as a system. Despite their feasibility, field flattening lenses are not widely used because bending the film generally causes no trouble. In section 14.2, we discuss field flatteners in more detail.

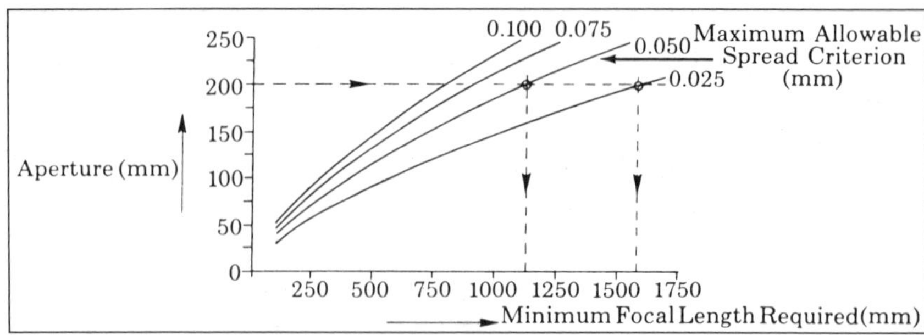

Fig. 8.9 *Performance of the Lensless Schmidt.*

8.7 The Lensless Schmidt

Many amateurs apparently hesitate to undertake a Schmidt camera because of the difficulty of making a Schmidt corrector. An alternative is the so-called "lensless Schmidt" in which the symmetry principle alone is used, and the corrector lens is omitted from the camera. The resulting system is a coma-free camera with spherical aberration. Since the amount of spherical aberration depends on the aperture and focal length, the designer can simply select a blur criterion, 0.025 mm, for example, and for a given aperture, the instrument is fully specified.

Fig. 8.9 gives a graph for determining lensless Schmidt blur figures. At our standard 200 mm aperture, meeting the 0.025 mm sharpness criterion requires a minimum focal length of 1600 mm. The focal ratio is *f*/8.0, and the tube of the system is an ungainly 3200 mm long. This length is necessary because the entrance pupil must be positioned at the center of curvature of the mirror, which is twice the focal length.

With a relaxed criterion of 0.050 mm—acceptable for 6 by 6 cm negatives—the focal length drops to 1120 mm, and the tube length is reduced to 2200 mm for the *f*/5.6 system. It is evident that to be fast, lensless Schmidts must have a small aperture. In lensless Schmidt cameras, the optimum focus is located slightly closer to the mirror than half its radius of curvature.

Chapter 9

The Schmidt-Cassegrain Telescope

9.1 General Classification

The SCT, or Schmidt-Cassegrain telescope, as offered nowadays by a variety of makers, is exceptionally popular with amateur astronomers. This is mainly due to the instrument's compactness, transportability, closed tube, and excellent color correction. Most systems offered have a focal ratio of *f*/10.

However, many users of these instruments are unaware that the standard commercial SCT is merely one member of an extensive family of Schmidt-Cassegrain systems with widely divergent characteristics. Although SCTs contain only three optical components—a primary mirror, a secondary mirror, and a Schmidt corrector—the number of configurations possible is rather large. Fig. 9.1 shows some of this potential diversity. The corrector, for example, may be placed near the focus of the primary mirror or near the center of curvature of this mirror. The first configuration is compact; the second has approximately twice the tube length. Of course, intermediate forms are also possible. The corrector may even be placed nearer the primary than the secondary mirror.

The other parameters also permit variety. The secondary may be attached to the corrector or supported by a spider. The focus may occur in front of the primary or behind that mirror. Furthermore, the mirrors may be spherical or have aspheric surfaces. Finally, there are systems designed to have a flat focal surface for photography, and those with a curved focal surface.

To a crude approximation, the design equations for the spacings and radii of Cassegrain systems also apply to the Schmidt-Cassegrain. Before continuing this chapter, therefore, the reader should review chapter 7. However, there are important differences between the two designs. The addition of the corrector lens to the two mirrors of the basic Cassegrain allows the designer to eliminate axial spherical aberration from any combination of mirrors. This is not at all the case for the off-axis aberrations. The correction of coma and astigmatism is possible only for specific combinations of power and position of the Schmidt corrector and certain shapes of the primary and secondary mirrors. We will explore the design procedure in detail in chapter 21.

However, it is pointless to discuss all possible configurations of the SCT, so we have restricted ourselves to those forms which are of practical importance to the amateur. Within the Schmidt-Cassegrain family, the most important distinc-

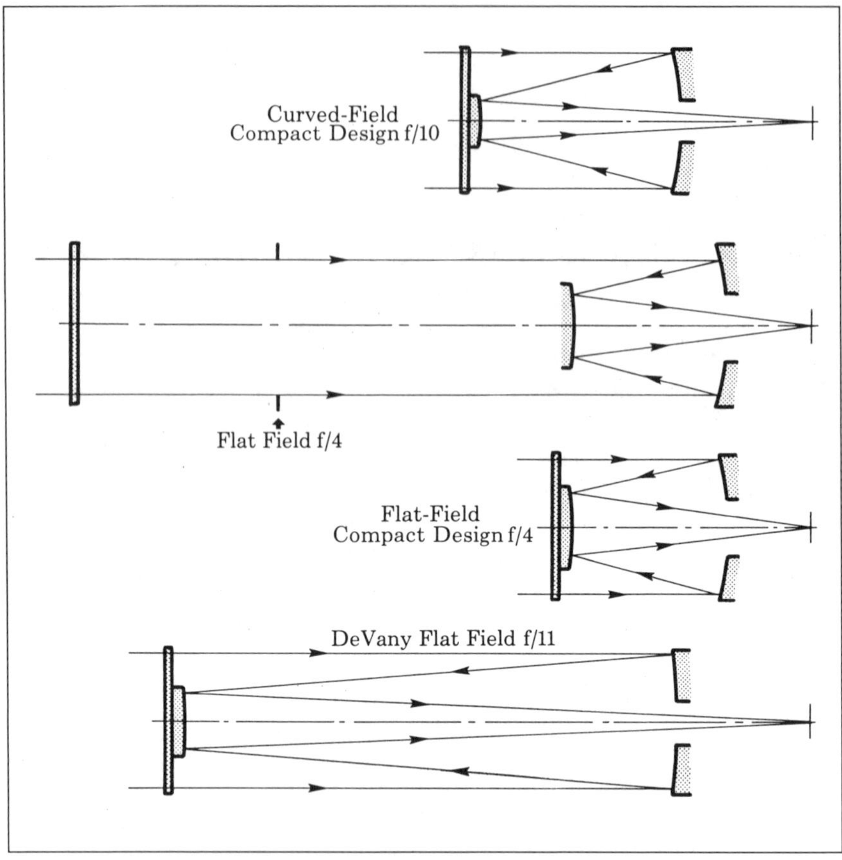

Fig. 9.1 *Layout of Four Schmidt-Cassegrain Systems.*

tion, and one that is most closely related to the needs of the amateur, lies between the systems with curved and those with flat focal surfaces.

Visual Schmidt-Cassegrain systems generally have strongly curved focal surfaces. This stems from a high secondary magnification which permits the secondary mirror to remain relatively small. Because these systems have been designed primarily for visual use, the designer must attempt to keep the diameter of the secondary mirror (or more precisely, the diameter of its holder) less than 30% the primary's diameter. Because of the curved field, these systems are not optimal for wide-field astrophotography unless a field flattener is used or the film is bent to fit the concave focal surface.

Flat-field Schmidt-Cassegrain systems are optimized for wide-field astrophotography on flat film. The flat focal surface is achieved, as it is in Cassegrains, by making the radii of curvature of the primary and secondary mirrors almost equal. As a result of this measure, the diameter of the secondary mirror becomes rather large—45% to 60% the diameter of the primary—when the focal surface

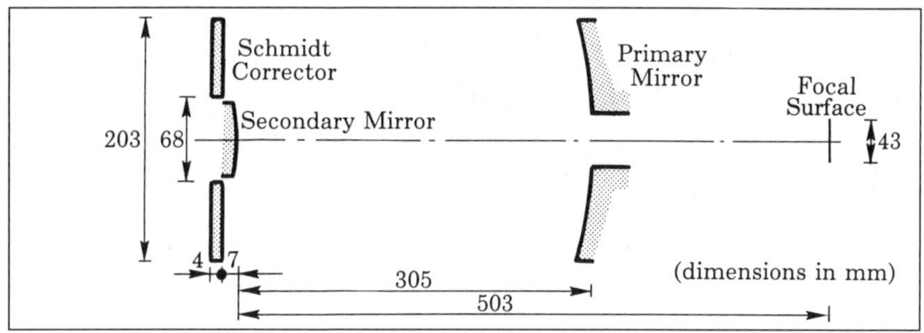

Fig. 9.2 *Dimensions of a Commercial 200 mm f/10 SCT.*

lies behind the primary mirror. These photographic systems are poor for visual use at high magnification.

9.2 Treatment of Systems

We first investigate a typical 200 mm *f*/10 system similar to well-known and popular commercial SCTs. These instruments have a curved focal surface, but offer a moderate size secondary mirror, making them suitable for visual use. In this compact design, the secondary mirror is mounted on the inside of the Schmidt corrector lens. The primary is *f*/2 and the secondary magnification is 5.

Next we discuss two typical flat-field systems with large secondary mirrors. Both have an aperture of 200 mm. The first system is a wide-field design with a focal ratio of *f*/4; the second has a slower focal ratio of *f*/11.

9.3 "Visual" Schmidt-Cassegrain Telescope

This instrument is similar to commercial 200 mm *f*/10 systems. Although the manufacturers do not supply accurate data on the positions, shapes, and radii of the optical components of their designs, it is possible to measure some of these dimensions from existing instruments. Fig. 9.2 shows the measured diameters and spacings of such a typical SCT focused at infinity. Although the exact shapes of the mirrors and Schmidt corrector are not determined, we have investigated combinations of parameters that produce a sharp axial image, and therefore have probably achieved a design close to that of the commercial systems.

We begin by investigating the performance of an SCT in which both mirrors are spherical. Fig. 9.3 shows green-light spot diagrams for such a system, on and off-axis, for flat and curved focal surfaces. The axial sharpness is excellent—the spot diagram is considerably smaller than the Airy-disk—but off-axis we find strong coma. This coma is approximately equal to that of a 200 mm *f*/5 Newtonian at the same off-axis distances (see fig. 5.6) and considerably worse for the same off-axis image angles.

It is clear that a compact Schmidt-Cassegrain with two spherical mirrors gives inadequate off-axis performance. How, then, does the designer confront the

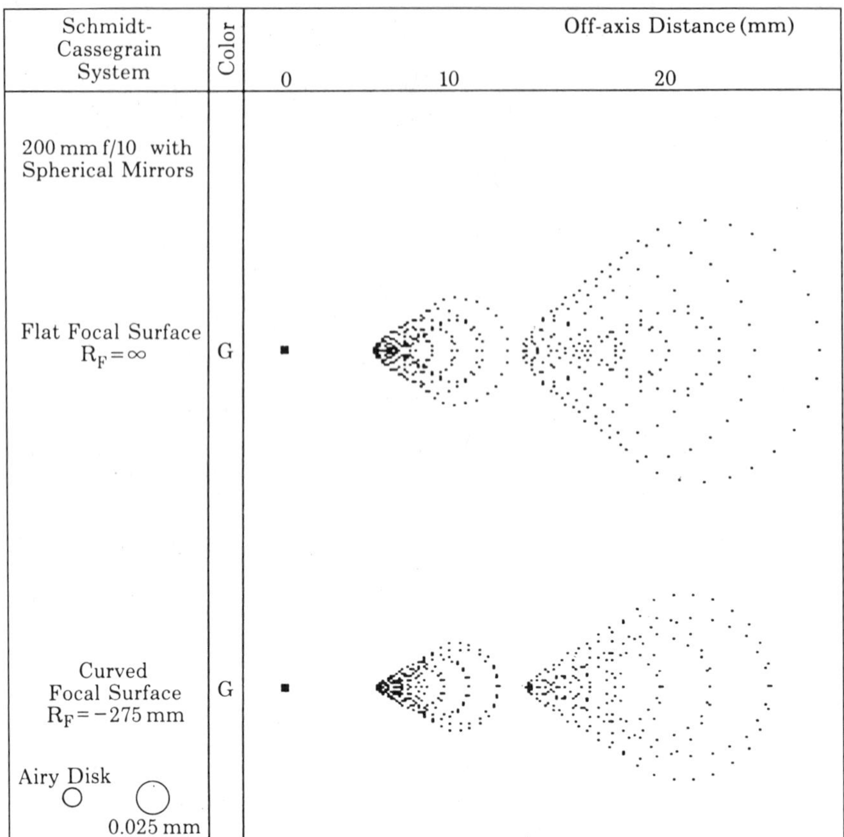

Fig. 9.3 *Spot Diagrams for a 200 mm f/10 All-Spherical Schmidt-Cassegrain Telescope.*

problem of eliminating coma while preserving good axial image sharpness? A number of possibilities exist: a) aspherizing the primary, b) aspherizing the secondary, c) both a) and b), and d) moving the Schmidt corrector, which is equivalent to abandoning the compact design. The most common solution is aspherizing the secondary. Aspherizing both mirrors has not been used in amateur instruments, at least as far as we know. Methods a) and d) will be discussed later.

Let us now examine the performance of a 200 mm *f*/10 Schmidt-Cassegrain optimized for both axial and off-axis performance by aspherizing the secondary mirror. Design data for the system are given in table 9.1; fig. 9.4 shows green light spot diagrams for both flat and curved focal surfaces. Because of the strongly curved focal surface, we felt it would be both instructive and interesting to show the off-axis image sharpness for both.

The upper row of spot diagrams (fig. 9.4) shows the image at the flat focal surface. It is evident that acceptable image sharpness is obtained only in the central 20 mm diameter if we apply the 0.025 mm criterion, and focus on the axial

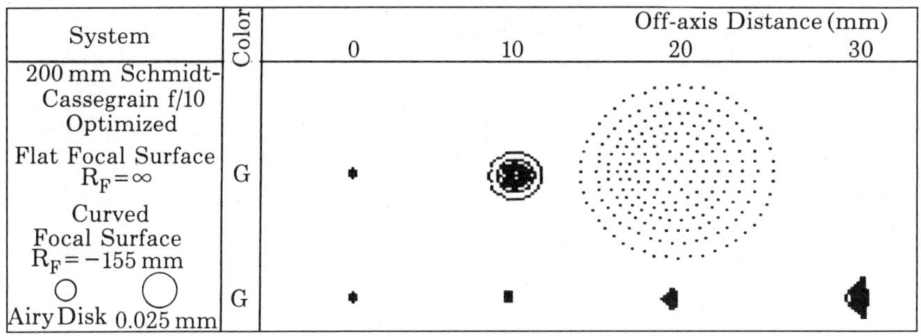

Fig. 9.4 *Spot Diagrams for an Optimized 200 mm f/10 Schmidt-Cassegrain Telescope.*

Table 9.1
Optimized 200mm *f*/10 Schmidt-Cassegrain
(all dimensions in millimeters)

Corrector	
Radius, first surface	flat
Thickness	4
Glass	517642
Paraxial Radius, second surface	–45029
Radius of Neutral Zone	86.6% of 100 mm
Relative Power	83.42%
Distance to Primary	312.5
Primary	
Radius of Curvature	–817.465
Distance to Secondary	–305.5
Secondary	
Radius of Curvature	–253.75
Deformation (sphere =0; parabola = 1)	–0.8857
Distance to focal surface	–503.06
Radius of Optimum Focal Surface	–157.6
Effective Focal Length	2027.0
1° Field	35.4

image. This diameter corresponds to the angular diameter of the full moon. Outside this circle the image swells up rapidly, causing fuzziness at the edge of the field.

The lower row of spot diagrams shows image quality at the curved focal surface. The image appears to be diffraction limited over a large part of the curved focal surface, so visual performance should be excellent. Using a film curved to match this focal surface or a field flattening lens would permit markedly better photographic performance.

In the optimized design, the power of the Schmidt corrector is less than

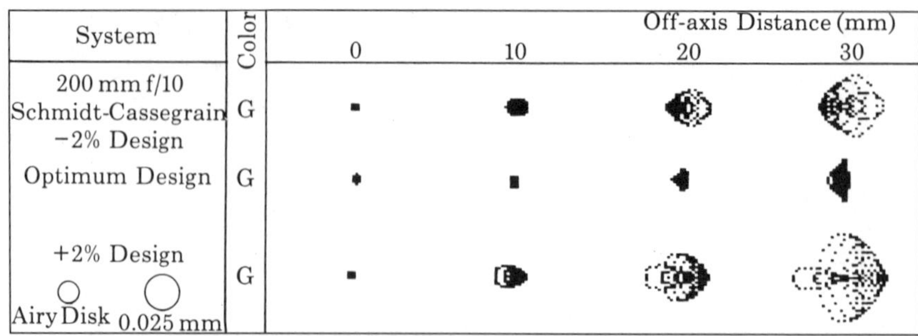

Fig. 9.5 *Spot Diagrams for Over- and Undercorrected 200 mm f/10 SCTs.*

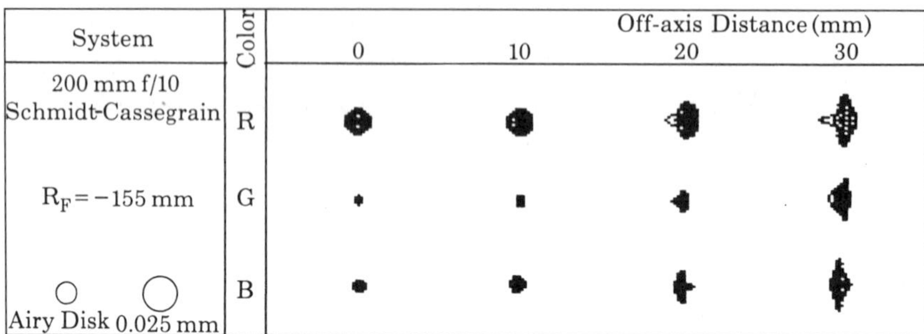

Fig. 9.6 *Color Correction for an Optimized 200 mm f/10 SCT.*

would be required to fully correct the primary mirror for spherical aberration. In this case, it is 83.42% corrected. The rest of the correction of the spherical aberration is done by the aspherical secondary mirror.

According to our analysis, the power of the corrector must be held within close tolerances because a relatively small deviation results in off-axis coma. This sensitivity is demonstrated in fig. 9.5., in which we varied the power of the Schmidt plate 2% with respect to the optimized design. The axial image sharpness appears the same in all cases since we also "refigured" the secondary to eliminate spherical aberration. The spot diagrams in fig. 9.5 are shown for the optimally curved focal surfaces. We chose the shape of the Schmidt corrector for minimum color aberration, placing the neutral zone at 86.6% of the radius.

Fig. 9.6. shows spot diagrams for red, green, and blue in a system optimized for green light. The remaining color aberration is a result of spherochromatism, i.e., the variation of spherical aberration with wavelength. This is nicely shown in fig. 9.7. All wavelengths come to the same focus at the neutral zone because at this zone the corrector has no power so the light is not refracted. At other places on the corrector, light is refracted, and the amount of refraction is different for each wavelength. In this example, color appears to be roughly five times smaller than for an *f*/15 Fraunhofer doublet refractor of the same aperture.

As we noted earlier, this instrument is intended primarily for visual use. Is there an SCT with better visual characteristics? The principal disadvantage of the

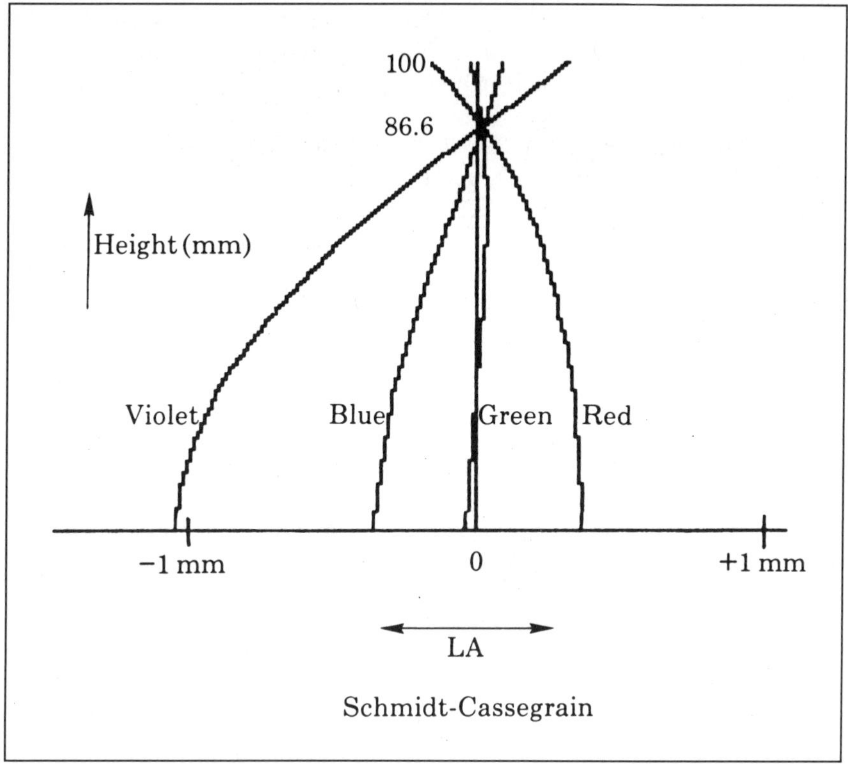

Fig. 9.7 *Spherochromatism in an Optimized 200 mm f/10 SCT.*

present design is the large central obstruction caused by the secondary mirror and its holder. In the telescope under consideration, this obstruction is 34% the diameter of the entrance pupil. This is larger than desirable, especially since observations would often involve objects with low inherent contrast, such as planetary surface details. The diameter of the secondary mirror can be reduced by making the system slower—*f*/15 or even *f*/20 instead of *f*/10. Unfortunately such changes reduce the angular field, so that the instrument becomes less versatile. The present instrument has a field of roughly one degree.

In chapter 19, we will see that the central obstruction of Cassegrain type instruments cannot be reduced arbitrarily even when a very small secondary mirror is chosen. The reason is that Cassegrain telescopes must have baffle tubes to prevent sky light from reaching the focal surface directly. Baffles make it difficult, in practice, to reduce the central obstruction of a 200 mm telescope much below 30%. In ref. 9.1, Ronald Willey presents an *f*/15 design in which the central obstruction is only slightly smaller than our design, namely 33%. When the obstruction is decreased to 25%, for example, baffling produces such strong vignetting that the useful field is substantially reduced. In chapter 19 we discuss vignetting in detail.

All in all, the optimized visual 200 mm *f*/10 SCT is not a bad compromise between the conflicting demands of small central obstruction, the focal ratio and the size of the usable field. Looking back at the lower row in fig. 9.4, we observe that the spot diagrams at off-axis distances greater than 20 mm swell somewhat as a result of residual astigmatism. This could be eliminated if the secondary magnification were increased from 5 to 6, a fact that will be explained in chapter 21. For this instrument, however, this has little practical significance because the image is already diffraction limited over a field 40 mm in diameter. This corresponds with the whole effective field diameter of this instrument, restricted as it is by the built-in rear baffle tube.

In comparing the off-axis performance of our optimized visual SCT with an equivalent *f*/10 two-mirror Classical Cassegrain, we find the latter displays off-axis spot diagrams roughly four times larger than those produced by the SCT (compare fig. 7.4 with the middle row in fig. 9.6). The reader may wonder why the SCT performs so much better. The improvement appears to be the combined effect of two changes. First, there is a slight improvement because the entrance pupil has been moved from the primary mirror to the Schmidt corrector. In very broad terms, the Schmidt corrector introduces an additional optical surface in the system. This offers the designer freedom to spread and balance the correction of aberrations over three surfaces instead of two, resulting in better overall performance.

Although we abandoned the compact all-spherical SCT earlier, it is worthwhile to explore what happens when the corrector is moved farther from the mirrors. As we will explain in chapter 21, the degree of aspheric deformation depends on the position of the Schmidt corrector. When the corrector is placed 550 mm from the primary and is given an appropriate power (72.5%), the resulting system has no coma even though both mirrors are spherical. Unfortunately we sacrifice a number of desirable properties when the corrector is moved. First, the system becomes longer. The secondary mirror, no longer attached to the corrector, must be supported with a spider, with all the problems involved. Another difficulty inherent in a longer design concerns vignetting of oblique bundles unless the primary is made larger.

We also investigated a coma-free compact system in which the primary is aspherized while the secondary is left spherical. This solution is more difficult than aspherizing the secondary because the primary is relatively large and the appropriate corrector must be stronger. This is apparently why manufacturers prefer to deform the secondary.

Earlier we mentioned mounting the secondary outside the Schmidt corrector. This would be advantageous during fabrication of the optics because the secondary mirror could be removed from its holder without disturbing the Schmidt corrector, especially for making one-of-a-kind systems. In mass-production, of course, optics are tested in permanent optical fixtures.

For the designer interested in SCTs, we recommend reading the comprehensive study by Sigler (refs. 9.2 and 9.3), which discusses a whole family of curved-

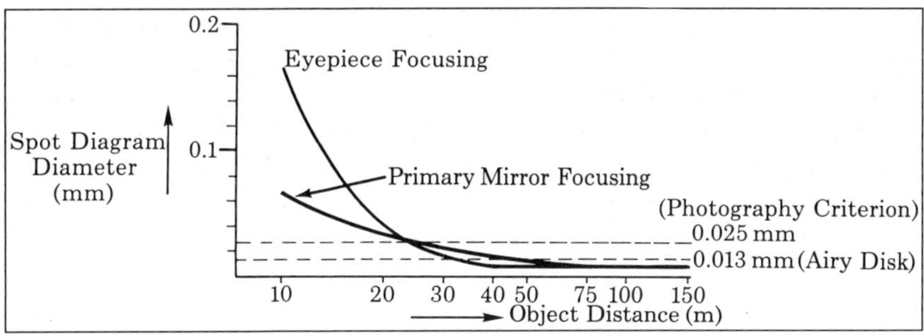

Fig. 9.8 *Comparison of Image Blur Diameters in Close-Focused SCTs.*

Table 9.2
Eyepiece Focusing versus Mirror Focusing

Obj. Distance (meters)	Eyepiece Focusing (mm)		Mirror Focusing (mm)	
	Eyepiece Movement	Equiv. Image Distance	Mirror Movement	Equiv. Image Distance
Infinity	0.00	2027	0.00	2027
1000	4.12	2040	0.163	2027
100	42.93	2178	1.675	2027
50	94.5	2336	3.354	2029
30	175	2680	5.598	2031
10	1200	5930	16.87	2037

focal surface SCTs. In this study, Sigler explores the optical characteristics as several parameters are varied. We have applied several of Sigler's formulae in chapter 21 in the section on Schmidt-Cassegrain Design.

9.4 Close Focusing in the SCT

With commercial Schmidt-Cassegrain telescopes, the ability to view objects at short range is a valuable asset. Studying and photographing birds, flowers, and insects at short range with a 200 mm telescope offers impressive potential for marketing and sales. While it is possible to correct spherical aberration for a given object distance, it will be corrected only for that distance. Since the instrument has been optimized for viewing objects at infinity, we have examined the loss in image sharpness at short range.

There are two ways of focusing on near objects. In the first, focusing is accomplished by moving the primary mirror; in the second, by moving the eyepiece. Fig. 9.8. graphs the image spread of a point source for the two methods. For object distances less than 20 meters, mirror focusing gives better images than eyepiece focusing. However, at such short ranges, neither system offers diffraction-limited images.

Table 9.2 indicates that there is another rather subtle disadvantage connected with eyepiece focusing when observing at short range: the equivalent image distance, which determines the image scale, grows considerably. This also means that the image brightness decreases. Moreover, the eyepiece movement becomes excessive, and vignetting of the light cone from the secondary by the baffle tube results in further light loss. It is clear that most manufacturers of SCTs have chosen the more expensive mirror focusing in order to make their instruments suitable for short range observing.

9.5 Flat-Field Schmidt-Cassegrain Systems

Baker, Linfoot, and others realized the possibility of designing flat-field Schmidt-Cassegrain systems (ref. 9.4) if the radii of curvature of both mirrors were made equal. In order to avoid coma and astigmatism, however, at least one of the mirrors had to be aspheric.

Slevogt developed an alternative in which both mirrors remain spherical (ref. 9.5). His design places the corrector somewhat outside the center of curvature of the primary, while giving the secondary slightly more power than the primary. The slight difference between their powers compensates the weak power of the Schmidt plate. In this way an astrocamera with a large, flat field having no coma and no noticeable residual astigmatism is obtained.

Fig. 9.1 shows the optical configuration, and table 9.3. contains optical data, for an instrument based on our optimization of the original Slevogt design. The instrument has an aperture of 200 mm and a focal ratio *f*/4. A major disadvantage of this design is the relatively long tube. Furthermore, when no preventive measures are taken, oblique bundles are seriously vignetted. Vignetting also depends on the position of the limiting stop in the instrument. When it coincides with the corrector, a large primary is necessary to intercept oblique bundles. A further disadvantage of this particular design is that the diameter of the secondary mirror is quite large.

In our version of the Slevogt instrument, we adopt a compromise: placing a 200 mm diameter stop 300 mm behind the corrector. This results in a primary diameter approximately equal to that of the corrector, though both are larger than the stop. An optical ray trace of the system shown in fig. 9.9 establishes that this design meets the 0.025 mm criterion over the entire 60 mm diameter flat field under consideration, and that lateral color is within the same limit as well.

To control vignetting in the Slevogt *f*/4 flat-field astrocamera with spherical mirrors, we balance the light loss due to the secondary obstruction against the vignetting caused by the large secondary mirror. The secondary's diameter is 110 mm, or 55% of the aperture. This causes a light loss in the center of the field of 30%. We chose the diameters of the mirrors in such a way that the light dropoff at the edge of a 6 by 6 cm negative is 35%, thereby illuminating the field almost uniformly. The effective focal ratio of the system is *f*/4.8, some 20% slower than the *f*/4 geometric focal ratio.

In a system designed for visual observation, such a large obstruction would

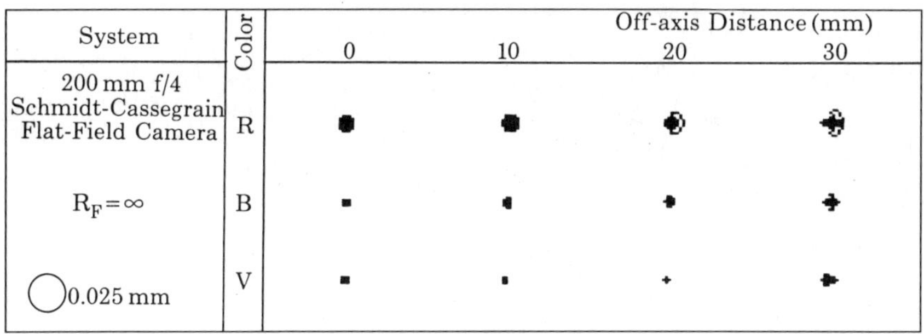

Fig. 9.9 *Spot Diagrams for a 200 mm f/4 Schmidt-Cassegrain Flat-Field Camera.*

Table 9.3
Data 200mm *f*/4 Schmidt-Cassegrain Flat-Field Camera
(All dimensions in millimeters)

Corrector	
Radius, first surface	flat
Thickness	4
Glass	517643
Paraxial Radius, second surface	–83250
Radius of Neutral Zone	86.6% of 100 mm
Relative Power	50.5%
Distance to Primary	977.2
Primary	
Radius of Curvature	–845.6
Distance to Secondary	–231.6
Secondary	
Radius of Curvature	–809.7
Distance to Focal Surface	358.285
Radius of Focal Surface	flat
Effective Focal Length	800
Geometric Focal Ratio	*f*/4
Diameters	
Schmidt Corrector	216
Stop	200
Primary Mirror	220
Secondary Mirror	110
Distance, stop to corrector	300
1° Field	14.0

be intolerable because too much light from the Airy disk would spread into the diffraction rings and cause excessive loss of contrast. For a photographic system, the situation is more favorable. A 55% central obstruction produces a star-image in

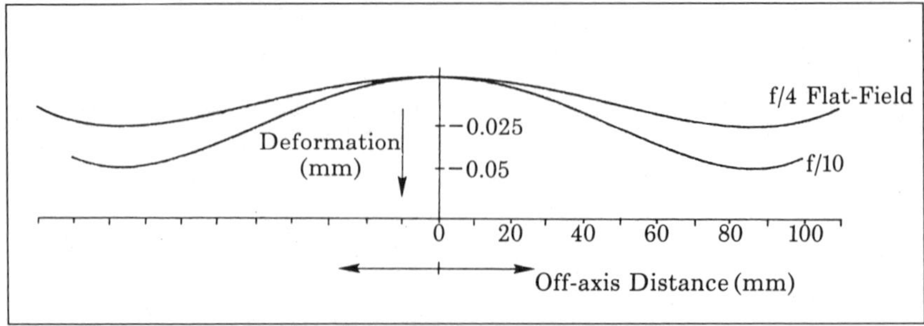

Fig. 9.10 *Profile Depths for Two Schmidt Correctors.*

which the Airy disk and the first diffraction ring together contain more than 80% of the light. The outside diameter of the first diffraction ring for an *f*/4 system is 0.011 mm. This figure is so much lower than our 0.025 mm photographic criterion that very nearly 100% of the light available will be concentrated within this circle. From a photographic point of view, a 55% obstruction cannot be considered detrimental to the sharpness of photographic images.

A significant advantage of the large secondary mirror is that this particular instrument can be built without a baffle system. This occurs because the large secondary, in conjunction with the long tube, prevents stray light from reaching the film.

From comparing the flat-field Schmidt-Cassegrain camera with the original Schmidt camera, it is evident that we must pay heavily for the comfort of having an easily accessible flat focal surface. The system is considerably more complex than the Schmidt camera.

Offsetting the complexity, however, is the possibility of using the flat-field camera as a coma-free visual telescope. By replacing the large secondary with a smaller, slightly aspheric secondary having stronger curvature, we convert it to visual use. The corrector and the primary mirror remain in their original positions; the new secondary is farther from the primary mirror. The focus moves back, leaving enough space for installing a focuser and eyepieces. Such a dual system can be designed rather quickly using computer programs, but, of course, the corrector must be smoothly aspherized to visual tolerances.

One final note on the practicality of the *f*/4 camera: in fig. 9.10 we compare the corrector profiles of the two Schmidt-Cassegrains discussed. Note that the diameter of the corrector for the *f*/4 system is somewhat larger than that of the *f*/10 system. Having seen good photographic images obtained over a flat field with a long Schmidt-Cassegrain camera, we next ask whether it is possible to design a compact camera, such as the one shown in fig. 9.1. The instrument would offer important advantages over the long design. The secondary would not be supported by a spider, and edge vignetting of oblique bundles would be less serious in a compact design. Unfortunately such a compact design does not appear to be feasible for most ATMs (see ref. 9.7). Although it is optically possible to make such a sys-

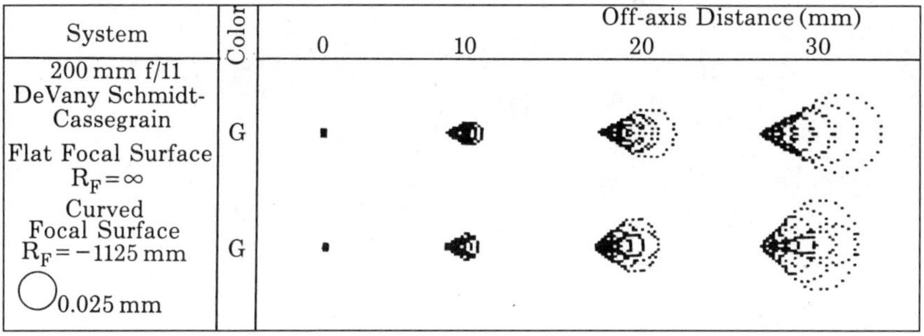

System	Color	Off-axis Distance (mm)			
		0	10	20	30
200 mm f/11 DeVany Schmidt-Cassegrain Flat Focal Surface $R_F = \infty$	G				
Curved Focal Surface $R_F = -1125$ mm	G				
0.025 mm					

Fig. 9.11 *Spot Diagrams for a 200 mm f/11 DeVany Schmidt-Cassegrain Flat-Field Camera.*

tem, it requires strongly deformed aspheric mirrors and a Schmidt plate with deep curves. These factors conspire to place the task beyond amateur skills, so we have omitted the optical data for this design.

In 1965, A. S. DeVany published the optical data of an *f*/11 compact design flat-field Schmidt-Cassegrain telescope with two spherical mirrors having the same radius of curvature (ref. 9.6), with the further desirable feature that the secondary mirror can be attached to the corrector. As ray tracing reveals, however, this system suffers from stronger coma than the equivalent Newtonian (compare figs. 5.6 and 9.11). On the positive side, the tube length of the DeVany system is less than 50% that of the Newtonian.

9.6 Computer-Aided Design

Computer-aided design of Schmidt-Cassegrain telescopes with both curved and flat focal surfaces is discussed in chapter 21. Computer programs are particularly useful if the designer wishes to investigate a variety of possible designs.

Chapter 10

The Maksutov Camera

10.1 Introduction

Soon after the success of the Schmidt camera became known, opticians tried to design a similar camera in which the aspheric Schmidt corrector was replaced by a corrector that could be more easily manufactured. By the beginning of the 1940s, researchers all over the world were working on the problem. The solution they found was a meniscus corrector with spherical surfaces---a strongly-bent negative lens of weak optical power.

A meniscus lens nearly eliminates the spherical aberration of a spherical mirror, or at least suppresses it, because it introduces spherical aberration opposite that of a spherical mirror. The idea occurred roughly at the same time, around 1940, to Maksutov in Russia, to Bouwers in the Netherlands, to Gabor in England, and to Penning in Germany. Because the invention was first published by Maksutov (ref. 10.1.), meniscus cameras are usually called Maksutov cameras.

As it has turned out, the manufacture of meniscus correctors for large astrocameras (i.e., those larger than 500 mm aperture) is no easier or less expensive than making a Schmidt corrector. This is one reason why meniscus correctors are not often used in large astronomical cameras; for relatively small amateur systems, however, the meniscus remains an important optical component.

10.2 Maksutov Camera Designs

Designers distinguish between two kinds of meniscus camera systems: those that are concentric, and those that are non-concentric. Fig. 10.1 shows a concentric system. All the radii, R_1, R_2, and R_3, as well as the radius of the focal surface, have the same center of curvature, *C*. The entrance pupil lies at the center of the system, in front of the meniscus. In this way every entering bundle, whether it is parallel to the mechanical axis or not, has its own optical axis of symmetry with respect to the mirror and the meniscus. Because of this symmetry coma and astigmatism do not occur. Furthermore, spherical aberration can be eliminated almost completely by a meniscus with the proper thickness, radii of curvature, and position.

The concentric system has one major disadvantage: because of the relatively thick meniscus, it suffers from longitudinal chromatic aberration. Bouwers, while emphasizing the advantages of the concentric system (ref. 10.2),suggests achromatizing the corrector by using two kinds of glass with the same refractive index

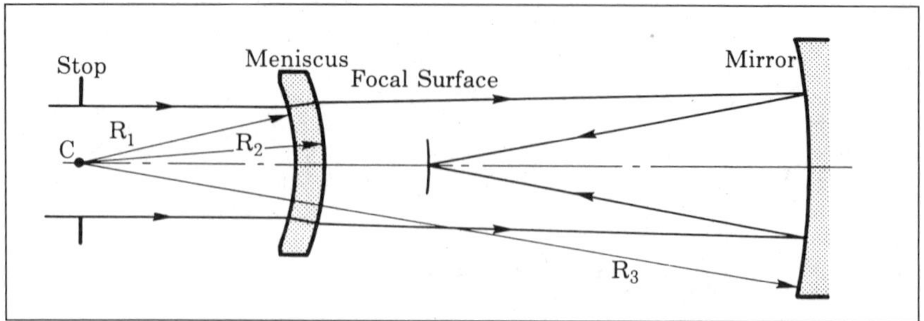

Fig. 10.1 *Layout of a Concentric Meniscus Camera.*

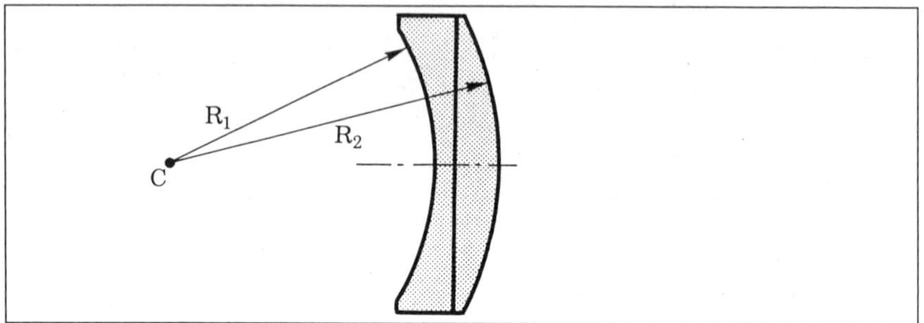

Fig. 10.2 *Achromatized Concentric Meniscus Corrector.*

at the design wavelength but different dispersions as shown in fig. 10.2. This method, unfortunately, is too cumbersome for most amateurs.

Maksutov found another method of achromatizing the corrector. He showed that the chromatic aberration of a meniscus for paraxial rays reaches a minimum when the axial thickness of the corrector equals:

$$t = \frac{n^2}{n^2 - 1} \cdot (R_2 - R_1) \tag{10.2.1}$$

where n is the refractive index of the color the corrector is designed for. For BK7 glass, with $n_F = 1.52237$, the axial thickness becomes:

$$t = 1.76(R_2 - R_1)\,. \tag{10.2.2}$$

But since:

$$t = (R_2 - R_1) \tag{10.2.3}$$

for a concentric meniscus, Maksutov's condition means that the meniscus is no longer concentric, and that its thickness is no longer equal at all places. Such a corrector is shown in fig. 10.3. Because the center of curvature is not at the same position for both optical surfaces, the system cannot be made concentric, so both coma and astigmatism occur, as well as lateral colors. Coma can be eliminated to

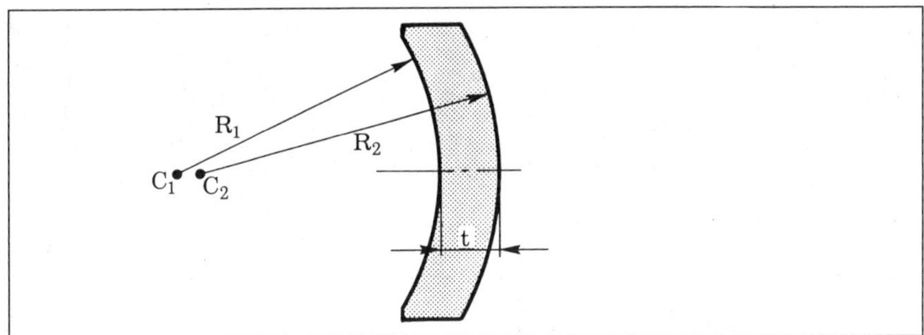

Fig. 10.3 *Non-concentric Meniscus Corrector.*

Table 10.1
Four 200mm Maksutov Cameras
(All dimensions in mm)

Focal Ratio	*f/2*	*f/2.5*	*f/3*	*f/4*
R_1 Radius of Curvature	−193.4	−224	−252.8	−305.6
T_1 Axial Distance	20	20	20	20
M_1 Medium	517643	517643	517643	517643
R_2	−205.4	−235.8	−264.6	−317.2
T_2	489	631	776	1078
M_2	Air	Air	Air	Air
R_3	−833	−1037	−1240.8	−1646.4
T_3	−427.5	−530.98	−634.65	−839.97
M_3	Air	Air	Air	Air
Effective Focal Length	400	500	600	800
1° Field	7.0	8.7	10.5	14.0
Axial Blur (Blue)	0.033	0.025	0.010	0.006
Axial Blur (Red/Violet)	0.035	0.030	0.010	0.006

a large extent by moving the corrector somewhat toward the mirror.

Maksutov even went one step further and placed the entrance pupil at the corrector instead of at the center of curvature of the mirror, but this leads to more coma and astigmatism. He could suppress this coma by moving the corrector still closer to the mirror. The result is a camera more compact than the concentric configuration, having a low chromatic aberration but some residual astigmatism. The Maksutov design has a tube length approximately 1.3 times the focal length, instead of twice the focal length as in the concentric design. It is also shorter than a comparable Schmidt camera.

A further advantage of the non-concentric system is that the diameters of the corrector and the mirror are smaller than they are for a concentric system of equal aperture. In addition, the corrector is relatively thick and strong. In amateur-size instruments of this design, it is possible to support the film holder on a bolt passing

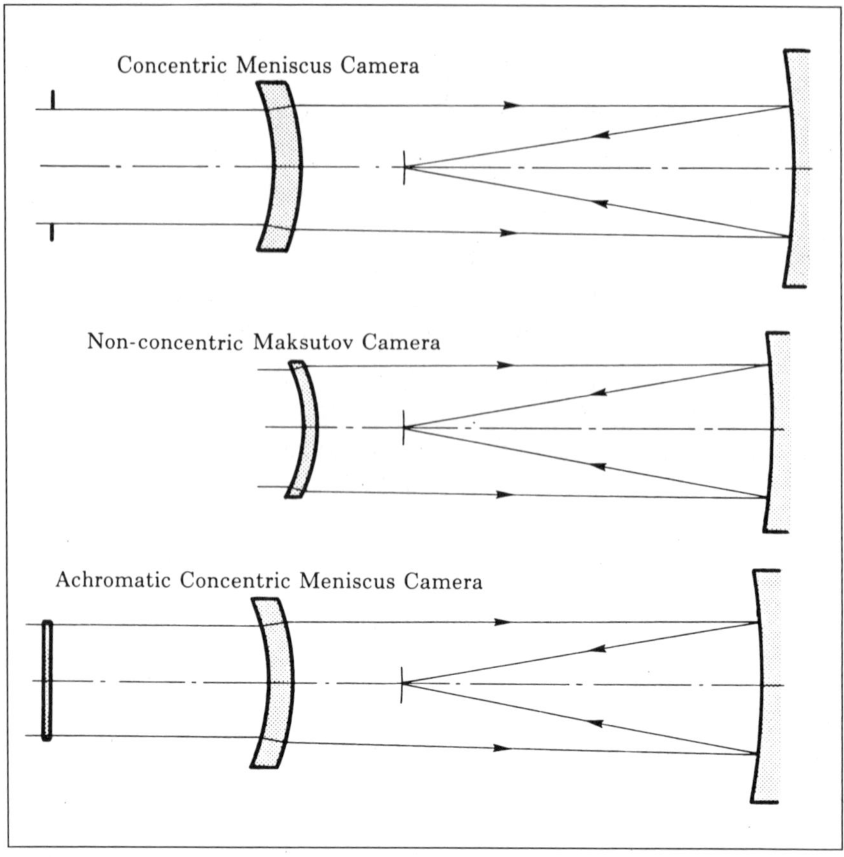

Fig. 10.4 *Layouts for Meniscus Cameras.*

through a hole in the center of the meniscus. This design, therefore, does not require a spider.

Table 10.1 gives data for 200 mm aperture *f*/2, *f*/2.5, *f*/3, and *f*/4 non-concentric cameras designed and optimized by Maksutov. For the corrector, he chose UBK7 glass; the axial thickness of the meniscus is $D/10$ (i.e., 20 mm for our standard 200 mm aperture) for all focal ratios. The table also lists the diameters of the axial image blur in blue light, an indication of the residual spherical aberration, and the combined image blurs for red and violet, a measure of axial color aberration, but gives no information on coma, astigmatism, or lateral color.

To determine these aberrations, we investigated the off-axis performance of three meniscus cameras, each having an aperture of 200 mm and a focal ratio of *f*/3. The first is the original Bouwers concentric system with single meniscus corrector; the second a non-concentric system of the type discussed above, by Maksutov; and the third, a concentric system color corrected by placing a weak positive lens at the entrance pupil.

Table 10.2
Optical Characteristics of
Three 200mm *f*/3 Meniscus Cameras
(All dimensions in mm)

System	Concentric Bouwers	Non-concentric Maksutov	Achromatic Concentric
Focal Ratio	*f*/3	*f*/3	*f*/3
R_1 Radius of Curvature	−382.686	−252.8	17311
T_1 Axial Distance	42.286	20	10.16
M_1 Medium	517643	517643	517643
R_2	−424.972	−264.6	−17311
T_2	843.601	776	380.97
M_2	Air	Air	Air
R_3	−1268.573	−1240.8	−380.98
T_3	−668.237	−634.65	40.637
M_3	Air	Air	517643
R_4			−421.607
T_4			797.489
M_4			Air
R_5			−1219.11
T_5			−619.224
M_5			Air
Effective Focal Length	600	600	600
1° Field	10.5	10.5	10.5

Table 10.2 lists the optical parameters and fig. 10.4 shows the three systems at the same scale. The correctors in the concentric systems are approximately 40 mm thick, in accord with Bouwers' proposal. The non-concentric system has a corrector with an axial thickness of 20 mm. It is evident that the concentric systems require longer tubes, and larger mirrors and correctors.

The optical ray traces shown in fig. 10.5 allow the amateur to decide whether it is worthwhile to construct the more complicated color-corrected concentric system or the non-concentric system by Maksutov. The spot diagrams were calculated for the optimum curved focal surface. Note the large color aberration present in the concentric Bouwers camera designed for the blue. The Maksutov has some astigmatism and significant lateral color. Only the achromatic concentric design meets our 0.025 mm criterion for photography over the whole field and in all colors. The optical performance of this system is roughly comparable with the equivalent *f*/3 Schmidt camera shown in fig.8.7.

10.3 The Optimum Meniscus Corrector

As we noted above, in his 1944 paper Maksutov points out that a corrector that

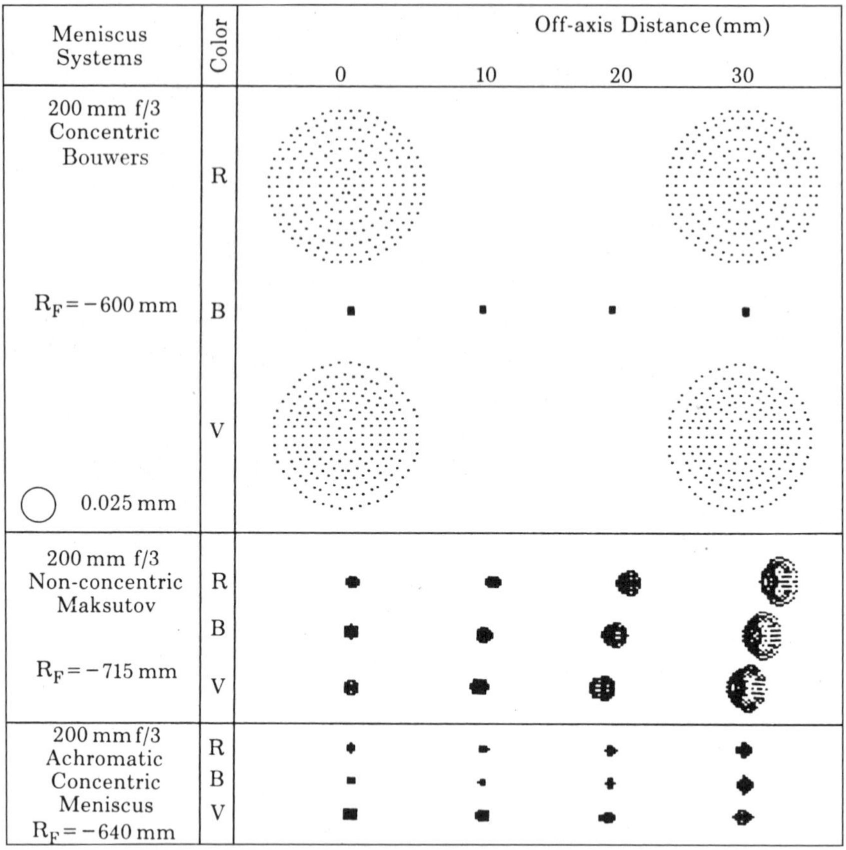

Meniscus Systems	Color	Off-axis Distance (mm) 0	10	20	30
200 mm f/3 Concentric Bouwers	R				
$R_F = -600$ mm	B				
0.025 mm	V				
200 mm f/3 Non-concentric Maksutov	R				
	B				
$R_F = -715$ mm	V				
200 mm f/3 Achromatic Concentric Meniscus	R				
	B				
$R_F = -640$ mm	V				

Fig. 10.5 *Spot Diagrams for Three 200 mm Meniscus Cameras.*

satisfies the condition:

$$t = \frac{n^2}{n^2 - 1} \cdot (R_2 - R_1) \tag{10.3.1}$$

will be paraxially achromatic. This means that rays close to the optical axis will have approximately the same focus for different colors. However, this is not the case for rays near the outer zones. As a result of spherochromatism, longitudinal color aberration occurs in the outer zones.

To minimize the overall influence of spherochromatism, the designer may wish to choose the axial thickness and the curvatures of the meniscus in such a way that the LA curves for red, blue, and violet intersect in the 70% zone. Table 10.1 indicates that Maksutov computed the thickness from:

$$t = 1.70 \cdot (R_2 - R_1) \tag{10.3.2}$$

rather than:

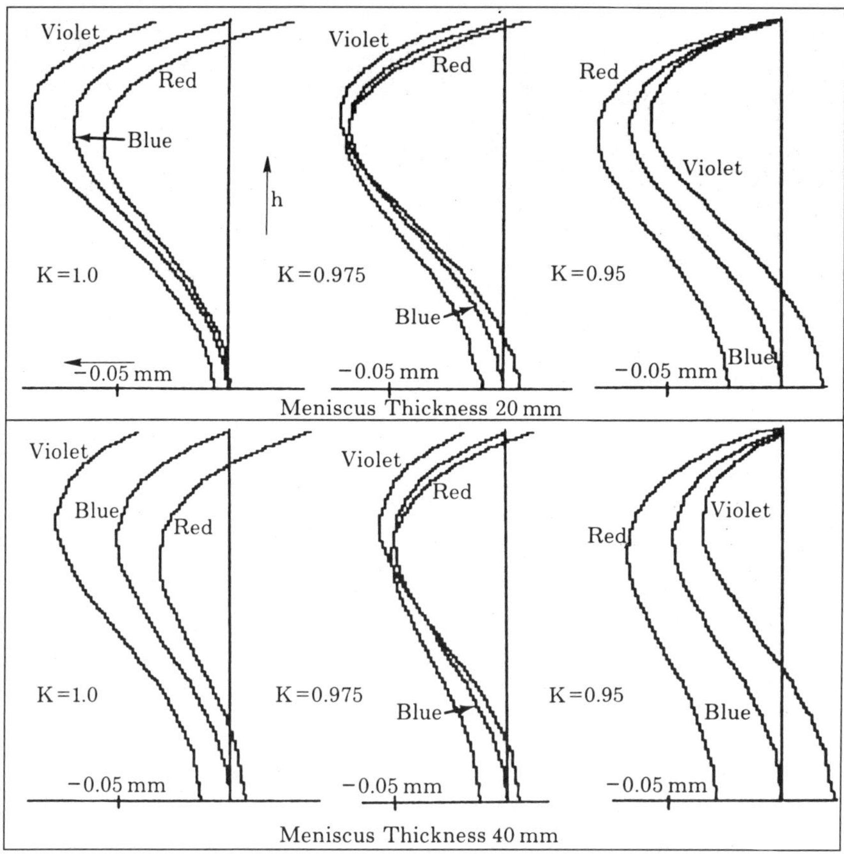

Fig. 10.6 *Spherochromatism of Non-concentric Meniscus Correctors.*

$$t = 1.76 \cdot (R_2 - R_1) . \quad \textbf{(10.3.3)}$$

Apparently he introduced a correction factor, k:

$$k = \frac{1.70}{1.76} = 0.97 \quad \textbf{(10.3.4)}$$

We investigated what correction factor is optimum for a 200 mm *f*/3 Maksutov camera. In order to determine how this factor depends on the thickness, we traced two thicknesses, 20 mm and 40 mm, and three values of K, 0.95, 0.975, and 1.00. Each of the resulting six correctors was then individually optimized. The position of each corrector with respect to the mirror was chosen to minimize coma.

For purposes of comparison, we chose radii of curvature for both corrector surfaces so that using UBK7 glass in blue light, the paraxial ray and the edge ray have the same focus. The radii were back-calculated from the assumed axial thickness using the relation:

Table 10.3
Five 200mm f/3 Maksutov Meniscus
(All dimensions in mm)

Thickness	10	20	30	40	50
R_1 RC	–213.48	–247.98	–270.59	–287.49	–301.07
T_1 Axial Dist.	10	20	30	40	50
M_1 Medium	517643	517643	517643	517643	517643
R_2	–219.34	–259.70	–288.17	–310.94	–330.38
T_2	825	788	750	710	685
M_2	Air	Air	Air	Air	Air
R_3	–1226.27	–1240.57	–1253.04	–1262.53	–1272.06
T_3	–623.101	–634.954	–644.88	–652.773	–660.377
M_3	Air	Air	Air	Air	Air

$$t = k \cdot \frac{n^2}{n^2 - 1} \cdot (R_2 - R_1) \qquad \textbf{(10.3.5)}$$

where k took values of 0.95, 0.975, and 1.00.

Fig. 10.6 shows the LA-curves for the six cases in three colors. We conclude that optimum color correction for both the 20 and the 40 mm thick correctors occurs for $k = 0.97$, in accord with Maksutov's results. It is clear that when $k = 1.0$, the paraxial focal distances of the three colors coincide, while for $k = 0.95$, the edge rays coincide.

Maksutov's choice for corrector thickness, *D*/10, has been followed by many telescope makers. This thickness is arbitrary and is certainly not optimum from an optical point of view. In later years various publications (refs. 10.4, 10.5, and 10.6) indicated that better correction of spherical aberration is obtained with thicker menisci, but that these suffer from worse lateral color.

In order to investigate the influence of the axial thickness on the residual spherical aberrations, we designed correctors 10 mm through 50 mm thick for a 200 mm *f*/3 Maksutov camera. Table 10.3 gives data for this camera with the different corrector thicknesses. These systems are not optimized for minimum blur. Correctors thinner than 10 mm are difficult to make without springing or cracking them, while thicknesses greater than 50 mm are too heavy. Each of the five correctors was individually optimized.

It appears that as the thickness increases, the corrector can be less strongly curved and closer to the mirror. From the graph of longitudinal spherical aberration shown in fig. 10.7, it is evident that a thin corrector exhibits considerable aberration, but that for thicknesses over 30 mm, the benefit from increasing the thickness is marginal.

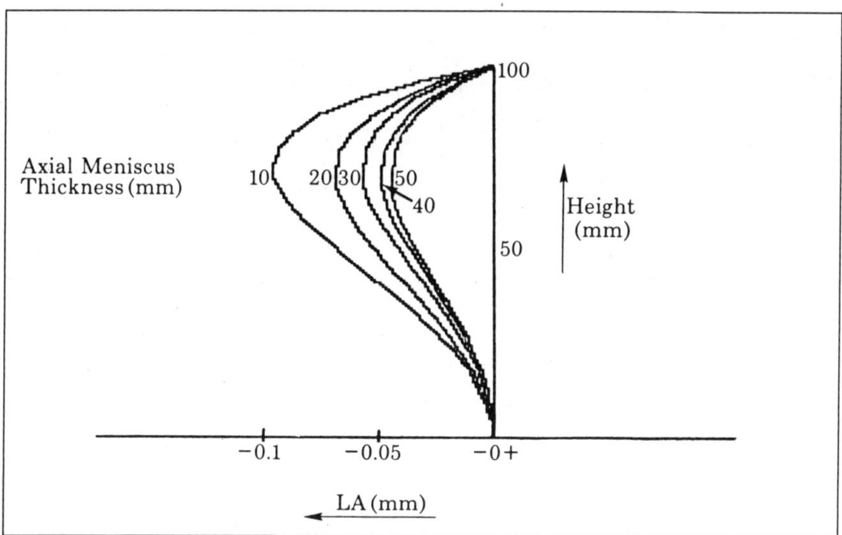

Fig. 10.7 *Longitudinal Spherical Aberration in a 200 mm f/3 Non-concentric Maksutov Camera for Various Meniscus Thicknesses.*

Chapter 11

The Maksutov-Cassegrain Telescope

11.1 Introduction

In their papers published in the forties, Maksutov and Bouwers suggested that the meniscus camera could be transformed into a Cassegrain system. But it was not until 1957, when John Gregory published his now-classic design for the *f*/23 meniscus telescope which bears his name, that these catadioptric instruments became available to amateurs (ref. 11.1). The primary mirror in the Gregory Maksutov-Cassegrain is spherical; the secondary is an aluminized spot on the back side of the meniscus corrector. It, too, is spherical. Like the Schmidt-Cassegrain, the Gregory Maksutov is short and closed, and offers good color correction and freedom from a spider.

In 1958, Gregory improved the color correction of his original design, also making possible an *f*/15 design (ref. 11.2). In the *f*/23 instrument, all the optical surfaces are spherical, but in the *f*/15 system, either the corrector or the primary mirror must be aspherized to eliminate residual spherical aberration. In the years following, modifications of the original Gregory design have been published.

This chapter describes the Gregory Maksutov and designs that have followed it, including systems optimized for astrophotography, and evaluates their optical performance in visual and photographic applications. As is our practice, all six instruments have an aperture of 200 mm.

11.2 Maksutov-Cassegrain Systems

Fig. 11.1 shows six representative Maksutov systems at the same scale. These six instruments offer an excellent opportunity to discuss an important optical concept: degrees of freedom. A degree of freedom means the ability to make a free choice with respect to an optical parameter such as a radius of curvature, a distance between two surfaces, a lens thickness, a glass type, an aspheric surface, and so forth. To correct the image aberrations in an optical system, the optical designer requires as many degrees of freedom as the number of image aberrations to be corrected, or more.

Consider now the Gregory Maksutov-Cassegrain system. Note that this instrument has strong coma and astigmatism, as shown in fig. 11.2. In this particular form, there is no possibility of correcting these aberrations because the secondary mirror is part of the back side of the corrector—and therefore necessarily has the

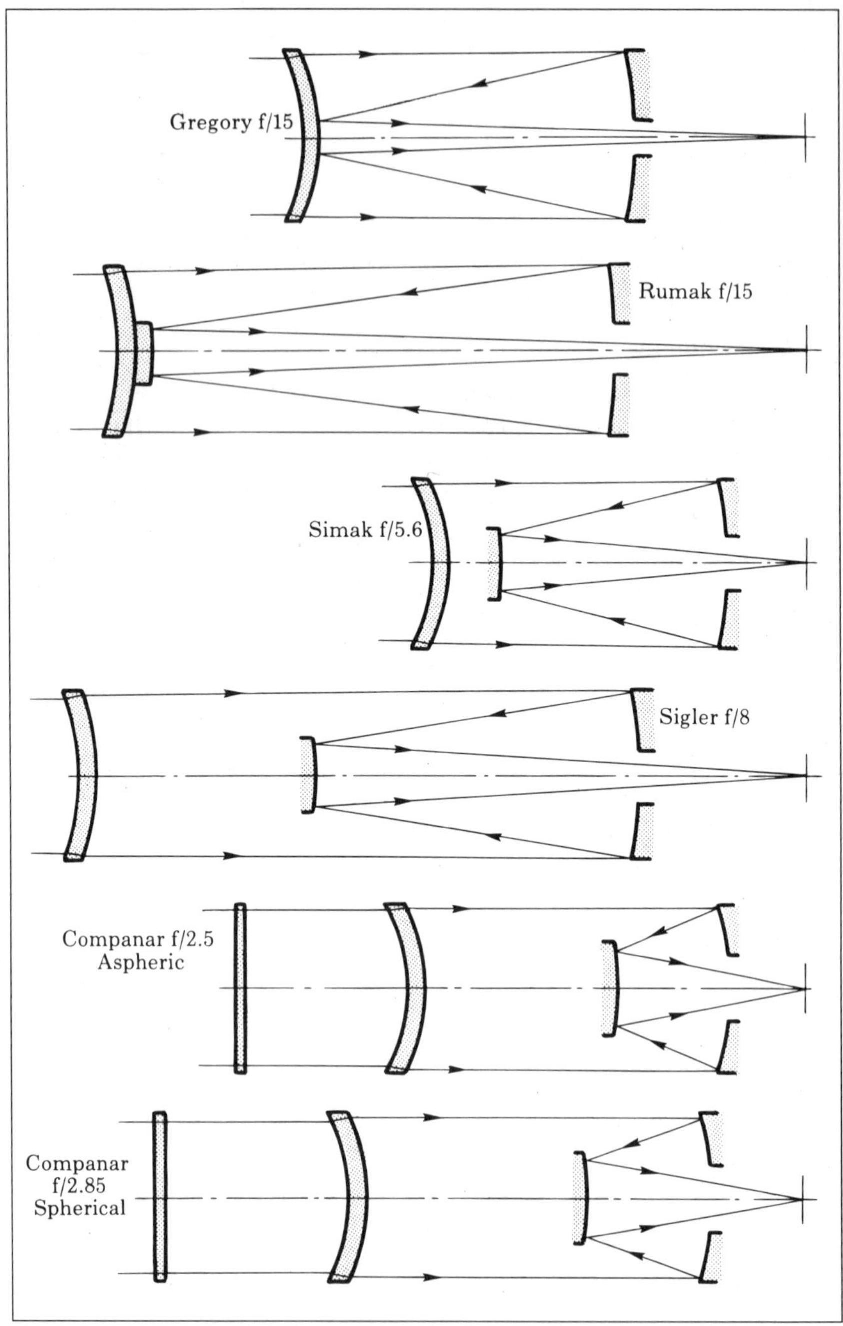

11.1 *Layout of Six Maksutov-Cassegrain Systems.*

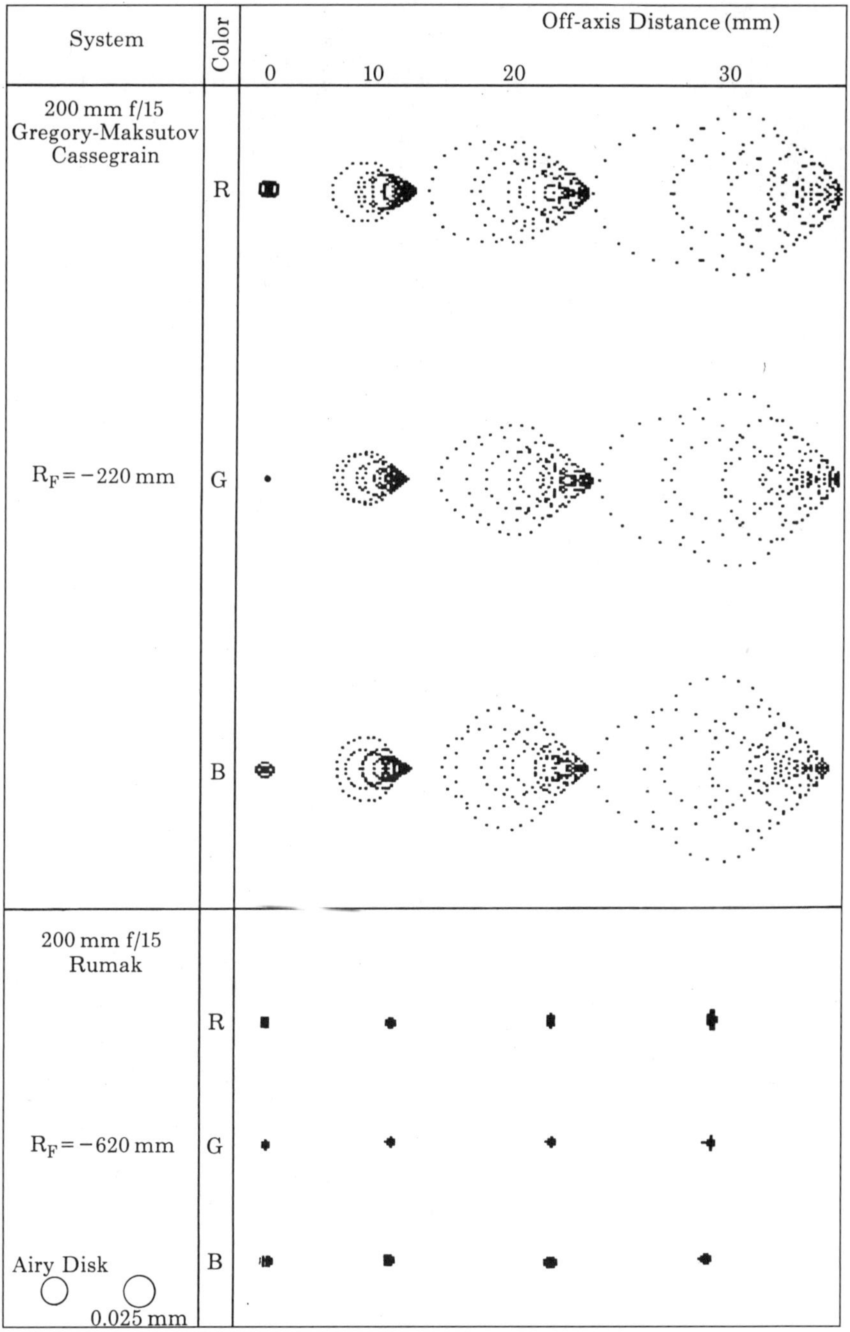

System	Color	Off-axis Distance (mm) 0	10	20	30
200 mm f/15 Gregory-Maksutov Cassegrain	R				
$R_F = -220$ mm	G				
	B				
200 mm f/15 Rumak	R				
$R_F = -620$ mm	G				
Airy Disk 0.025 mm	B				

Fig. 11.2 *Spot Diagrams for the Gregory-Maksutov and Rumak.*

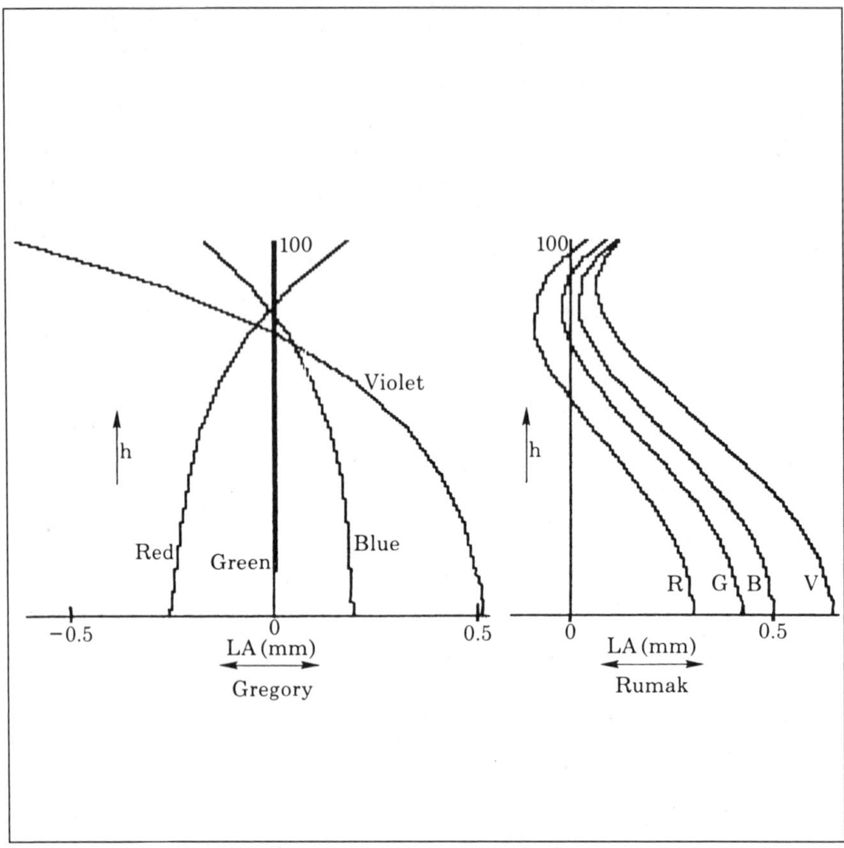

Fig. 11.3 *Spherochromatism in Two 200 mm f/15 Maksutov-Cassegrain Systems.*

same radius of curvature. In order to improve the off-axis aberrations of this system, the designer must have at least one additional degree of freedom, such as independently changing the radius of curvature of the secondary mirror. Although one manufacturer of Maksutov-Cassegrain telescopes uses a design of this sort—grinding the center of the corrector's back surface to a longer radius than the corrector itself—this method is too difficult for most amateurs. A simpler solution is to attach a separate convex mirror to the corrector.

Fig. 11.1 shows such a system, the Rumak, designed by Harrie Rutten. The Rumak is an *f*/15 system, like the Gregory Maksutov, so we can make a direct comparison of their optical performance (see fig. 11.2). The all-spherical Rumak, with one extra degree of freedom, gives considerably better off-axis images, improved color correction, and a flatter focal surface.

Because the Gregory and the Rumak are primarily visual instruments, the spot diagrams are shown for red, green, and blue only. We see that both instruments appear to be diffraction limited on-axis. However, fig. 11.3 shows curves

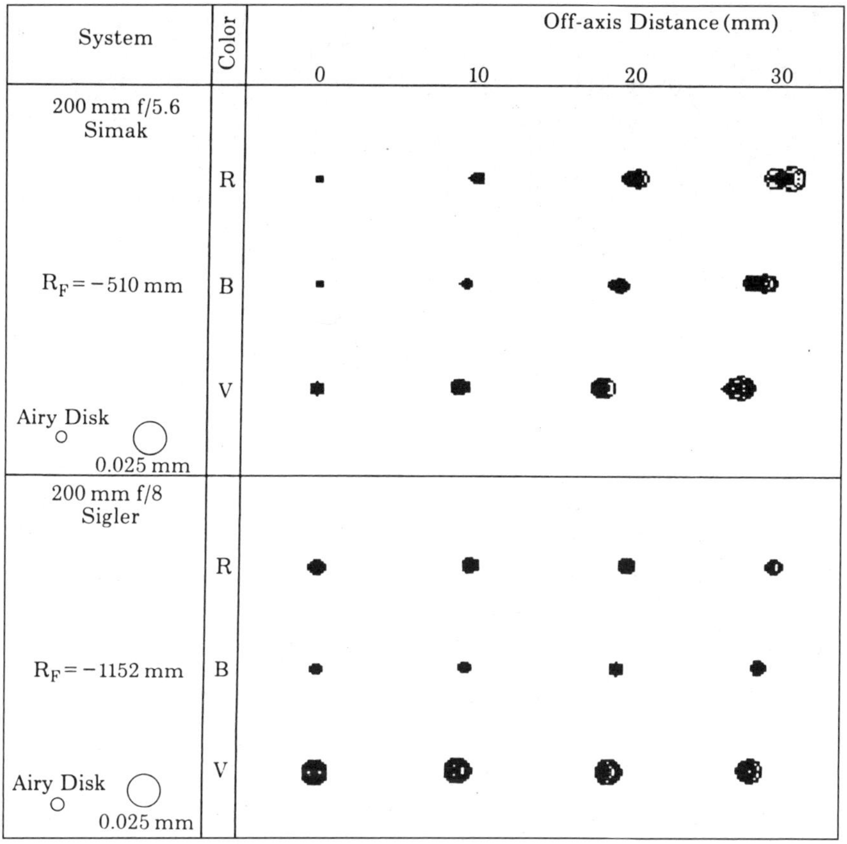

System	Color	Off-axis Distance (mm) 0	10	20	30
200 mm f/5.6 Simak	R				
$R_F = -510$ mm	B				
Airy Disk 0.025 mm	V				
200 mm f/8 Sigler	R				
$R_F = -1152$ mm	B				
Airy Disk 0.025 mm	V				

Fig. 11.4 *Spot Diagrams for the Simak and Sigler Maksutov Systems.*

for the spherical aberration in four colors, including violet light which is necessary for photography. The Gregory design suffers from considerable spherochromatism. The Rumak's aberration curves lie closer together, particularly in the outer zones, where it most matters. The improvement of the Rumak is largely due to the less steeply-curved meniscus.

Of course, the Rumak also has its disadvantages: the tube is longer and the secondary mirror is larger than those of the Gregory. Would it be possible to design a Rumak-like system with the same tube length as the Gregory? It is possible, but the designer needs still another degree of freedom. Aspherizing one optical surface—corrector, primary, or secondary mirror—is sufficient, but the difficulty of figuring aspheric optical surfaces renders this possibility unattractive.

However, other degrees of freedom may be exploited. In specifying a fast Maksutov-Cassegrain (i.e., faster than *f/*8) in which the optical surfaces must be left spherical, the designer may alter the distance between the corrector and the secondary mirror. Two examples of this are the Simak (ref. 11.3) and the Sigler

Table 11.1a
Three 200 mm Maksutov Cassegrain Systems
(All dimensions in millimeters)

System	Gregory	Rumak	Simak
Focal Ratio	*f*/15	*f*/15	*f*/5.6
R_1 Radius of Curvature	–219.413[a]	–334.956	–217.465[a]
T_1 Axial Distance	17.347	20	19.425
M_1 Medium	517642	517642	517642
R_2	–229.6	–346.673	–229.025
T_2	403.653	605.8	355.575
M_2	Air	Air	Air
R_3	–980.453	–1592.2	–847.2
T_3	–403.653	–589.7	–289.675
M_3	Air	Air	Air
R_4	–229.6	–598.072	–451.75
T_4	620.679	832.12	388.222
M_4	Air	Air	Air
R_5			
T_5			
M_5			
R_6			
T_6			
M_6			
Effective Focal Length	3000	2964	1122
1° Field	52.4	51.7	19.6

a. aspheric deformation approx. 1/4-wavelength

Maksutov (ref. 11.4). Both are shown in fig. 11.1. In these instruments, the corrector was moved with respect to the secondary mirror.

The limiting focal ratio for a Maksutov-Cassegrain with three spherical optical components appears to be approximately *f*/4. The aberration of faster systems becomes intolerable when only spherical surfaces are used.

The designer of the *f*/2.5 Companar, Klaas Compaan, suppressed color aberrations in his concentric meniscus system by placing a weak positive lens in front of the corrector. In order to eliminate residual spherical aberration, this lens was made slightly aspheric. An entirely spherical *f*/2.5 design would have an axial image blur of 0.045 mm; aspherizing the lens is unnecessary for focal ratios slower than *f*/2.85.

The optical characteristics of the six Maksutov systems are given in table 11.1a and b, while their spot diagrams are shown in figs. 11.2, 11.4, and 11.5.

Table 11.1b
Three 200 mm Maksutov Cassegrain Systems
(All dimensions in millimeters)

System	Sigler	Companar	Companar
Focal Ratio	*f*/8.0	*f*/2.5	*f*/2.85
		(aspheric)	(spherical)
R_1 Radius of Curvature	–297.709	11070[a]	12300
T_1 Axial Distance	20.6	10.0	11.1
M_1 Medium	517642	517642	517642
R_2	–309.455	–11070	–12300
T_2	693.582	210.0	233.3
M_2	Air	Air	Air
R_3	–1311.818	–210.0	–233.3
T_3	–413.473	18	20.1
M_3	Air	517643	517643
R_4	–847.091	–228.0	–253.4
T_4	631.75	388.175	439.88
M_4	Air	Air	Air
R_5		–520.84	–584.13
T_5		–140.0	–155.56
M_5		Air	Air
R_6		–500.68	–562.44
T_6		240.283	275.19
M_6		Air	Air
Effective Focal Length	1595	504	572
1° Field	27.8	8.8	10.0

a. aspheric deformation approx. 1/4-wavelength

11.3 Meniscus Correctors

The meniscus correctors used in the Gregory, Rumak, Simak, and Sigler are all designed for low color aberration. As we saw in chapter 10, for an optimal color correction of paraxial rays, the corrector thickness is:

$$t = \frac{n^2}{n^2 - 1} \cdot (R_2 - R_1) . \tag{11.3.1}$$

For BK7 glass this becomes:

$$t = 1.76(R_2 - R_1) . \tag{11.3.2}$$

Gregory's original 1957 design and Sigler's instrument conform, at least approximately, to this value. In Gregory's improved 1958 design, and for the Rumak

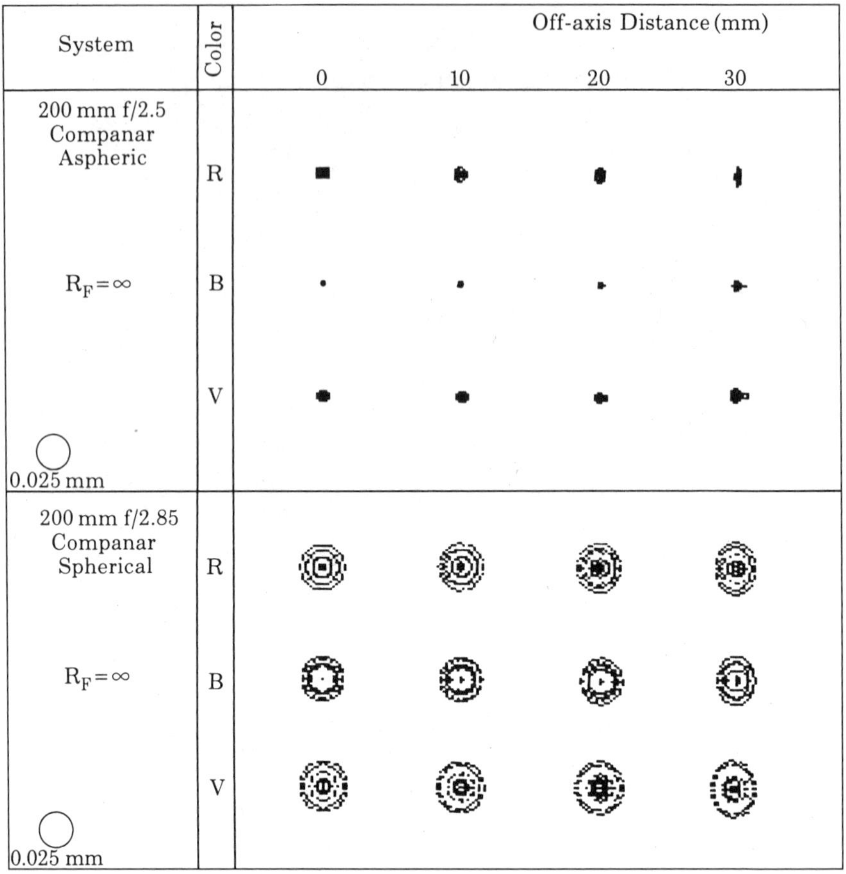

System	Color	Off-axis Distance (mm)			
		0	10	20	30
200 mm f/2.5 Companar Aspheric	R				
$R_F = \infty$	B				
0.025 mm	V				
200 mm f/2.85 Companar Spherical	R				
$R_F = \infty$	B				
0.025 mm	V				

Fig. 11.5 *Spot Diagrams for Two Companar Flat-Field Cameras.*

and Simak, however, the corrector thickness is:

$$t = 1.70(R_2 - R_1)\,. \tag{11.3.3}$$

The slightly thinner corrector offers better overall color correction than the paraxially designed corrector, as explained in the previous chapter. In order to minimize coma and lateral color, the meniscus of the Companar has been made fully concentric with respect to the entrance pupil, so that this meniscus satisfies:

$$t = (R_2 - R_1)\,. \tag{11.3.4}$$

Although the meniscus thickness causes longitudinal color aberration, this is eliminated by placing a weak positive lens in the entrance pupil.

11.4 Curved- and Flat-Field Maksutov-Cassegrain

Of the Maksutov-Cassegrain telescopes we have examined, all except the Companar have a curved focal surface. The Companar was designed solely for photographic use as an astrocamera; the other instruments can also be used visually. For visual use, the diameter of the secondary should be kept as small as possible. Table 11.1 does not list the diameters of the primary and secondary mirrors for the visual Maksutov-Cassegrain telescopes because they depend not only on the size of the desired field and the baffling employed, but also on the amount of vignetting and diameter of central obstruction that the user will accept. The relationships between these parameters are discussed in detail in chapter 19. With the guidelines given there, the amateur can readily determine the dimensions required for his own design.

As we said above, the Companar has been designed exclusively for photographic use. Its flat field results from making the radii of curvature of both mirrors nearly equal. The design unfortunately requires a secondary mirror 120 mm in diameter—60% of the entrance pupil. At the center of the field, approximately 36% of the light is blocked, so the effective focal ratio is some 20% slower than the geometric focal ratio. In order to prevent excessive vignetting of oblique bundles of rays in the extreme corners of a 6 by 6 cm negative, the meniscus must be slightly larger and the primary mirror considerably larger than the entrance pupil. For zero edge vignetting, the secondary should actually have a larger diameter. When the secondary mirror is restricted to 60%, a light dropoff of 35% occurs at the extreme edge.

A 60% central obstruction is far less detrimental to the photographic image quality than it would be in a visual system. It produces a star image in which the Airy disk and the first diffraction ring together contain approximately 80% of the light reaching the focal surface. At *f*/2.5, the outside diameter of the first diffraction ring is 0.007 mm, so much smaller than our photographic criterion that nearly 100% of the light available will fall within our 0.025 mm blur criterion. From the point of view of diffraction, the instrument's 60% obstruction is not detrimental to the photographic image sharpness of this instrument.

When we compare this flat-field Maksutov-Cassegrain design with the original Maksutov camera, it is evident that the convenience of an accessible, flat focal surface necessitates a considerably more complex system. In the Sigler and Companar designs, the secondary mirrors must be supported with a spider; in the other instruments, the secondary is attached to the corrector. In one respect, however, the Companar is simpler: it can be built without a baffle system because the large secondary mirror prevents stray light from reaching the film directly.

How do the Maksutov and Schmidt-Cassegrain systems compare? In the Maksutov, the elements must be slightly larger than in the Schmidt. Because the Maksutov corrector is a negative lens, rays passing through it are bent outward, so the primary mirror, in order to catch all rays, must have a larger diameter than the corrector, even for the axial bundle. The outer rays in a Schmidt corrector are bent slightly outward, too, but in a Maksutov corrector this effect is stronger.

How does the Maksutov-Cassegrain stack up against the Schmidt-Cassegrain? We may compare the image quality of the flat-field Maksutov-Cassegrain with an *f*/4 Schmidt-Cassegrain (described in section 9.5, spot diagrams in fig. 9.9) with the spot diagrams for the aspheric Companar shown in fig. 11.5. Both are exceptional in their performance, but neither instrument is easy to build. Only experienced workers will be able to overcome the difficulties involved in making these instruments to the required accuracy. These astrocameras are intended for people who wish a wide-field instrument with a truly flat focal surface that is easy of access, and who demand the very best photographic image quality over a wide spectral range.

Chapter 12

The Schiefspiegler

12.1 Introduction

The simplest Schiefspiegler is a two-mirror reflecting telescope that offers, as its most desirable feature, an obstruction-free light path. This is accomplished by tilting the system's primary mirror so that the secondary mirror does not block incoming light. Although tilting a single mirror results in strong coma and astigmatism, by the proper choice of radii and figure, one aberration can be corrected and the other reduced to an acceptable low value in small telescope systems.

Conventional reflecting telescopes such as the Newtonian and Cassegrain, and all the various catadioptrics, have a central obstruction in their light path due to the secondary mirror. In the Newtonian and the Cassegrain these mirrors are supported with a spider. In the Schmidt- and Maksutov-Cassegrain designs, the secondary may be attached to the corrector. The obstruction of the mirror and spider diffracts light from the Airy disc into the diffraction rings, resulting in loss of image contrast. This can be observed particularly in objects having a low inherent contrast, such as planetary surface details. In chapter 18, these effects will be discussed in detail.

In the 1950s, Anton Kutter, a German, designed various mirror-systems which do not have these obstructions. He named them schiefspieglers, or "oblique reflectors." Kutter was not the first to investigate unobstructed mirror systems. William Herschel had, over 150 years earlier, tilted the mirror of his largest telescope in order to be able to observe outside the entering bundle of light. Although Herschel's motives had mainly to do with avoiding light loss due to the low reflectivity of his speculum-metal mirror, tilted primary systems are still called Herschelian telescopes.

The first two-mirror telescope without an obstruction was the Brachyt, or "broken," telescope. This system consists of an off-axis part of a parabolic primary and an off-axis part of a hyperbolic secondary. Both mirrors have the same optical axis. The Brachyt is an eccentric part of a conventional Cassegrain telescope. Because the mirrors are difficult to make, these instruments have never become popular.

Schiefspiegler systems are called tilted-component telescopes, or TCTs for short. Owners of the Kutter schiefspieglers and the later designs by Buchroeder (ref. 12.8) are enthusiastic about the contrast and image sharpness of their instru-

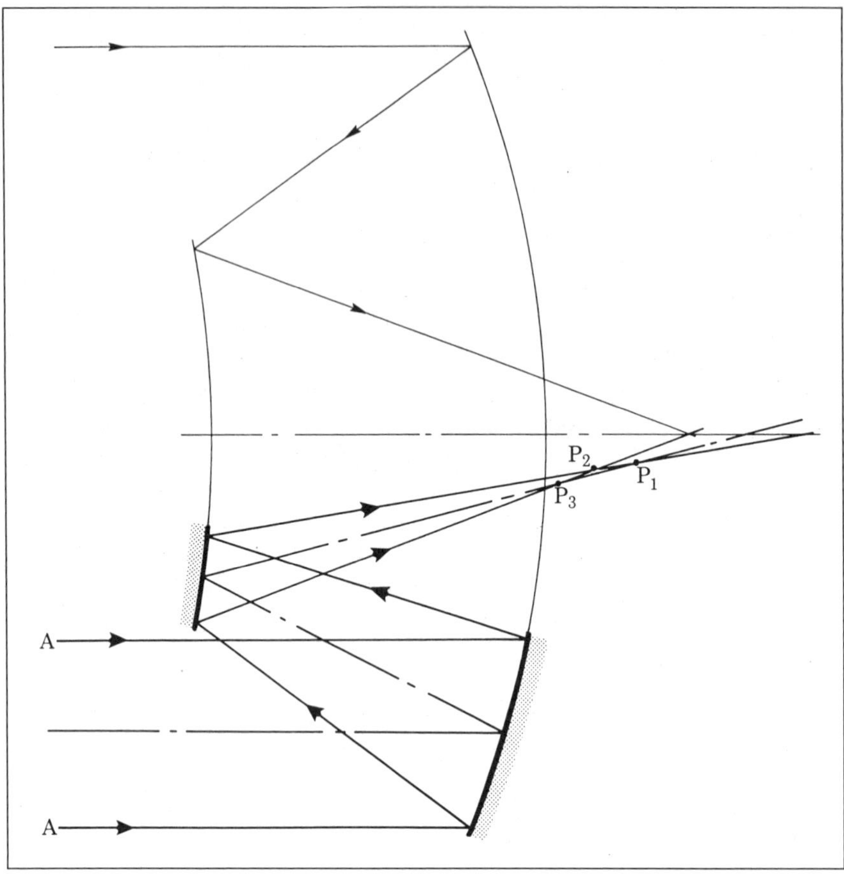

Fig. 12.1 *Derivation of the Schiefspiegler Telescope.*

ments. In this chapter, we investigate the design principles of schiefspieglers, then evaluate the optical performance of three forms of the TCT.

12.2 Optical Principles of Schiefspieglers

When designing his schiefspieglers, Anton Kutter took the Cassegrain reflector as his starting point. To avoid hard-to-make aspheric surfaces, Kutter began with two spherical mirrors, and cut an eccentric pupil from the incoming light, as shown in fig. 12.1. In this initial form, however, the convergence of rays at the focal plane is rather bad. Coma can be clearly recognized in this figure: the ray intersection points P_1, P_2, and P_3 do not coincide. Astigmatism, which is also present, cannot be seen in this particular drawing, because this aberration results from rays in the sagittal plane, outside the plane of the drawing.

Anton Kutter's merit as an optical designer is that he examined methods to eliminate or suppress sufficiently coma and astigmatism, and thereby found vari-

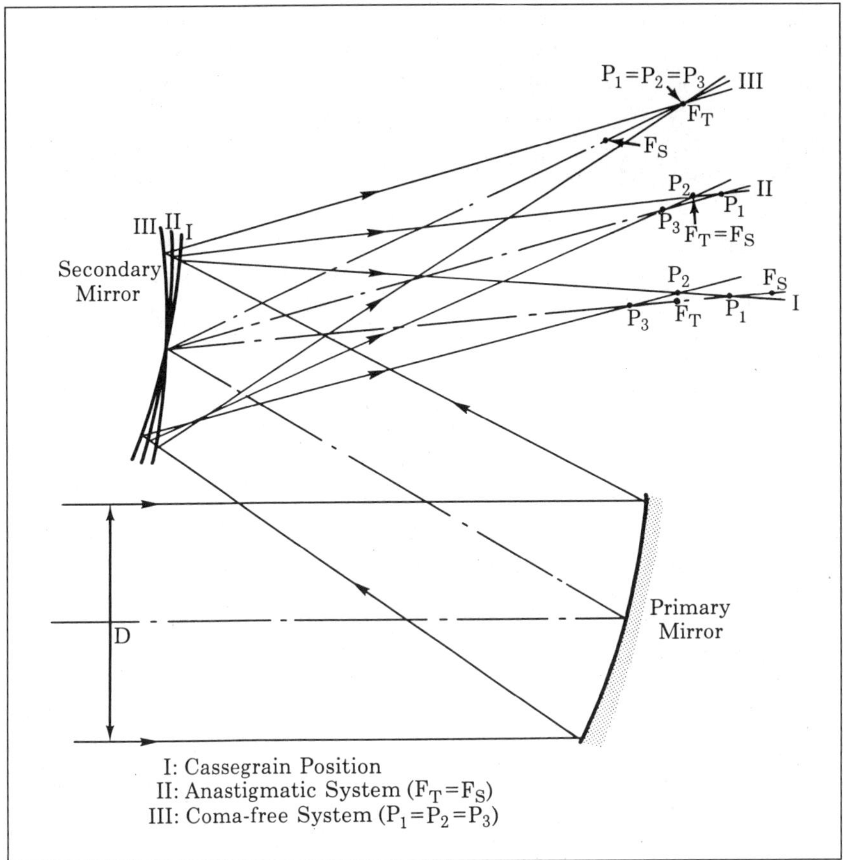

Fig. 12.2 *The Schiefspiegler for Various Tilt Angles (after A. Kutter).*

ous solutions. His first important decision was to set the radii of curvature of both mirrors equal, or roughly equal, to facilitate fabrication.

His second decision was to make the radii of curvature of both mirrors large. In a normal Cassegrain telescope, in order to obtain a short tube length, the focal ratio of the primary mirror is normally between 3 and 5. In a schiefspiegler, the primary's focal ratio may be 12. In an ordinary Cassegrain, the secondary magnification lies between 2 and 5; in a schiefspiegler, this number lies near 1.7. The focal ratio of the whole system normally comes out between 20 and 30. This makes the schiefspiegler less suitable for observation or photography of large star-fields and nebulae, but preeminently suitable for observing lunar and planetary details and close double stars. This is consistent with the high contrast available from an unobstructed aperture.

Anton Kutter emphasized the great importance of a small secondary magnification. When a high secondary magnification is used, he felt that great demands are placed on the surface accuracy of the primary mirror because the irregularities are magnified by the secondary. In this notion, he was wrong, because the optical path differences caused by surface irregularities do not depend on secondary magnification. Nevertheless, for the observation of low contrast objects, smooth mirror surfaces are of the utmost importance. Moreover, the difficulty of producing the rather "shallow" mirrors for a schiefspiegler should not be underestimated by amateur opticians.

In considering the performance of schiefspieglers, the importance of the long focal length for obtaining high performance from eyepieces should not be ignored. Image aberrations, particularly the off-axis aberrations of an eyepiece, are strongly dependent on the focal ratio of the telescope objective. At focal ratios between 20 and 30, eyepieces deliver sharper images than they do at focal ratios around 5. This point, one that is often forgotten, will be elucidated more fully in chapter 16. With schiefspieglers, relatively cheap long focal length Huygenian eyepieces can be used to advantage. This type of eyepiece, which the amateur can easily construct, is normally rejected by users of fast Newtonians because of intolerable aberrations.

Kutter's third and most important decision was to minimize image aberrations by tilting the secondary mirror. Fig. 12.2 shows a secondary in three positions: I, original position based on the configuration of the original Cassegrain system with coma and astigmatism; II, the position giving an anastigmatic system, corrected for astigmatism but suffering from residual coma; and III, a coma-free system, with residual astigmatism.

In an anastigmatic system, the rays in a plane perpendicular to the plane of the drawing will intersect at the same point F_s as the rays in the plane of the drawing itself, F_t, so $F_t = F_s$. In the coma-free system, the points P_1, P_2, and P_3 coincide, so $P_1 = P_2 = P_3$. Kutter established that it is impossible to find a tilt angle whereby coma and astigmatism are corrected simultaneously. However, he developed various methods to solve, or at least minimize, this problem.

In anastigmatic systems, coma can be suppressed below the size of the Airy disc by choosing a long focal length. The focal ratio required depends on the aperture as follows:

1. 80 mm aperture: *f*/20, with a tube length of 700 mm
2. 110 mm aperture: *f*/24.7, with a tube length of 1100 mm
3. 150 mm aperture: *f*/29, with a tube length of 1780 mm.

A 200 mm aperture anastigmatic schiefspiegler would require a tube some 2500 mm long, which would be impractical.

Coma-free systems can be freed from astigmatism by giving the secondary mirror a special deformation in the sagittal plane. Since the deformation has a saddle form, it is very difficult to make with the required accuracy. In order to restrict the tube length for apertures up to 400 mm, Kutter proposed a different solution, one which he called the "golden mean," of tilting the secondary mirror to an angle

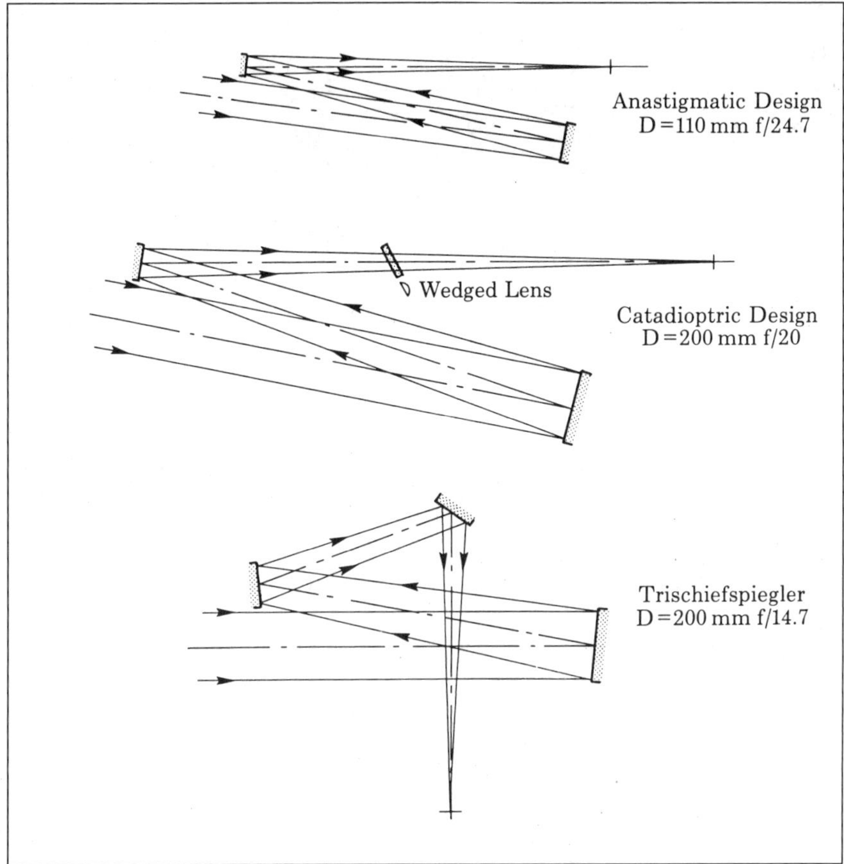

Fig. 12.3 *Three Forms of Tilted-Component Telescopes.*

between those of the anastigmatic and the coma-free systems.

Such a two-mirror system has both residual astigmatism and coma, but these can be nearly eliminated by placing a weak, tilted plano-convex lens in the exit bundle of the secondary mirror. The color aberration of this weak lens is small, and may be further decreased by making the lens very slightly wedged. Kutter calls this compound system the catadioptric schiefspiegler system.

According to the literature, for apertures greater than 220 mm, the primary mirror must be elliptically deformed to 60% of a parabola, to suppress the residual spherical aberration. Calculations show, however, that even for an aperture of 200 mm, the aspheric primary leads to better results. The overall length of a 200 mm catadioptric schiefspiegler is approximately the same as a 150 mm anastigmatic system. Because of these barely-suppressed aberrations and larger geometric aberrations in a larger system, Schiefspieglers may not always be scaled up; proportional downscaling is, of course, always allowed.

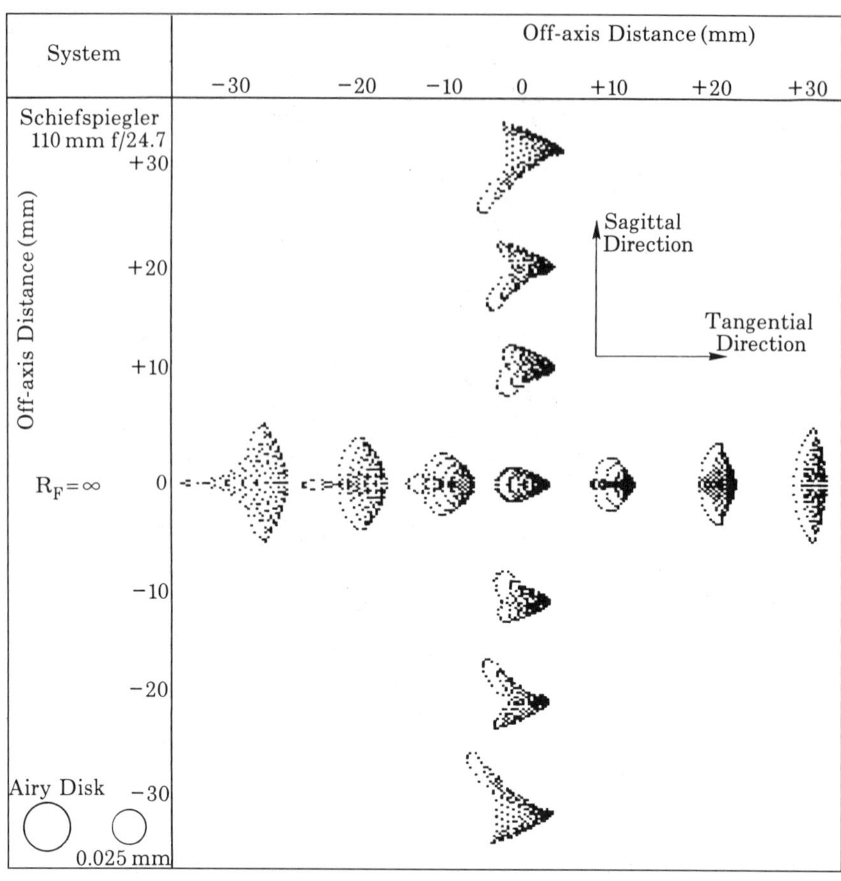

Fig. 12.4 *Spot Diagrams for the 110 mm f/24.7 Anastigmatic Schiefspiegler.*

In the 1960s, Kutter also developed a trischiefspiegler, a system with three mirrors. This system is much more compact than the two-mirror schiefspieglers, but looks rather exotic. (Fig. 12.3 compares the optical layout of the three telescopes.) The secondary and tertiary mirrors of the trischiefspiegler are spherical; the primary mirror is elliptical.

12.3 Results of Optical Ray Tracing

It would be cumbersome to give ray-tracing results for all the TCT designs proposed by Anton Kutter, Arthur Leonard, Richard Buchroeder, and others. Instead, we have restricted ourselves to three systems popular among amateurs (see refs. 12.2, 12.3, and 12.5). We have based our analyses on Kutter's original designs.

The three systems are: an anastigmatic 110 mm *f*/24.7 two-mirror system, a 200 mm *f*/20 catadioptric two-mirror system, and a 200 mm *f*/14.7 trischiefspiegler. Table 12.1 gives the construction data; fig. 12.7 shows their layouts.

Before discussing the ray-tracing results, we must explain the unusual spot diagrams. When an optical system is rotationally symmetric, the image characteristics on the axis in the tangential and sagittal directions are the same, as are the focal surfaces. On the optical axis, such systems produce round, rotationally-symmetric image blurs, so the calculation of spot diagrams can be restricted to a tangential calculation only.

For a tilted system, a different situation occurs. Because the optical power of the various tilted elements is different in the tangential and sagittal planes, the positions of the tangential and sagittal foci on the principal axis of a schiefspiegler can be different. If only the tangential calculation is carried out, the designer may draw false conclusions, because the blur on the sagittal axial focus can have a considerable size.

Off-axis imaging characteristics for the two types of systems differ. In rotationally symmetric systems, the calculations can be restricted to one direction because the spot diagrams are the same no matter what direction they are from the axis. This is not the case for a schiefspiegler; the calculations must be carried out for more directions. Here, we give spot diagrams for four directions: two in the tangential direction and two in the sagittal direction, in each case calculated for off-axis distances of 10, 20 and 30 mm. Note that for schiefspieglers, spot diagrams are mirror-symmetric with respect to the tangential direction, while for a rotationally symmetric system these are mirror-symmetrical about the optical axis.

In studying the spot diagrams, note that in the 110 mm *f*/24.7 anastig-

Table 12.1
Schiefspieglers
(All dimensions in millimeters)

110 mm *f*/24.7 Anastigmatic Schiefspiegler	
Primary	
Radius of Curvature	–3240
Angle of Deviation α_1	5.234°
Distance to Secondary	–965
Secondary	
Radius of Curvature	–3240
Angle of Deviation α_2	13.666°
Distance to Focal Plane	1108
Angle of Image Plane α_3	5.16°
Effective Focal Length	2720
1° Field	47.5
200 mm *f*/20 Catadioptric Schiefspiegler	
Primary	
Radius of Curvature	–4800
Deformation	–0.6
Angle of Deviation α_1	6.3°
Distance to Secondary	–1365
Secondary	
Radius of Curvature	–5060
Angle of Deviation α_2	18.133°
Distance to Corrector Lens	745
Corrector	
Radius of Curvature	–15000
Thickness at Center	7
Medium	517642
Angle of Wedge	0.0382°
Angle of Tilted Plane Surface α_3	28°
Distance, Secondary to Focus	1725
Angle of Image Plane α_4	8.05°
Effective Focal Length	4000
1° Field	69.8
200 mm *f*/14.7 Trischiefspiegler	
Primary	
Radius of Curvature	–4065
Deformation	–0.53
Angle of Deviation α_1	10.313°
Distance of Secondary	–1012.5
Secondary	
Radius of Curvature	–5930.25
Angle of Deformation α_2	30.419°
Distance to Tertiary	613.5
Tertiary	
Radius of Curvature	–32745
Angle of Deviation α_3	68.893°
Distance to Focus	879
Angle of Image Plane α_4	9.15°
Effective Focal Length	2930
1° Field	51.5

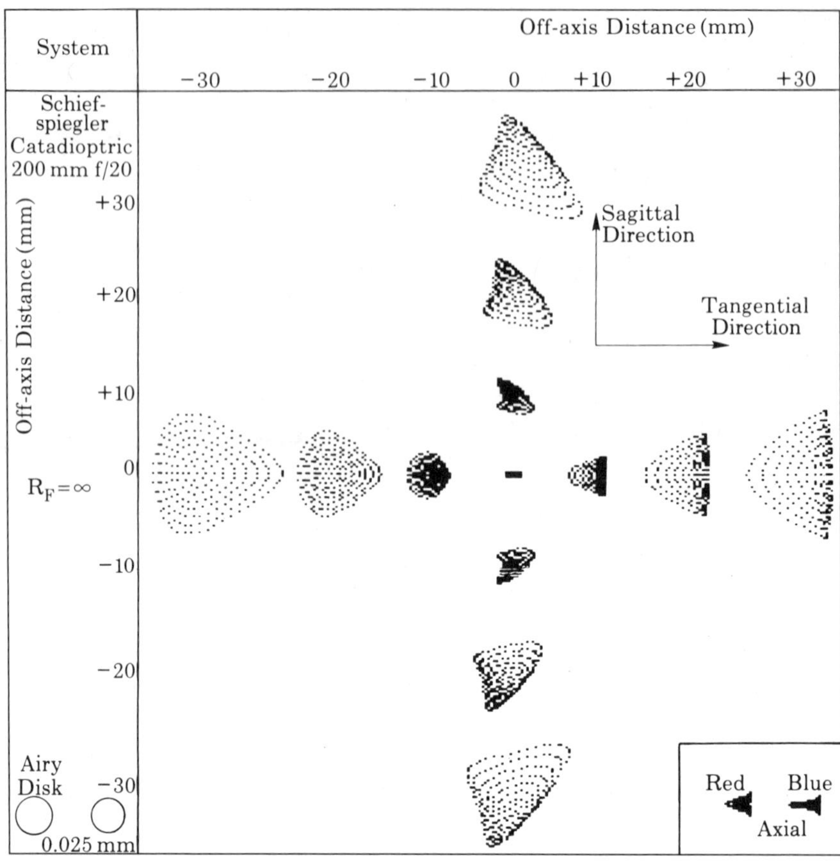

Fig. 12.5 *Spot Diagrams for the 200 mm f/20 Catadioptric Schiefspiegler.*

matic system (fig. 12.4), residual coma is clearly visible but does not, on the main axis, exceed the size of the Airy disk. The 200 mm *f*/20 catadioptric system (fig. 12.5) gives excellent sharpness on the main axis, and is diffraction-limited over a field approximately 15 mm in diameter. The main-axis spot diagrams for blue and red light do not exceed the Airy disk either. Although there is a slight amount of lateral color axially, caused by the oblique rays on the tilted corrector lens, this could probably be eliminated by slightly changing the wedged shape of the lens.

The 200 mm *f*/14.7 trischiefspiegler (fig. 12.6) appears to be a special case. For it, the spot diagram does not have its smallest size on the main axis. The diameter of the axial spot diagram is approximately 67 microns diameter, while about 14 mm off-axis to the left, the smallest spot diagram is some 30 microns diameter. This is still somewhat larger than the Airy disk, 20 microns diameter. Note that *in all cases* the optimum focal plane, for which all spot diagrams were calculated, is not perpendicular to the axis, but is slightly tilted (fig. 12.7).

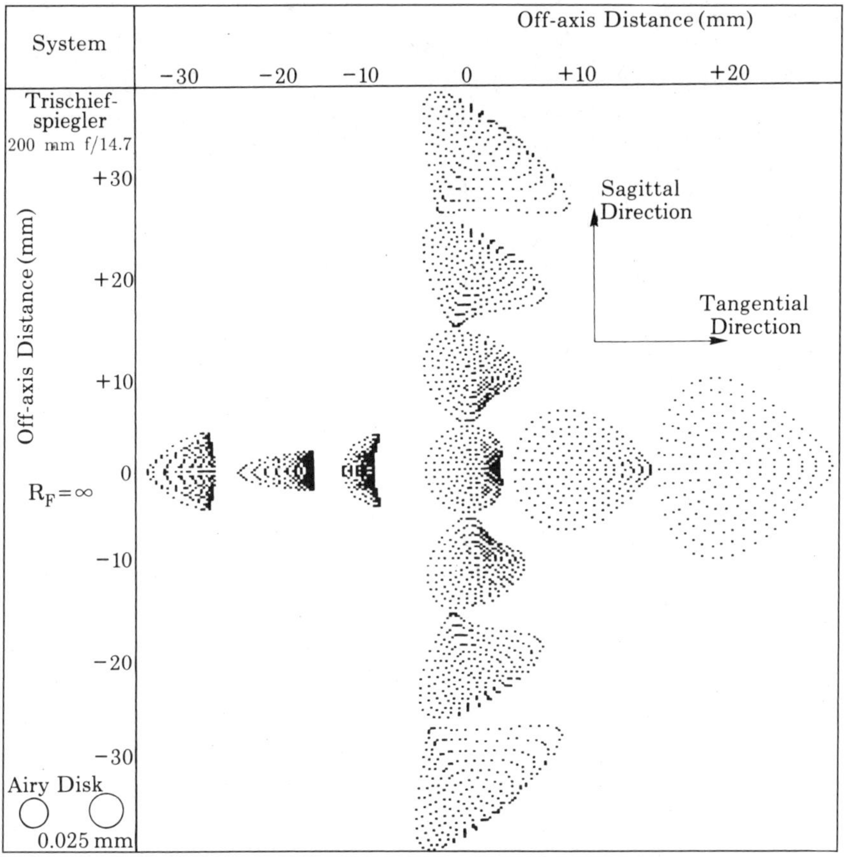

Fig. 12.6 *Spot Diagrams for the 200 mm f/14.7 Trischiefspiegler.*

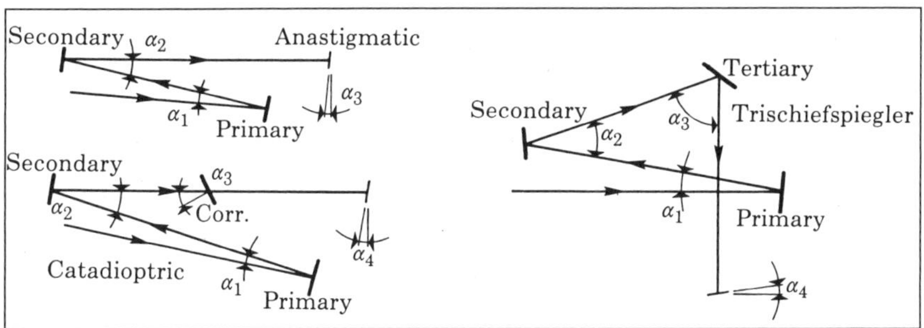

Fig. 12.7 *Tilted-Component-Telescopes*

Chapter 13

Other Compound Systems

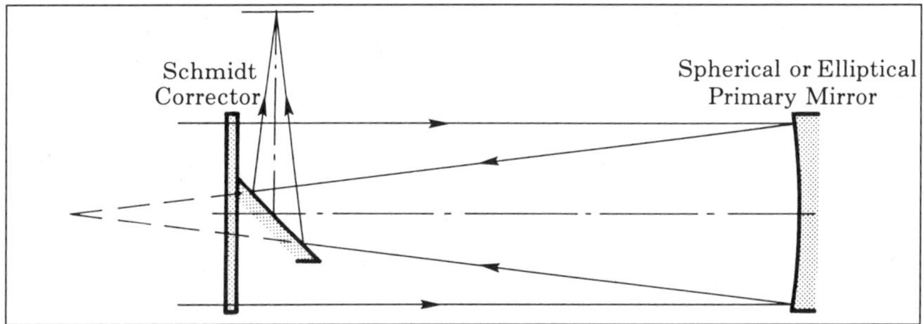

Fig. 13.1 *Layout of the Schmidt-Newtonian and Wright Systems.*

13.1 Introduction

Each of the telescopes treated in the previous chapters belongs to a particular family of systems. This is not the case with the systems to be discussed below. Three groups are treated:

1. Full-aperture correctors, mirror/lens systems in which a corrector is placed in the parallel entering bundle.

2. Focal correctors, mirror/lens systems in which a corrector is placed in the converging light cone in front of the focal plane.

3. Miscellaneous systems that can be useful for the amateur.

13.2 Full-Aperture Correctors: Schmidt Derivatives

As we saw in chapter 8, the Schmidt camera consists of a spherical mirror and a Schmidt corrector. The special characteristic of the Schmidt camera is that the corrector lies at the center of curvature of the mirror. In this way a symmetrical configuration is obtained, that is, there is no preferential direction for the entering bundles. All bundles, whether these enter parallel to the optical axis or obliquely, have, at least in principle, the same image sharpness in the focal plane. Such a system should be free of coma and astigmatism.

A disadvantage of the Schmidt camera, however, is its long tube, which must be twice the focal length of the mirror. In order to avoid this disadvantage, the

Table 13.1
Two 200 mm Schmidt-Derived Systems
(All dimensions in mm)

System	Schmidt-Newtonian	Wright
Focal Ratio	*f*/4	*f*/4
R_1 Radius of Curvature	flat	flat
T_1 Axial Distance	4	4
M_1 Medium	517642	517642
R_2 Paraxial Radius	–282040	–108016
T_2	617	617
M_2	Air	Air
Relative Power of Corrector	100%	259.319%
R_3	–1600	–1600
T_3 Back Focal Length	–798.825	–796.988
Deformation of Mirror	0	1.561
M_3	Air	Air
Effective Focal Length	799.8	797.0
1° Field	14.0	13.9

Schmidt corrector may be placed nearer to the mirror, for instance, at or near the focal plane. The resulting instrument is the Schmidt-Newtonian. The mirror remains spherical (fig. 13.1).

In the Schmidt-Newtonian, the symmetry with respect to entering bundles is lost so that coma and astigmatism are not completely corrected. The system is, nevertheless, better corrected than a Newtonian of the same aperture and focal length. The amount of coma and astigmatism depends on the position of the corrector. The shorter the distance between corrector and mirror, the stronger these aberrations will be.

Coma correction can be obtained by deforming the mirror to an ellipsoid. The resulting system is called the Wright telescope. The ellipsoidal deformation must be carried out in such a way that the degree of curvature at the edges of the mirror is stronger than in the center, unlike the normally-used conic aspherical deformations in which the curvature of the center is greatest. The Wright ellipsoid is the figure of revolution produced by an ellipse where the focal points of the ellipsoid are not located on the optical axis, but on both sides of the axis. This is the so-called oblate ellipsoid. (The more familiar ellipsoid, having greatest curvature in the middle of the mirror, is called a prolate ellipsoid.)

Table 13.1 shows the strong aspherical deformation of the Wright mirror. This is necessary in order to eliminate coma for the particular chosen position of the Schmidt corrector. In order to compensate this deformation and bring all axis-parallel rays to one point on the axis, the power of the Schmidt corrector must be approximately 2.6 times stronger than for the Schmidt-Newtonian of the same fo-

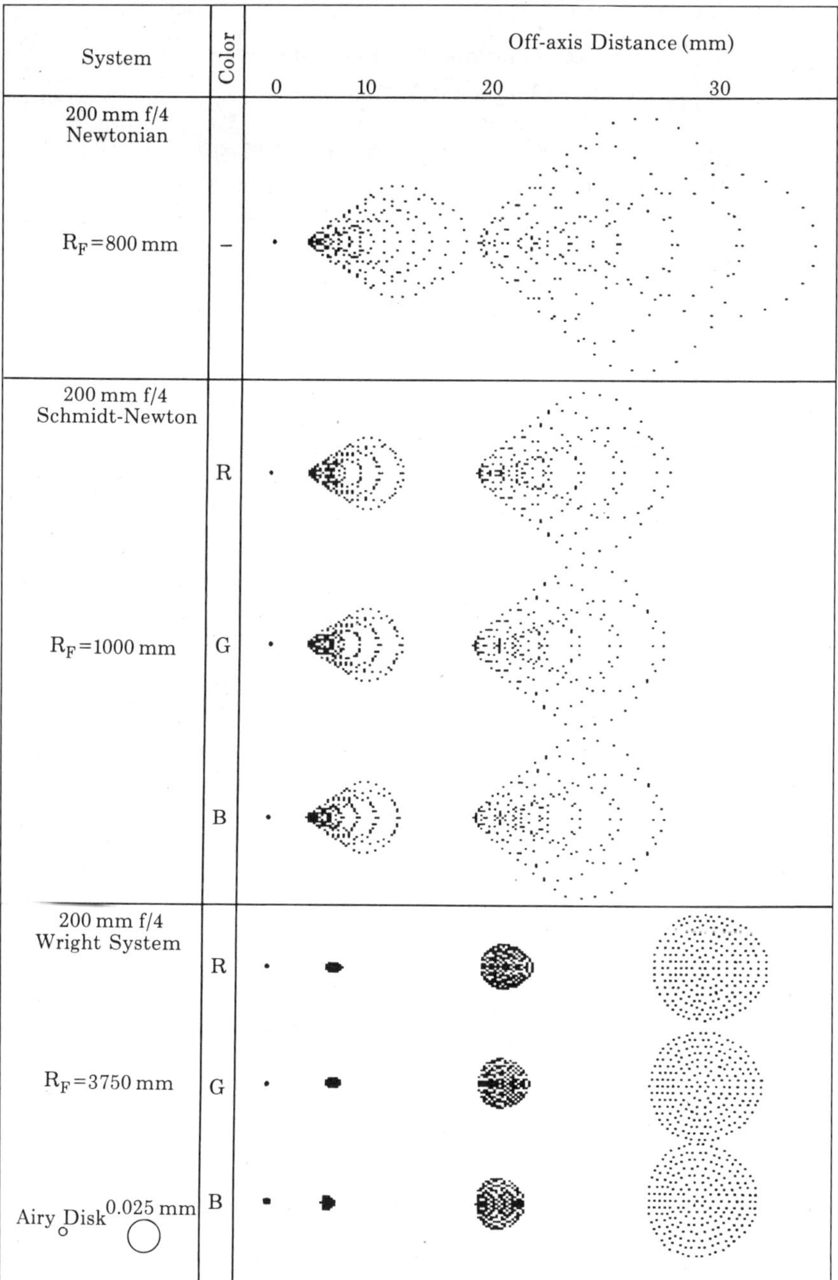

Fig. 13.2 *Comparison of Newtonian, Schmidt-Newtonian, and Wright Systems.*

Table 13.2
Four 200 mm Houghton-Derived Designs
(All dimensions in millimeters)

System	Buchroeder Houghton	Lurie Houghton	Houghton Cassegrain	Houghton Cassegrain
Focal Ratio	*f*/3	*f*/4	*f*/10	*f*/5.3
R_1 Radius of Curvature	1143.73	1286.81	939.8	566.6
T_1 Axial Distance	16.67	16	14.5	22.5
M_1 Medium	517642	517642	517642	517642
R_2	flat	–4810.34	–1196.6	–585.5
T_2	10	3.0	1.5	0.6
M_2	Air	Air	Air	Air
R_3	–1143.73	–1286.81	–939.8	–566.6
T_3	8.33	12	5	5
M_3	517642	517642	517642	517642
R_4	1143.73	4810.34	1196.6	585.5
T_4	10	617.67	435	1112.3
M_4	Air	Air	Air	Air
R_5	flat	–1590.97	–1200	–1200
T_5	16.67	–793.84	–415.38	–338.5
M_5	517642	Air	Air	Air
R_6	–1143.73		–527.47	–1200
T_6	1182.67		595.48	443.78
M_6	Air		Air	Air
R_7	–1200			
T_7	–597.3			
M_7	Air			
Effective Focal Length	604.26	800.4	1968.97	1072.3
1° Field	10.5	14.0	34.4	18.7

cal ratio. The color aberration of the Wright design is higher than in the Schmidt-Newtonian, but it remains within acceptable limits.

In fig. 13.2, we show a direct comparison between the 200 mm aperture *f*/4 Newtonian, Schmidt-Newtonian, and Wright telescopes. In the Schmidt-Newtonian, coma is about 40% less than it is in the *f*/4 Newtonian. Far better off-axis images are achieved in the Wright design because coma can be fully corrected, despite the astigmatism remaining. Note that in both Schmidt-derived designs, the entrance pupil has been moved from the mirror to the corrector, which means that the mirror should have a larger diameter to avoid vignetting off-axis beams. The mirror remains smaller in both cases than in the case of the Schmidt camera. An advantage of the configurations shown is that the diagonal mirror can be attached to the Schmidt corrector so that no spider problems exist.

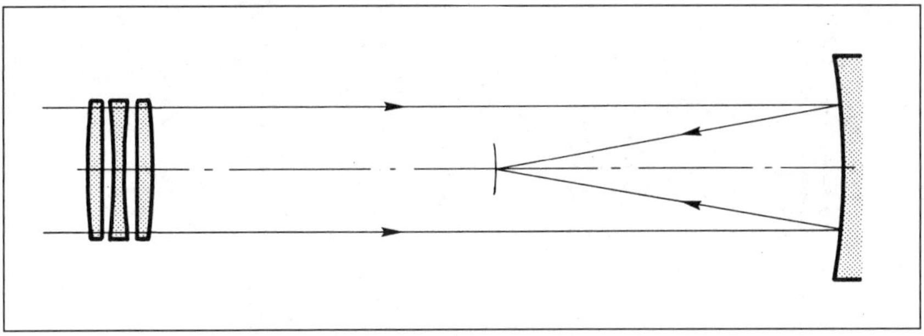

Fig. 13.3 *The Buchroeder-Houghton Camera.*

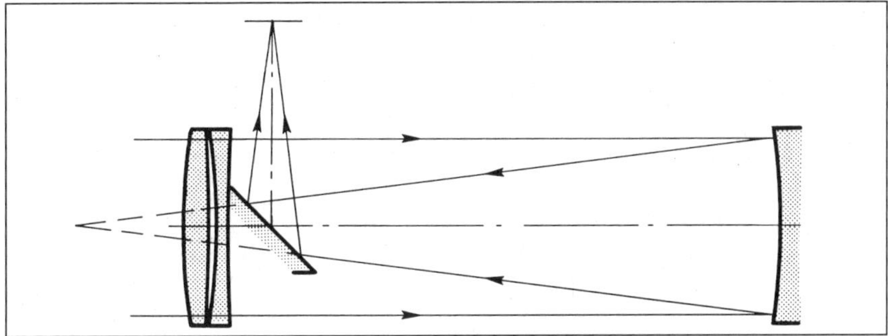

Fig. 13.4 *Layout for Lurie's Houghton Telescope.*

13.3 Full-Aperture Correctors: Houghton Derivatives

In 1944, Houghton suggested a new type of full-aperture corrector for spherical mirrors as an alternative to the hard-to-make Schmidt corrector (ref. 13.1). These correctors consist of a combination of two or three positive and negative spherical lens elements that can be considered afocal (i.e., having no optical power) and achromatic. They are over-corrected for spherical aberration in order to compensate for that of the mirror.

One special advantage of the Houghton corrector is that its spherical aberration depends on the spacing between the lenses. This means that a residual spherical aberration in the assembled system can be corrected by decreasing or increasing the spacing between the lenses.

Houghton pointed out the possibility of using a single type of glass for all of the lenses. Other designers were inspired by the Houghton systems to design variations; a few of these systems are analyzed here. We have deliberately chosen those systems having lenses with the same kind of glass and, in pairs, the same radii of curvature of the lens surfaces, because this facilitates making the optics. Furthermore, in the four cases discussed, all of the optical surfaces are spherical.

The first system we analyzed was designed by Buchroeder (ref. 13.2). In it, the corrector lies at the center of curvature of the mirror, as is the case in a Schmidt camera. This system is shown in fig. 13.3. A much shorter configuration was pre-

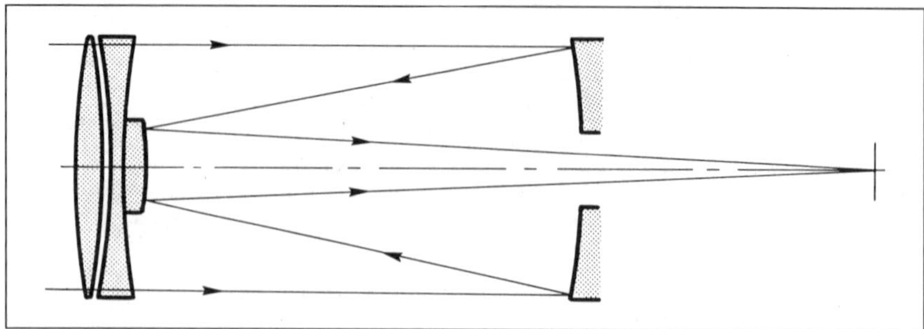

Fig. 13.5 *Layout for an f/10 Houghton-Cassegrain Telescope.*

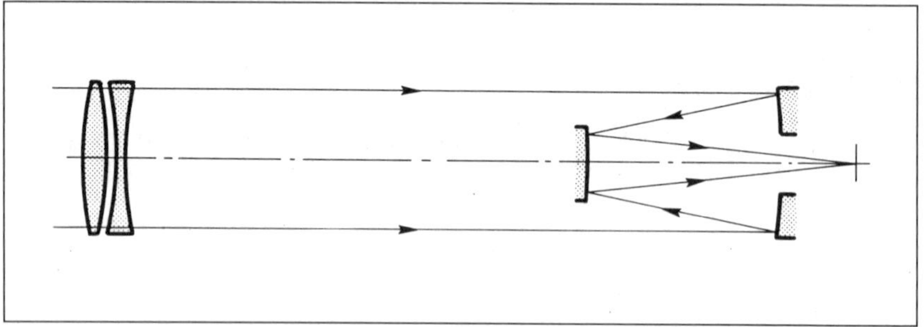

Fig. 13.6 *Layout for an f/5.3 Houghton-Cassegrain Flat-Field Camera.*

sented by Lurie (ref. 13.3). Lurie placed a two-lens corrector close to the mirror (fig. 13.4), even closer than the focus. This system has the same configuration as the Schmidt-Newtonian and the Wright, so that a direct comparison of the performance of these systems can be made.

The Houghton-based corrector lens can also be placed at the entrance pupil of a catadioptric Cassegrain, following the example of the Schmidt or Maksutov corrector. Two designs are presented here: an *f*/10 and an *f*/5.3 system, the last one being a flat-field camera (shown in figs. 13.5 and 13.6). These were designed by a method given by Sigler in ref. 13.4 and are described in chapter 21.

Fig. 13.7 summarizes the ray-tracing results for these four Houghton-derived designs. All systems are coma-free. This may be surprising, especially in the fast system designed by Lurie. The reason is that, compared with Schmidt-derived designs, Lurie's design contains three (rather than two) elements, giving the designer an additional degree of freedom to correct this aberration. The mirror can be left spherical.

The Buchroeder design appears to be excellent. On the whole curved focal surface the photographic image sharpness is sufficient over a wide spectral range. The system designed by Lurie is suitable for both visual and photographic use over a large flat field. Note that although the Lurie has astigmatism, it is only half that of the Wright camera, and the system is much easier to build. The Lurie instrument seems to be an almost ideal rich field telescope for the demanding ama-

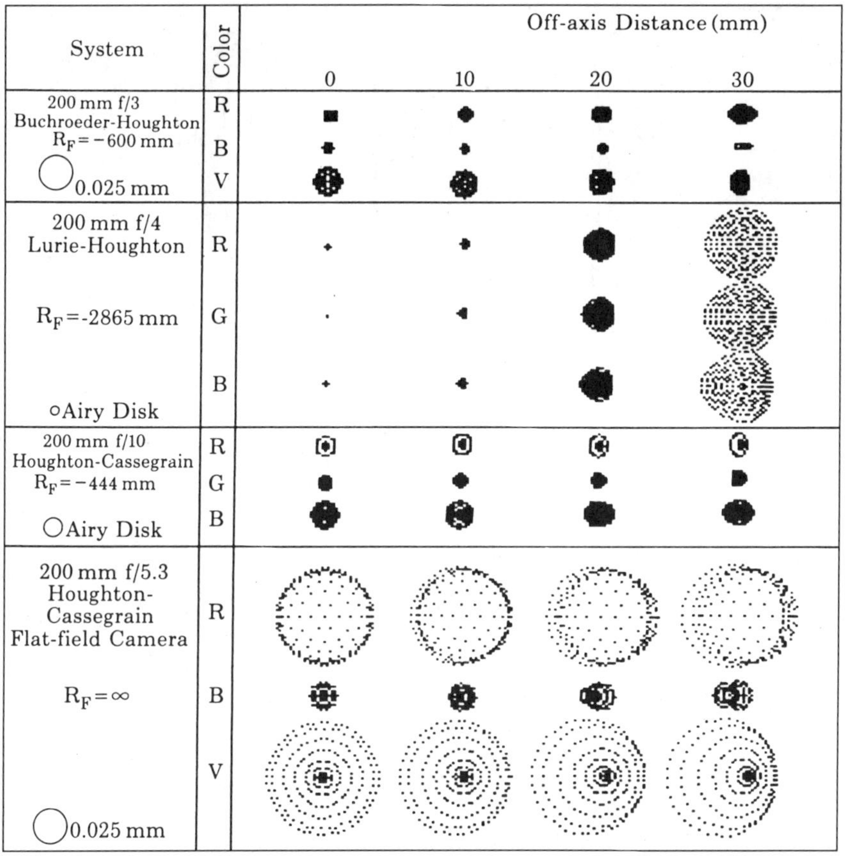

System	Color	Off-axis Distance (mm)			
		0	10	20	30
200 mm f/3 Buchroeder-Houghton $R_F = -600$ mm	R				
	B				
○0.025 mm	V				
200 mm f/4 Lurie-Houghton	R				
R_F =-2865 mm	G				
○Airy Disk	B				
200 mm f/10 Houghton-Cassegrain $R_F = -444$ mm	R				
	G				
○Airy Disk	B				
200 mm f/5.3 Houghton-Cassegrain Flat-field Camera	R				
$R_F = \infty$	B				
○0.025 mm	V				

Fig. 13.7 *Spot Diagrams for Four 200 mm Houghton-Derived Designs.*

teur.

The *f*/10 Houghton-Cassegrain was designed with the computer program described in chapter 22. Its performance for visual use is comparable with the compact *f*/10 Schmidt-Cassegrain described in chapter 9. The *f*/5.3 flat-field Houghton-Cassegrain was designed likewise. It must be emphasized that both systems were the result of a single computer run—just to show what can be achieved with the design program—and have not been optimized. The resulting *f*/10 system is satisfactory as designed, but the *f*/5.3 should be optimized in order to reduce the color aberration. This can be accomplished by using two different glasses or changing the radii of curvature. Of course, the ease of manufacturing would then be partially lost. We conclude that designs derived from Houghton have favorable characteristics and deserve more attention from amateurs (see ref. 13.18).

13.4 Focal Correctors: Jones, Bird, and Brixner

By placing a combination of lenses in the converging light cone of a spherical mir-

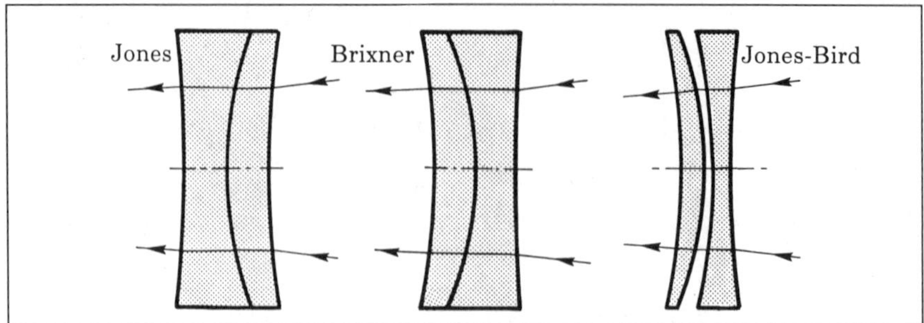

Fig. 13.8 *Focal Correctors for Spherical Mirrors.*

Table 13.3
Three Focal Correctors for 200 mm Spherical Mirrors
(All dimensions in millimeters)

System	Jones	Brixner	Jones-Bird
Focal Ratio	*f*/10	*f*/10	*f*/6
R_1 Radius of Curvature	–1600	–1600	–1562.96
T_1 Axial Distance	–586.87	–579.4	–610.06
M_1 Medium	Air	Air	Air
R_2	252.67	587.833	331.66
T_2	–11	–11	–5.02
M_2	648338	517642	517642
R_3	110.67	–93.933	–174.14
T_3	–11	–11	–0.6
M_3	517642	617366	Air
R_4	–350	–202.033	–104.04
T_4	–512.9	–518.8	–5.64
M_4	Air	Air	620364
R_5			–178.12
T_5			–247.54
M_5			Air
Effective Focal Length	1999.46	2000	1200
1° Field	34.9	34.9	20.9

ror, at some distance from the focal plane, we can correct the spherical aberration of the spherical mirror and as much coma as possible, without introducing much chromatic aberration.

In 1957, Jones presented such a design (ref. 13.5), while Brixner (ref. 13.6) and Bird (ref. 13.7) published improved designs in later years. These are shown in fig. 13.8. These correctors have negative power, so that the focal length of the mirror is increased. Jones, Brixner, and Bird all used an *f*/4 spherical primary mirror. The focal length amplification factor for the Jones and Brixner correctors is 2.5

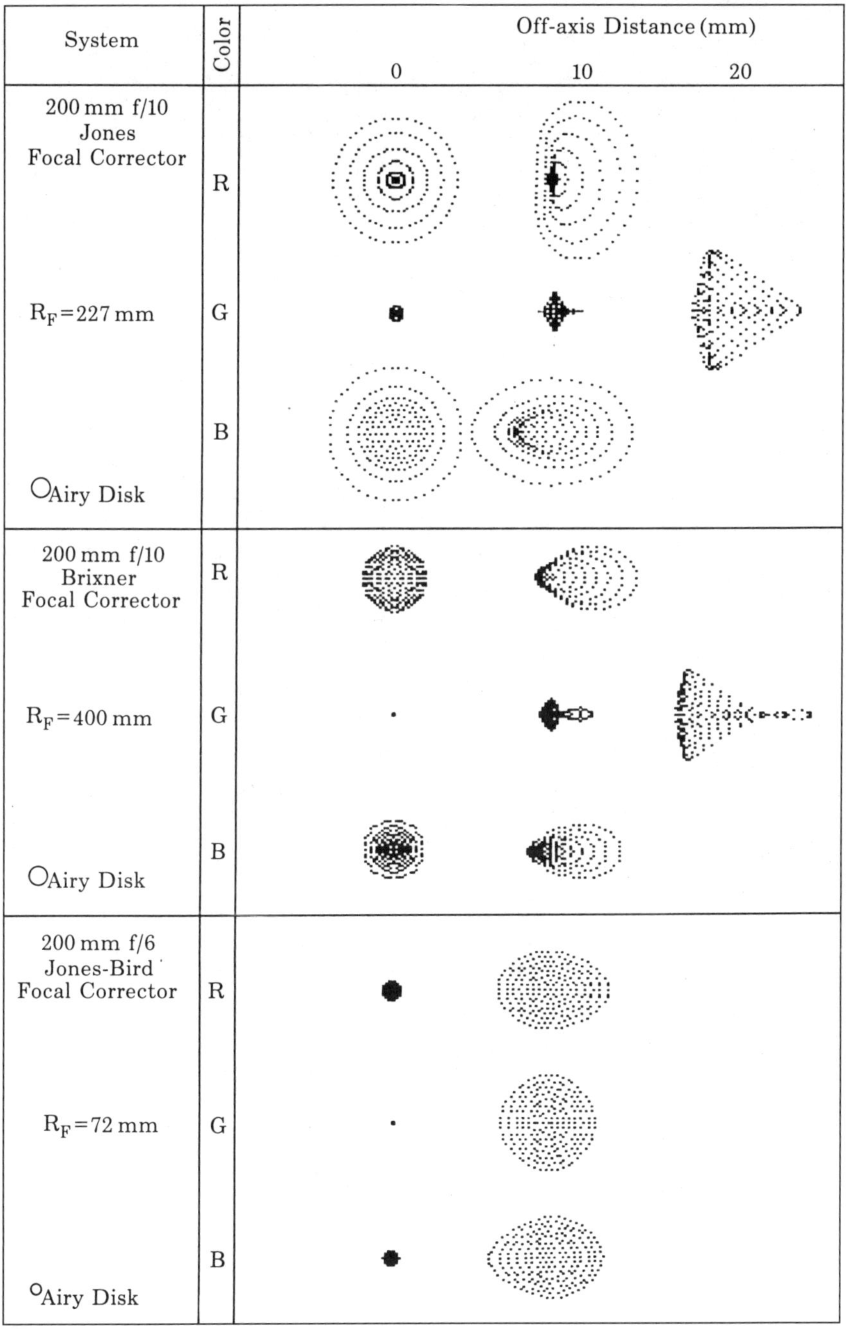

System	Color	Off-axis Distance (mm) 0	10	20
200 mm f/10 Jones Focal Corrector	R			
R_F=227 mm	G			
○Airy Disk	B			
200 mm f/10 Brixner Focal Corrector	R			
R_F=400 mm	G			
○Airy Disk	B			
200 mm f/6 Jones-Bird Focal Corrector	R			
R_F=72 mm	G			
○Airy Disk	B			

Fig13.9 *Spot Diagrams for Corrected Spherical Primaries.*

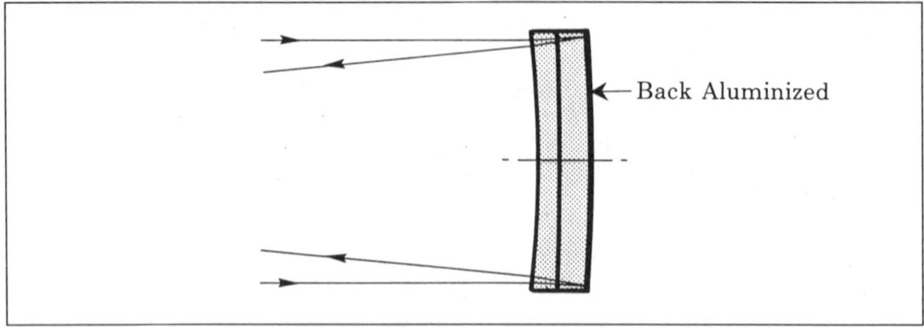

Fig. 13.10 *An Achromatic Mangin Mirror.*

times, while Bird used 1.5 times, producing *f*/10 and *f*/6 systems, respectively. Unlike the arrangement of a normal refractor objective, but similar to all negative achromats, the negative element is made from crown glass and the positive lens consists of flint glass. Brixner's paper shows that the best results are obtained when the first element is the negative lens. An advantage of these systems is their relatively short tube length in comparison with a Newtonian of equal focal length.

The ray-tracing results, presented in fig. 13.9, reveal that it is difficult to achieve a good performance with these systems. The reason for this is that three aberrations (spherical aberration, coma, and color) must be corrected simultaneously with a simple doublet. The axial color correction of the third system, the Jones-Bird, appears to be good, but the system exhibits both astigmatism and field curvature. Note that coma could be eliminated by separating the two lenses. Focal correctors of this type are sometimes mistakenly called Barlow lenses. This is incorrect because a Barlow lens does not change spherical aberration (see section 15.1); rather, a Barlow lens is designed for use in systems in which spherical aberration is already corrected.

13.5 Unusual Compound Systems

A Mangin mirror is a negative meniscus with the back aluminized to produce a catadioptric mirror, as shown in fig. 13.10. The system bears the name of its designer, Mangin, a French officer who applied this form of mirror instead of a paraboloidal mirror in searchlights. The meniscus suppresses the spherical aberration of the spherical mirror. Mangin mirrors are still in use today in, for example, modern catadioptric telephoto lenses.

When the Mangin mirror consists of a single element, the system suffers from chromatic aberration. This can be corrected when two kinds of glass are used. Since most of the optical power is supplied by the mirror, the lens need only correct spherical aberration, and thus has a relatively low power. Despite the fact that the light passes the lens twice (quite unusual in astronomical units) the secondary spectrum amounts to only 0.0002f with two kinds of normal glasses. Compare section 6.2: with a Fraunhofer doublet with the same kinds of glass the secondary spectrum amounts to 0.0005f. Table 13.4 lists the characteristics of a

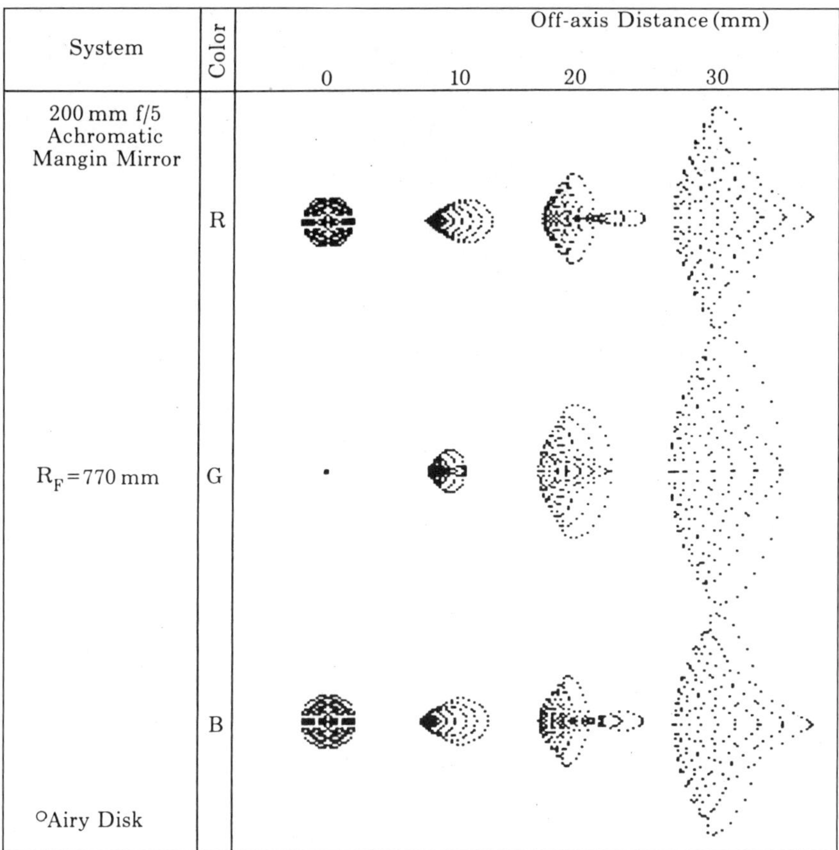

Fig. 13.11 *Spot Diagrams for a 200 mm f/5 Achromatic Mangin Mirror.*

200 mm *f*/5 achromatic Mangin mirror.

Note that the positive part of the lens, contrary to a refractor, consists of flint glass and the negative lens of crown. Despite its relatively low secondary spectrum, a 200 mm *f*/5 system would not be satisfactory. A focal ratio of *f*/10 would be necessary to reduce the visual secondary spectrum sufficiently, But this telescope would lose much of its attraction because of its long tube length. Coma of this system appears to be somewhat lower than half of the coma of an equivalent Newtonian (see fig. 13.11) but may be reduced by choosing matching glasses and even eliminated by air spacing.

An intriguing design that uses a Mangin relay lens to correct the chromatic aberration of a simple primary lens is known as the Schupmann, or medial telescope (ref. 13.20). There are many variations, including an all spherical "super-Schupmann" design.

As we saw in chapter 6, good color correction of a 200 mm doublet refractor objective can only be obtained with special glasses or by increasing the focal

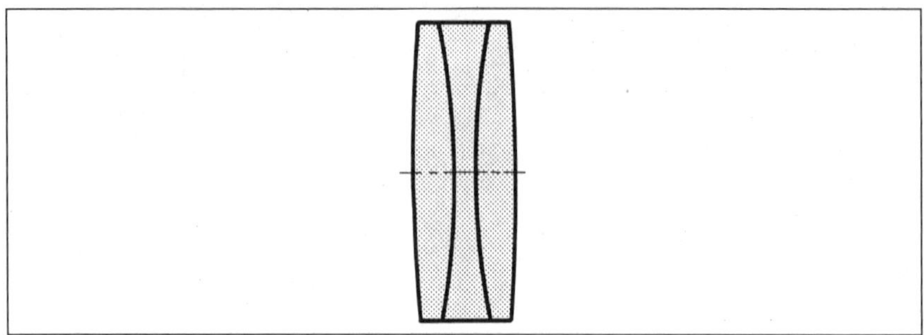

Fig. 13.12 *The Christen Triplet Objective.*

System	Color	Off-axis Distance (mm)			
		0	10	20	30
200 mm f/10 Christen Immersion Three Element Objective	R				
	G				
$R_F = \infty$	B				
Airy Disk 0.025 mm	V				

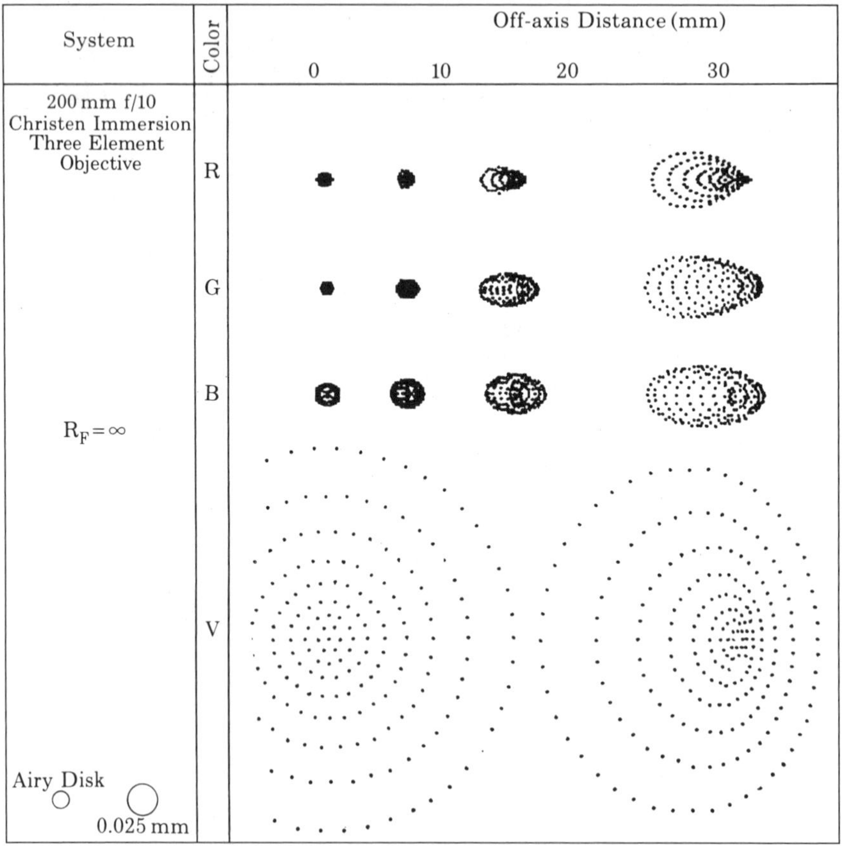

Fig. 13.13 *Spot Diagrams for the f/10 Christen Triplet Objective.*

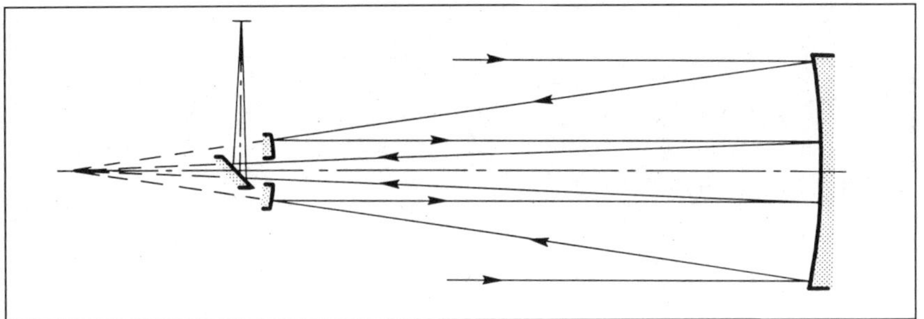

Fig. 13.14 *Layout for the Loveday Telescope.*

Table 13.4
Four 200 mm Optical Systems
(All dimensions in millimeters)

System	Achromatic Mangin	Christen Triplet	Loveday	Maksutov-Newtonian
Focal Ratio	*f*/5	*f*/10	*f*/24	*f*/4
R_1 Radius of Curvature	–922	2080	–2400[a]	–308.76
T_1 Axial Distance	15	22.5	–900	20
M_1 Medium	517642	517642	Air	517642
R_2	–3502.42	–507.5	–600[a]	–320.439
T_2	20	12.5	900	617
M_2	613370	613443	Air	Air
R_3	–1515.27	476.25	–2400[a]	–1627
T_3	–20	22.5	–1200	–829.9
M_3	613370	670471	Air	Air
R_4	–3502.42	–2080		
T_4	–15	1998.9		
M_4	517642	Air		
R_5	–922			
T_5	–985.5			
M_5	Air			
Effective Focal Length	1000.7	2012.5	4800	799.81
1° Field	17.5	35.1	83.8	14.9

a. Deformation: SC = –1 (parabola)

length enormously. By adding a third element of an abnormal dispersion glass, however, we can readily obtain excellent color correction. Various multi-lens refractor objective designs have been published in amateur-oriented literature; one very special design is the "oiled" triplet objective. Four of six surfaces have matching radii of curvature, which facilitates their manufacture. The thin spaces

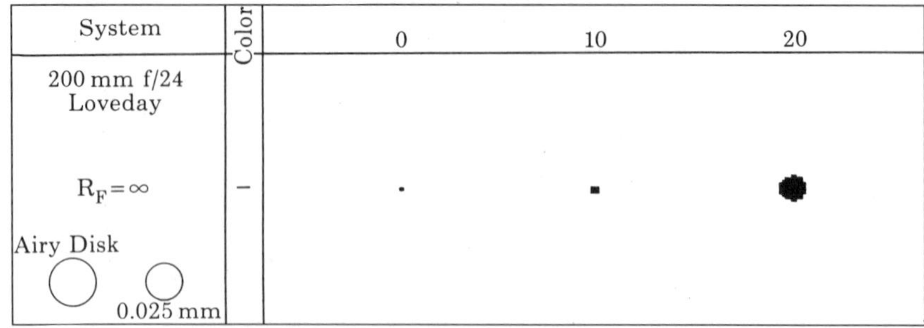

Fig. 13.15 *Spot Diagrams for the 200 mm f/24 Loveday Telescope.*

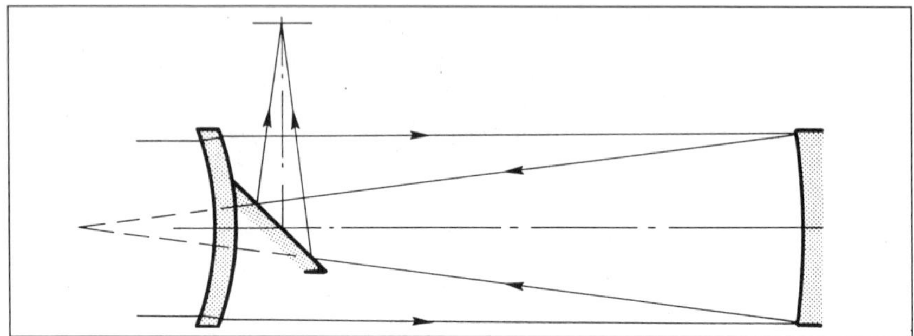

Fig. 13.16 *The Maksutov-Newtonian Telescope.*

between the lenses are filled with an oil which compensates small surface irregularities so that the intermediate surfaces need not be fully corrected (fig. 13.12). Only the front and back surfaces need be fully corrected. Table 13.4 lists the construction data of an *f*/10 design made by Roland Christen (ref. 13.8). Fig. 13.13 shows that this triplet has been optimized for visual use. It has a very good color correction. Unfortunately there is some coma. For photographic use a filter should be used to avoid unsharpness. We discuss several optimized three-lens designs in section 21.16.4.

A further interesting and unconventional design is the system discussed by Loveday (ref. 13.9), which consists of a concave and a convex parabolic mirror which are placed confocally, i.e., their focal points coincide. This results in a situation where the converging beam of the primary, after being reflected by the secondary, is a parallel beam. This reflects again from the primary (which is a tertiary mirror now) and comes to focus at the original focal plane of the primary (fig. 13.14). This is a dual telescope design: when the secondary mirror is removed, the system is a fast Newtonian; with the secondary in place, the telescope is a slow system. In order to limit the central obstruction, the secondary mirror and its hole should be made as small as possible. Therefore this kind of instrument has a strongly restricted field, but an excellent diffraction limited image, mainly the result of the small image (fig. 13.15).

The system with two confocal paraboloids was originally proposed by

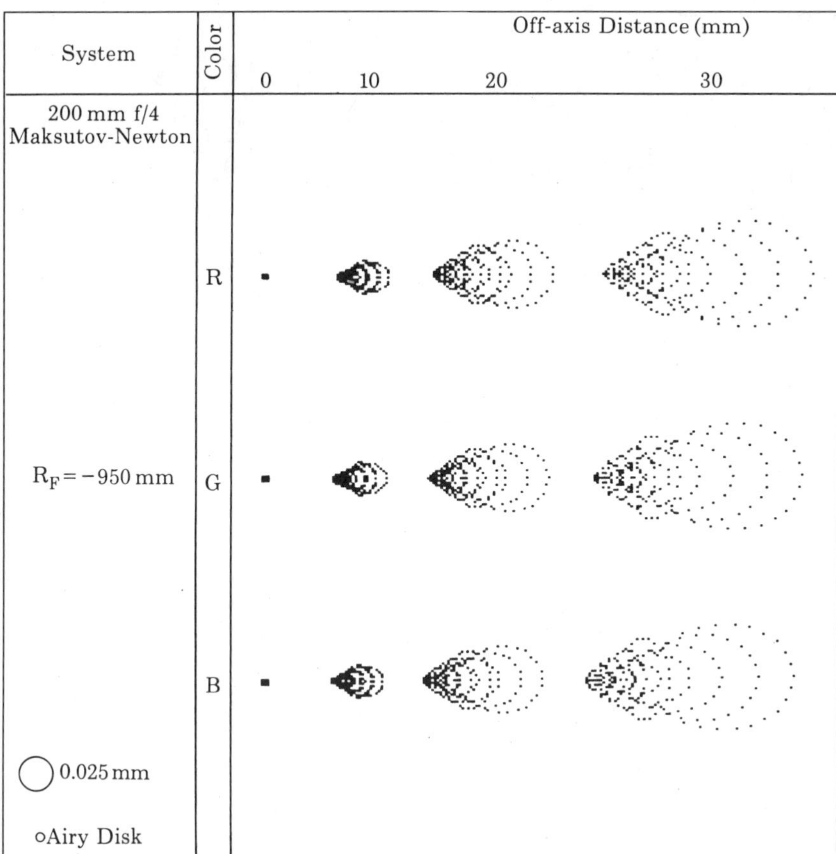

Fig. 13.17 *Spot Diagrams for the 200 mm f/4 Maksutov-Newtonian.*

Mersenne in the 17th century. Unlike Loveday's design, the Mersenne telescope delivers a parallel exit beam of smaller diameter than the aperture, so this system can be used in combination with another imaging device (ref. 13.10).

Another possibility is the Maksutov-Newtonian. In it, the Maksutov meniscus corrector is placed closer to the mirror than in the original Maksutov camera so the diagonal can be easily attached to the meniscus. In the design shown, the corrector has been optimized for its particular position and is slightly flatter than the corrector for the original *f*/4 Maksutov camera. Fig. 13.16 shows an *f*/4 system which has approximately the same main dimensions and configurations as the Schmidt-Newtonian and Wright telescopes, so that a direct comparison of their spot diagrams can be made. The Maksutov Newtonian is not free of coma, but the aberrations are about half those of a comparable Schmidt-Newtonian (compare figs. 13.17 and 13.2). Because the Maksutov-Newtonian should also be somewhat easier to make, this instrument seems to be a more favorable rich-field telescope

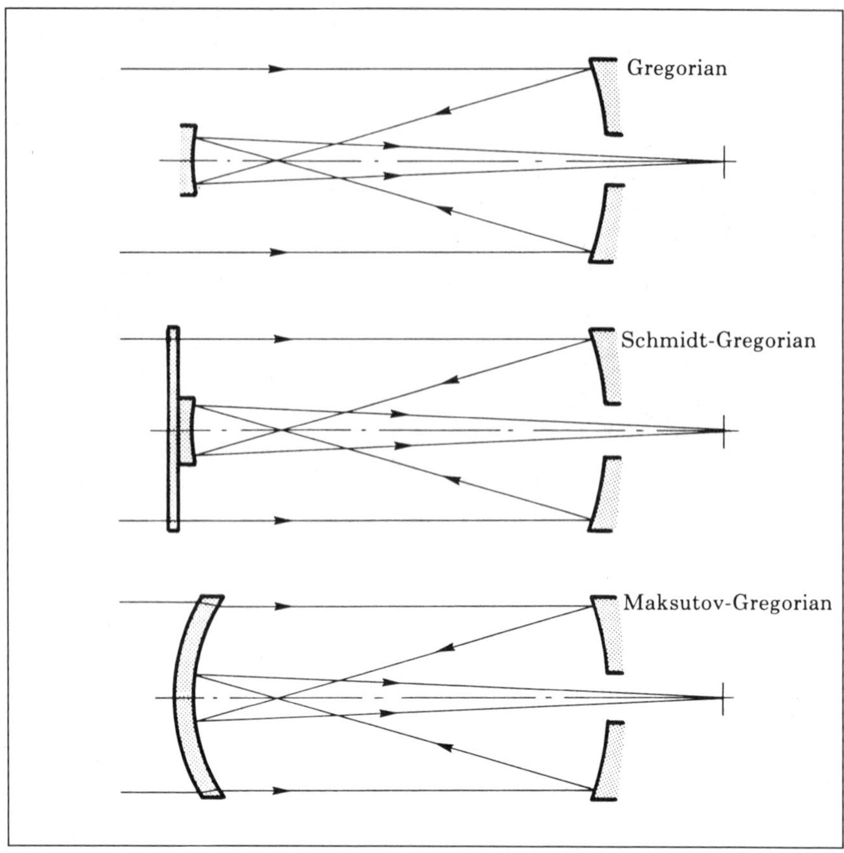

Fig. 13.18 *Layouts for Gregorian Systems.*

than the Schmidt-Newtonian.

13.6 Gregorians, Relay Telescopes, and Wright's Off-Axis Catadioptric

Many more possibilities exist. We present just a few of these, but without spot diagrams or detailed analysis. Advanced amateur designers may wish to explore these systems further.

A significant and interesting variation on the Cassegrain is the Gregorian telescope. Instead of a convex secondary inside the focus of the primary, this system has a concave secondary outside the focus. As can be seen from fig. 13.18, this leads to a longer tube length. The distance between the mirrors is slightly more than the sum of their focal lengths. The image is upright; before the invention of the achromatic objective, this design was often used for terrestrial observation.

The classical Gregorian has a parabolic primary and an elliptical secondary.

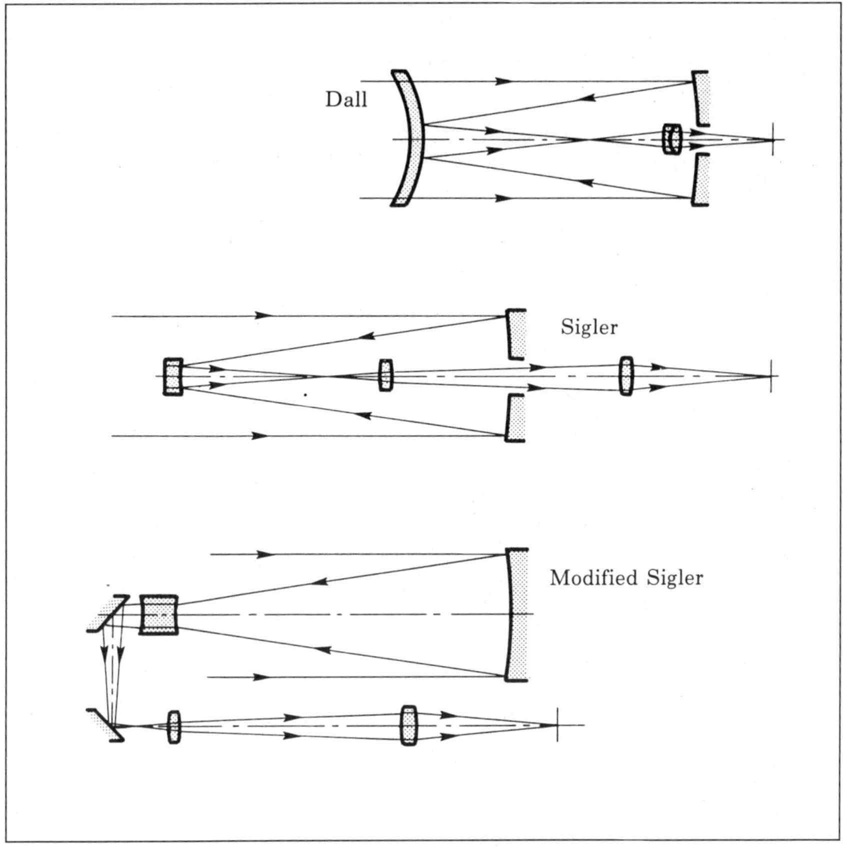

Fig. 13.19 *Layout for Three Relay Telescopes.*

Spherical aberration is eliminated, but the off-axis images suffer from coma, astigmatism, and field curvature. As is the case with the Cassegrain, other mirror shapes are possible; likewise, although both spherical aberration and coma can be corrected, astigmatism remains present. Fig. 13.18 also shows some derivations of the Gregorian with a Schmidt and a meniscus corrector. The correctors close the tube, and, of course, the secondary can be attached to the corrector. A meniscus system of this design would be called a Maksutov-Gregorian, and should not be confused with the Gregory Maksutov described in chapter 11. Bouwers (ref. 13.11) reports a good axial sharpness when the secondary is an aluminized spot on the spherical back side of the meniscus, but the off-axis performance leaves much to be desired. Better correction of coma is possible when the secondary is separate and aspheric.

Unlike the Cassegrain and most other telescopes, the Gregorian has an outward curving focal surface. This suggests the possibility of making an almost aberration-free visual telescope by matching the focal surface to the focal surface of

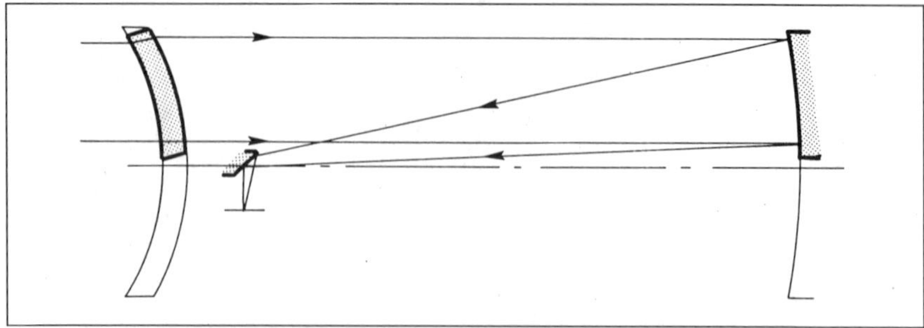

Fig. 13.20 *The Off-axis Catadioptric Wright-Newtonian.*

the eyepiece. In a Schmidt-Gregorian system (ref. 13.12) in which coma has been fully corrected by choosing the proper aspherically deformed secondary mirror, we find astigmatism still present (fig. 4.11, case 4). This means that the tangential focal surface is farther away from the objective than is the sagittal focal surface, or that astigmatism is overcorrected.

Suppose that we use this telescope with an eyepiece which has its tangential and sagittal focal surfaces coinciding exactly with the corresponding focal surfaces of the objective. In this way, the astigmatism of the eyepiece, which is mainly responsible for the off-axis unsharpness in telescopes, could be fully eliminated, resulting in an equally sharp image in the center and the field edge. Of course, this can only be achieved when the objective and the eyepiece are designed as an integrated system, and objective and eyepieces are fully matched.

Another group of systems are the relay telescopes. These are Cassegrain-like instruments in which the focus lies between the two mirrors, inside the system, resulting in a smaller secondary mirror and a simpler baffle system. In order to be viewed or photographed, the image is transferred out of the system by relay lenses to a location behind the primary mirror. Outstanding examples are the Dilworth (ref. 13.13), Sigler (ref. 13.14), Buchroeder (refs. 13.21 and 13.22), and Dall (ref. 13.15) systems.

A special aspect of the Sigler relay telescope, shown in fig. 13.19, is the flat backside of the Mangin secondary mirror. This allows an alternate configuration (also shown) in which a double-thickness secondary transmits the beam to the relay elements rather than reflecting it to them. The Dilworth is not shown because it is quite similar to the Sigler system. The main difference between them is that the back side of the Dilworth secondary is curved rather than flat.

Relay systems are not easy for the amateur telescope maker to fabricate because of their close centering and mounting tolerances. In most cases, their field of view is limited, making such systems mainly useful for visual observing.

Many unobstructed catadioptric systems other than the Schiefspiegler have been proposed. One such design, proposed by Wright (ref. 13.16), is a mixture of the Herschelian TCT and an eccentric section of a Maksutov corrector (fig. 13.20). Because of its closed tube, this unobstructed system may reasonably be expected to give excellent definition, making it particularly suitable for lunar and planetary observation. Note, however, that Wright's design incorporates a prismatic lens. This element would be difficult to make because of the close tolerances. An alternative to this design could employ a tilted concentric meniscus, which would be equally thick in all places, but it would suffer from color aberration.

Chapter 14

Field Correctors

14.1 Introduction

Field correctors are lenses, or combinations of lenses, placed in the converging light cone of the telescope in front of focus. Their purpose is to enlarge the usable field of the instrument by correcting image aberrations in an existing system without introducing other serious aberrations. Field correctors are used mainly in professional telescopes, but seldom in amateur instruments. The reason is that professional astronomers place much higher demands on off-axis image quality than most amateur astronomers do. However, amateurs should be aware of the possibilities created by field correctors, especially for fast Newtonians. In this chapter we will examine two kinds of correctors: field flatteners and field correctors for Newtonians.

14.2 The Single-Lens Field Flattener

As we have seen in previous chapters, many telescopes and astrocameras have a curved focal surface. For visual use, this is not detrimental because the eye accommodates when viewing off-axis images. If the telescope is used photographically, particularly when high demands are made on off-axis sharpness, however, it is desirable that best focus coincide with the photographic emulsion.

One way to accomplish this is to bend the film to the same curvature as the focal surface. Refractors, Newtonians, and Cassegrain-like telescopes all have an inward curving, or concave, field. In order to bend it properly, a piece of the film must be held against the inside of a concave cassette by suction from behind. It should be obvious that a precise focus would be difficult with such a system.

Another method is to use a field flattening lens.

The simplest possibility is a single plano-concave lens placed as close as possible in front of the film, with its flat side to the film. Fig. 14.1 shows this system with three converging bundles of light. Since the light path in the lens is longer at the edge than near the axis, the originally curved focal surface is flattened. When the radius of curvature of this focal surface is r, the radius of curvature of the lens, R, must be:

$$R = r \cdot \frac{n-1}{n} \tag{14.2.1}$$

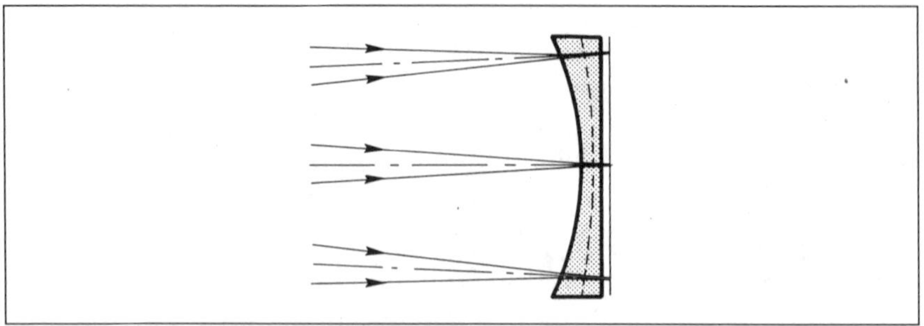

Fig. 14.1 *Field Flattener for a Cassegrain-like Telescope.*

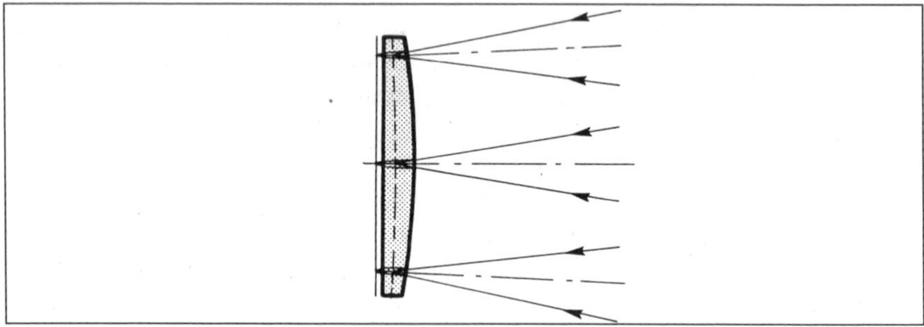

Fig. 14.2 *Field Flattener for a Schmidt Camera.*

where n is the refractive index of the glass in the color light for which the lens is designed. It must be emphasized that this formula is valid only when the original system has no astigmatism. Furthermore, in order to limit the aberrations, especially the color aberrations, such a lens must be placed very close to the film. If astigmatism is present, R must be optimized for the particular system to minimize astigmatism for the combined system.

This can be nicely illustrated with three examples. First, consider a focal field flattener that was designed for the 200 mm *f*/3 Schmidt camera, as described in section 8.6. This system has very little astigmatism, so we expect that the curvature R of the field flattener will follow the formula quite accurately. The radius of curvature of the convex focal surface, r, is = –600 mm. When we use a BK7 glass field flattener in blue light (n = 1.52237), the value of R turns out to be:

$$R = -600 \cdot \frac{0.52237}{1.52237} = -205.9 \text{ mm}. \tag{14.2.2}$$

This also appears to be the best practical value of the first surface of this plano-convex lens (fig. 14.2). The spot diagrams shown in fig. 14.3 were calculated with this value. In order to achieve this result, the Schmidt plate had to be moved 25 mm toward the mirror to minimize coma. Compare this with section 8.6.

Next, consider a field flattener that was designed for the 200 mm *f*/10 curved field Schmidt-Cassegrain described in section 9.3. Note that this instrument has a concave field, so in this case the field flattener is a plano-concave lens. The

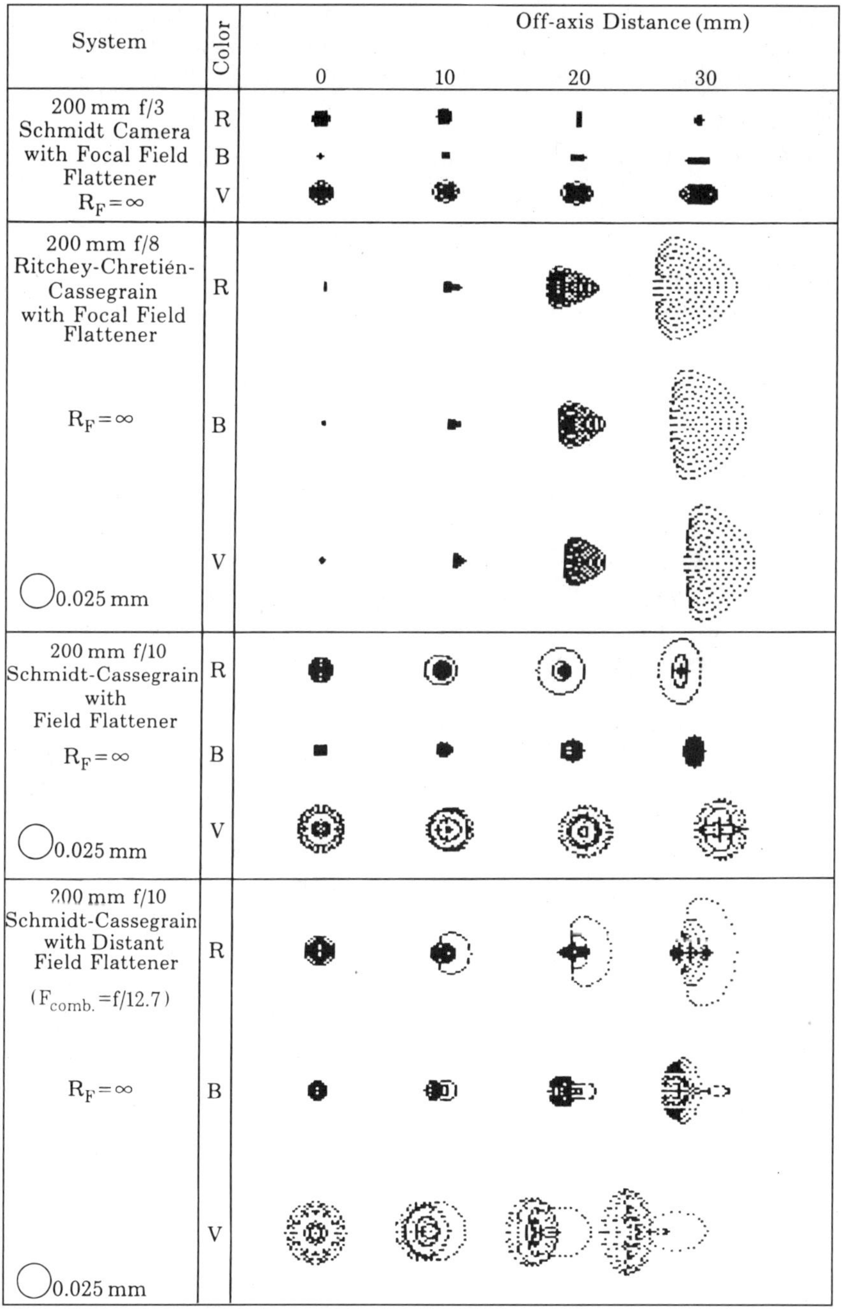

System	Color	Off-axis Distance (mm) 0	10	20	30
200 mm f/3 Schmidt Camera with Focal Field Flattener $R_F=\infty$	R				
	B				
	V				
200 mm f/8 Ritchey-Chretién-Cassegrain with Focal Field Flattener	R				
$R_F=\infty$	B				
◯0.025 mm	V				
200 mm f/10 Schmidt-Cassegrain with Field Flattener	R				
$R_F=\infty$	B				
◯0.025 mm	V				
200 mm f/10 Schmidt-Cassegrain with Distant Field Flattener ($F_{comb.}$=f/12.7)	R				
$R_F=\infty$	B				
◯0.025 mm	V				

Fig. 14.3 *Spot Diagrams for Four Optical Systems with Field Flatteners.*

Schmidt-Cassegrain has some residual astigmatism, so we expect that the first-order and optimized values of R will differ. The radius of curvature of the Schmidt-Cassegrain's focal surface is –155 mm (see fig. 9.4). Using BK7 glass again, we find R:

$$R = -155 \cdot \frac{0.52237}{1.52237} = -53.18 \text{ mm}. \quad (14.2.3)$$

However, when we optimize the radius for this particular system, we find it should be –73.95 mm.

Our third example is a telescope with pure and rather strong astigmatism: the 200 mm *f/*8 Ritchey-Chrétien described in section 7.3. Its optimum curved focal surface has a radius of –199 mm, so that the theoretical value of R is:

$$R = -199 \cdot \frac{0.52237}{1.52237} = -68.28 \text{ mm}. \quad (14.2.4)$$

The optimized value of the radius of the field flattener was found to be –82.8 mm. As it appears from fig. 14.3, the original astigmatism of the Ritchey-Chrétien could be reduced by 30% with the introduction of a field flattener though some coma was introduced (i.e., compare fig. 14.3 with fig. 7.3, upper row).

Note that while field flatteners with negative power exhibit some lateral color, this is opposite and much less important for the positive field flatteners. In some cases this effect can be favorable, particularly when the original system has lateral color of the opposite sign. This is the case, for instance, with the Simak described in section 4.4. When this instrument is combined with a single lens plano-concave field flattener of the correct dispersion, the combined system can be made free of lateral color.

The single-lens focal field flattener, having the advantage of a simple construction, has also its disadvantages. Mounting such a lens close to the film in a conventional 35 mm SLR or 6 by 6 cm camera body is difficult. Focusing is also a problem. Because the lens must be very close to the film plane, small surface irregularities (scratches) and dust particles will be visible as shadows on the film. Moreover, this lens is often a source of internal reflections (ghosts). These can be reduced by a good antireflection coating.

14.3 The Distant Field Flattener

To avoid the problems inherent in single-lens field flatteners close to the focal surface, we can call upon another design class: the distant field flattener. Because of its greater distance from the focal surface, this type of field corrector causes far fewer focusing and mounting problems, and may be used with 35 mm SLR and 6 by 6 cm reflex camera bodies. A representative of this design class is shown fig. 14.4. In order to suppress color aberrations, the design must be achromatized, i.e., it must have at least two elements. As is the case with the single-lens field flattener, the distant field flattener is a negative lens. As a result of the divergence of the

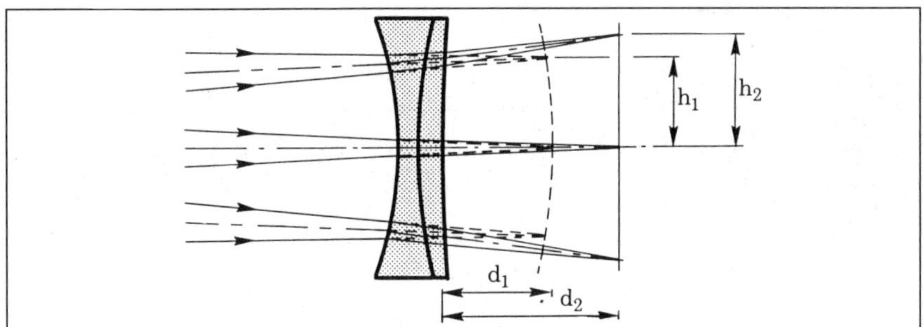

Fig. 14.4 *Distant Field-Flattener for a 200 mm f/10 Schmidt-Cassegrain Telescope.*

light bundles, the image scale, and also the effective focal length and focal ratio of the system will be increased by the ratio h_2/h_1 (defined in the figure), which is, of course, a disadvantage. In fact, a distant field flattener works like a focal extender (see sec. 15.1).

An important point, especially with respect to the two-element achromatic field flattener, is that this lens must be designed for a specific telescope. When a field flattener designed and optimized for one telescope is combined with another telescope, even when both telescopes have the same curvature of field, optimal functioning is not guaranteed. This is especially the case when the incident angles of the light cones from the objective (i.e., the objective's focal length, ratio, and field curvature) are different from those the lens was designed for.

The lens data in table 14.1 specifically refer to a field flattener for the 200 mm *f*/10 Schmidt-Cassegrain telescope described in section 9.3. The field flattener is placed 48 mm ahead of the film in an SLR camera, increasing the focal length of the combined system from 2027 mm to 2543 mm. Note that to achieve proper correction, the negative element must be crown glass and the positive element flint. In the example given, the leading element is the negative lens, which results in a biconcave lens combination. When the positive lens is placed in front, it will be a meniscus, and the resulting color aberrations will be slightly greater than those of the biconcave crown-first design, assuming the same glasses are used.

In fig. 14.3, we show spot diagrams for various off-axis distances and colors for a flat focal surface. Some of the color aberration in red and violet results from the Schmidt-Cassegrain telescope itself, because this instrument has some spherochromatism (see fig. 9.7).

14.4 Field Correctors for Newtonians

Over the years, many proposals have been made for increasing the useful field of a Newtonian telescope. The best-known example is that of the Ross corrector (ref. 14.1). Designed by Frank Ross in 1935, at the request of the director of Mount Wilson Observatory, for the 2.5-meter reflector, the system consisted of three lenses placed close to the focal plane in the converging light cone of the telescope. The corrector had no optical power, but it effectively corrected coma, expanding

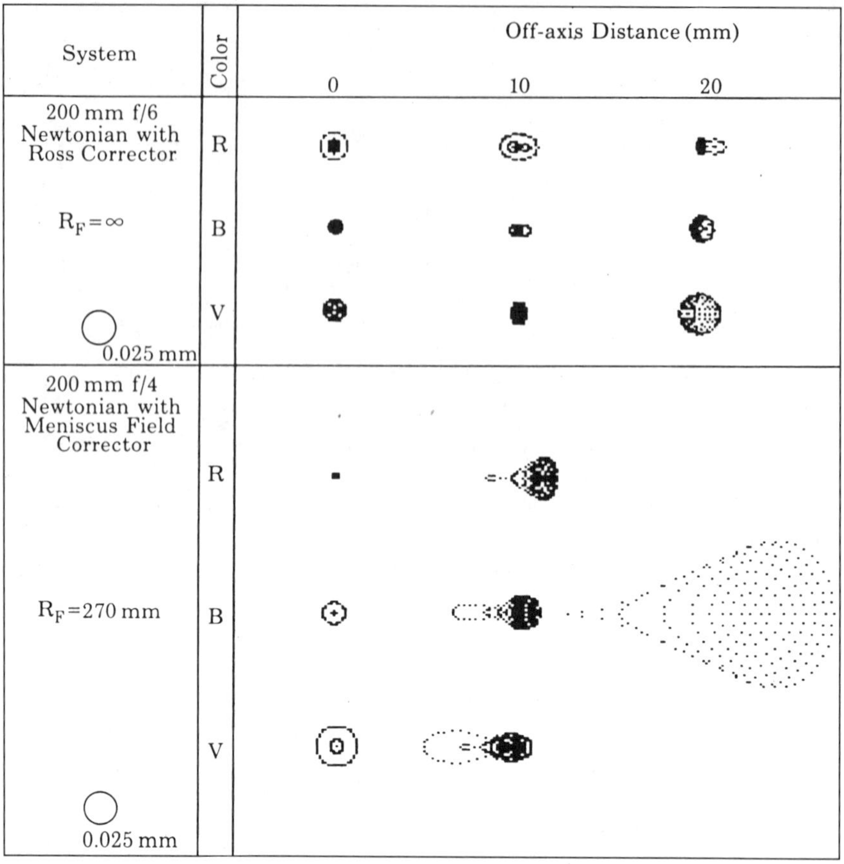

System	Color	Off-axis Distance (mm) 0	10	20
200 mm f/6 Newtonian with Ross Corrector	R			
$R_F = \infty$	B			
0.025 mm	V			
200 mm f/4 Newtonian with Meniscus Field Corrector	R			
$R_F = 270$ mm	B			
0.025 mm	V			

Fig. 14.5 *Spot Diagrams for Two Newtonians with Field Correctors.*

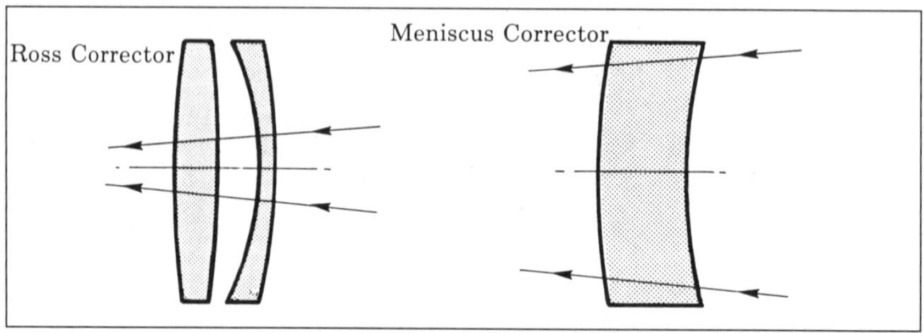

Fig. 14.6 *Field Correctors for Newtonians.*

Table 14.1
Field Correctors
(All dimensions in millimeters)

System	Dist. Field Flatnr.	Ross Corrector	Meniscus Corrector
System Corrected	200 mm *f*/10 Schmidt-Cassegrain	200 mm *f*/6 Parabolic Mirror	200 mm *f*/4 Parabolic Mirror
Combined Focal Ratio	*f*/12.7	*f*/5.8	*f*/4
Light Travels	left to right	right to left	right to left
Distance from Mirror	460.5	–1122	–530
R_1 Radius of Curvature	–102.41	–123.44	146.64
T_1 Axial Distance	4	–3.048	–25.04
M_1 Medium	517642	517642	517642
R_2	157.99	–60.45	146.64
T_2	5	–8.13	–270
M_2	762270	Air	Air
R_3	570.69	–217.42	
T_3	47.86	–8.89	
M_3	Air	517642	
R_4		217.42	
T_4		–64.53	
M_4		Air	
Effective Focal Length	2543.2	1196.4	800
1° Field	44.4	20.9	14.0

the photographically useful field from a few arcminutes to roughly half a degree. Without such correctors, large Newtonian telescopes would have been impractical.

The Ross corrector we analyzed was a combination with a 200 mm *f*/6 paraboloidal mirror, employing two lenses (ref. 14.3). This combination would be too poorly corrected for visual use—the axial spot diagrams are larger than the Airy disk (fig. 14.5)—but for photography, the combination satisfies the 0.025 mm criterion over a 24 mm by 36 mm field.

However, an even simpler solution (particularly suitable for amateur-size instruments) for reducing coma of a parabolic mirror was invented by Maksutov (ref. 14.2). Maksutov placed a relatively thick meniscus lens ahead of focus, with its concave side toward the mirror. A recent modification (ref. 14.4) is shown in fig. 14.6. Although coma is considerably less than that of an *f*/4 Newtonian, the sharpness in the outer field leaves much to be desired (compare figs. 14.5 and 13.2).

Chapter 15

Focal Extenders and Reducers

15.1 Focal Extenders

A focal extender is a system of lenses with negative power placed in the converging light cone of an objective to increase the effective focal length of this objective. Focal extenders—or "tele-X-tenders"—are often used in combination with photographic objectives as an easy and cheap way to produce a greater focal length. Those focal extenders typically consist of three or more lenses in order to maintain good off-axis sharpness.

The focal extenders used by amateur astronomers with telescopes are, in contrast, simpler. Most consist of two lens elements, and are called Barlow lenses. The focal length amplification factor usually lies between 1.5 and 3.0. A Barlow lens is an extremely useful accessory for the following reasons:

1. when it is used with a fast objective, for instance an *f*/5 Newtonian, high powers can be obtained without resorting to eyepieces having extremely short focal length and low eye relief;
2. when it is used for photographing lunar and planetary details, an increase of the image scale is easily attained; and,
3. the application of a Barlow lens usually improves the off-axis sharpness of the eyepiece used.

This last occurs because the image surface of most telescope objectives is inward curving; the Barlow tends to flatten the focal plane. (This is explained in greater detail in section 14.2, on field flatteners.) In addition, the greater focal ratio of the system reduces astigmatism present in the eyepiece.

Some manufacturers supply a set of eyepieces with a matched Barlow lens. The Barlow may be designed so that its inherent astigmatism partially compensates the eyepiece astigmatism, thus providing considerable improvement in the off-axis definition of the eyepieces used with it. Furthermore, a Barlow lens often permits the use of relatively poorly corrected eyepieces with Newtonians and other telescopes with fast focal ratios. These eyepieces are normally rejected by users of fast Newtonians because of the intolerable off-axis aberrations.

An important point that should be kept in mind is that the Barlow's amplification factor depends on the position of this lens with respect to the original focal

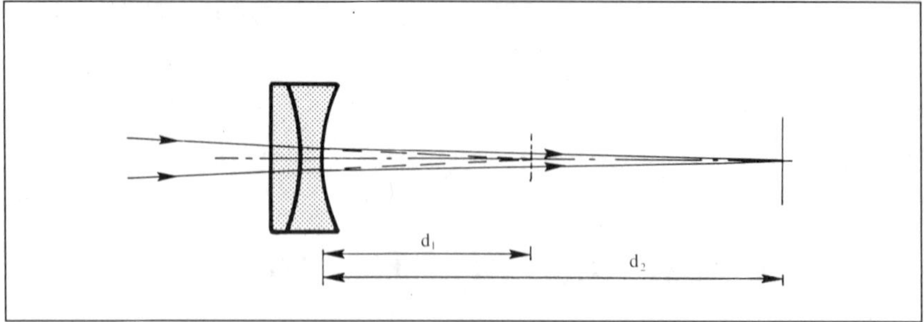

Fig. 15.1 *Barlow Lens for a 200 mm f/10 Schmidt-Cassegrain Telescope.*

plane. The amplification factor, M_B, and the effective focal length of the system used with a Barlow, F_{comb}, are computed from:

$$F_{comb} = \frac{F_o \cdot F_B}{F_B - d_1} \tag{15.1.1}$$

$$M_B = \frac{F_{comb}}{F_o} = \frac{F_B}{F_B - d_1} \tag{15.1.2}$$

where F_o is the focal length of the objective, F_B is the focal length of the Barlow lens, and d is distance the Barlow lens lies inside the original focus. In developing these equations, the thickness of the Barlow was ignored. The image height and the position of the new focal surface are thus:

$$h_2 = M_B \cdot h_1$$

and

$$d_2 = M_B \cdot d_1 .$$

Thus the amplification factor is independent of the focal length and focal ratio of the objective. When $d_1 = 0$, then $M_B = 1$, meaning that the Barlow has no effect. This phenomenon, in fact, also happens with a single-lens field flattener. When $d_1 = F_B$, the magnification goes to infinity, meaning that the Barlow has converted the converging beam into a parallel beam, and the telescope becomes a Galilean, or Dutch, telescope.

However, in the design of a Barlow, the optical characteristics of the objective must be taken into account in order to obtain optimum performance. Because coma, astigmatism, and field curvature are different in Newtonians, Cassegrains, and refractors, Barlow designs should be different too. Unfortunately, this is often ignored in practice. Few Barlow lens construction data have been published in the literature. In table 15.1, we give data for a Barlow designed for use with the 200 mm *f*/10 Schmidt-Cassegrain described in section 9.3. This Barlow design is shown in fig. 15.1.

Table 15.1
Focal Extenders and Reducers
(All dimensions in millimeters)

System	Focal Extender	Focal Reducer
Used With	200 mm *f*/10 Schmidt-Cassegrain	200 mm *f*/10 Schmidt-Cassegrain
Combined Focal Ratio	*f*/20	*f*/5.5
Distance from Secondary Mirror	432	390
R_1 Radius of Curvature	799.34	71.913
T_1 Axial Distance	8	20
M1 Medium	613370	517642
R_2	–58.038	–85.316
T_2	5	5
M_2	517642	762270
R_3	48.048	–225.59
T_3	124.75	45.91
M_3	Air	Air
Combined Effective Focal Length	4000	1100
1° Field	69.8	19.2

Common glasses were used in this case. Note that the positive element (made of flint glass) leads, while the negative lens (made of crown) comes second. This arrangement gave better performance. When the elements are reversed, the inner radii of the Barlow must be strongly curved, leading to greater off-axis aberrations. This is confirmed in the literature (ref. 15.1).

The amplification factor of the particular Barlow we designed is 2.0, but the aberrations are still acceptable when it is used at an amplification of 3.0. For greater amplification, the Barlow must be moved toward the telescope. The distance between a Barlow and its new focal plane is:

$$d_2 = F_B \cdot (M_B - 1) . \qquad \textbf{(15.1.3)}$$

The difficulty of designing a Barlow increases as

1. the Barlow amplification factor increases,
2. the focal ratio of the objective decreases, and
3. off-axis aberration present in the objective increases.

When designing a Barlow for use with an eyepiece, a rather subtle effect must be kept in mind. Normally telescope eyepieces are designed to work best when the principal (i.e., central) rays of the entering off-axis light cones are parallel to the optical axis. A Barlow bends off-axis light beams outward, as shown in fig. 14.4. When off-axis light enters the eyepiece at an angle it was not designed for, the exit pupil is moved back, and additional off-axis aberrations occur. The

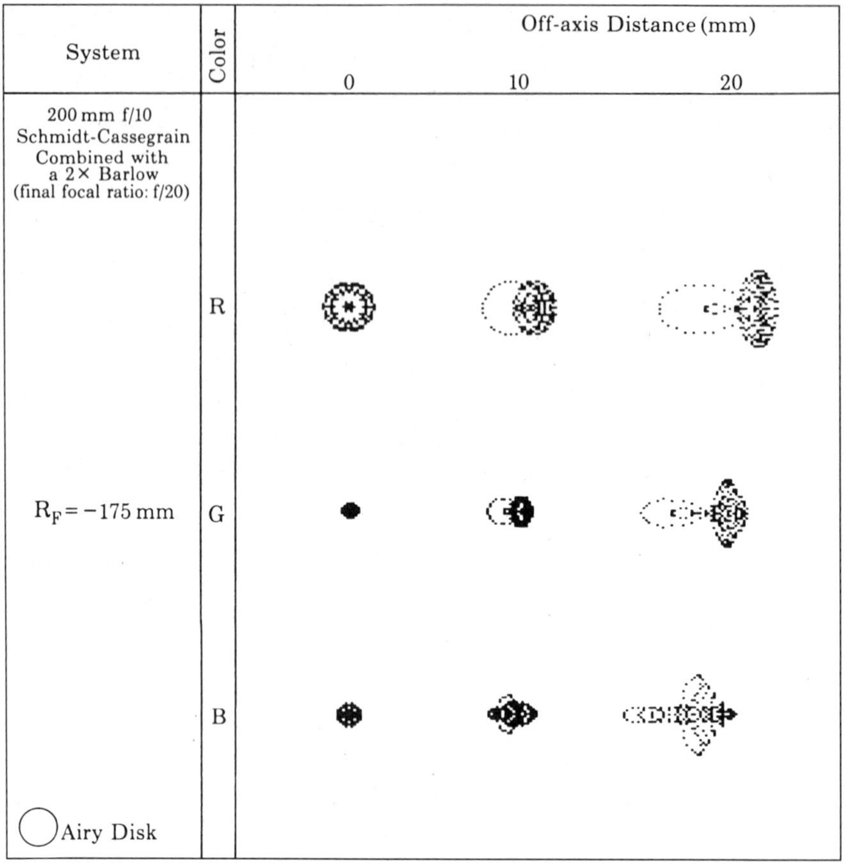

System	Color	Off-axis Distance (mm)		
		0	10	20
200 mm f/10 Schmidt-Cassegrain Combined with a 2× Barlow (final focal ratio: f/20)				
	R			
$R_F = -175$ mm	G			
	B			
Airy Disk				

Fig. 15.2 *Spot Diagrams for a 2 × Barlow Lens and 200mm f/10 Schmidt-Cassegrain.*

first effect is unavoidable, but the second can be suppressed, providing the eyepiece manufacturer designs his "matched" Barlow in such a way that this effect is taken into account.

We would like point out the possibility of using a Barlow lens attached to the diagonal in a Newtonian telescope. Because the Barlow extends the light cone, the diagonal can be made smaller than it would be without using the Barlow. This is favorable because the diagonal will then cause less diffraction (ref. 15.2).

15.2 Focal Reducers

A focal reducer (sometimes called a Shapley lens) is a positive system of lenses placed in the converging beam of an objective with the aim of decreasing the effective focal length of this objective. Focal reducers are used to increase the speed of slow systems (from, for instance, *f*/10 to *f*/5) in order to obtain shorter exposure times for astrophotography.

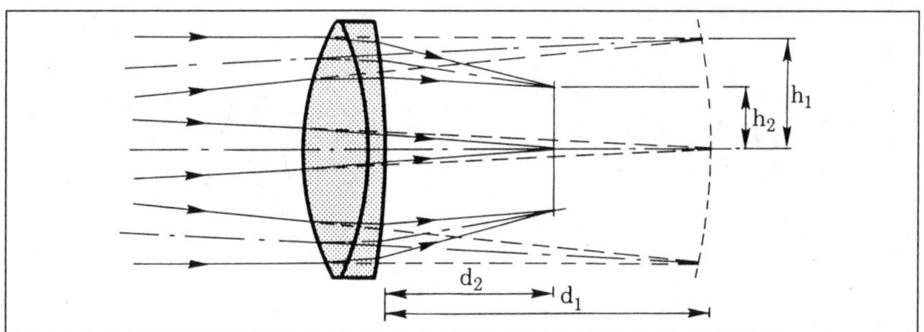

Fig. 15.3 *Focal Reducer for a 200 mm f/10 Schmidt-Cassegrain Telescope.*

Sometimes focal reducers are employed in hopes of increasing the angular field. This is often problematic in small telescopes because of limitations introduced by vignetting, either from the inner diameter of the baffle tube in Cassegrain-like instruments or from the size of the rack and pinion mount, which in fact prohibits increasing the angular field. Therefore, in many cases there is a reduction of the linear field coverage; often only part of the 24 by 36 mm negative is covered.

A focal reducer is a positive lens system and works in the opposite sense from a focal extender. This means there is a tendency to produce a focal surface more strongly curved than it was initially. Unless care is taken, off-axis image sharpness may suffer.

When a focal reducer is used, the focal length of the system and the amplification factor are computed as follows:

$$F_{\text{comb}} = \frac{F_o \cdot F_R}{F_R + d_1}$$

$$M_R = \frac{F_{\text{comb}}}{F_o} = \frac{F_R}{F_R + d_1}$$

the image height and the position of the new focal surface are thus:

$$h_2 = M_R \cdot h_1$$

and

$$d_2 = M_R \cdot d_1$$

where F_o is the focal length of the objective, F_R is the focal length of the focal reducer and d_1 is the distance the focal reducer lies inside the original focus.

In order to flatten the field it is necessary to design the focal reducer overcorrected for astigmatism, so that the situation shown in the third panel in fig. 4.11 occurs. Table 15.1 lists the optical data of a focal reducer designed especially for our 200 mm *f*/10 Schmidt-Cassegrain telescope (see fig. 15.3). The focal length

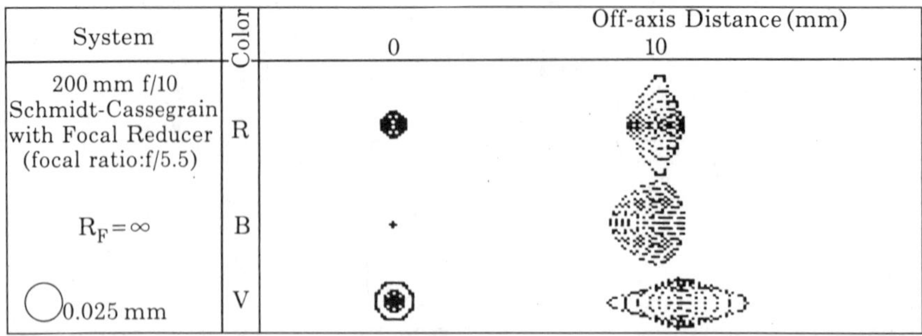

Fig. 15.4 *Spot Diagrams for a 200 mm Schmidt-Cassegrain with an f/5.5 Photographic Focal Reducer.*

reduction factor is 0.55, producing an overall focal length of 1100 mm.

Fig. 15.4 shows spot diagrams for the combination. The design has a useful field of about 25 mm, and is restricted by astigmatism and light drop-off. This lens turned out to be quite difficult to design. Because of its short focal length and resultant strong radii of curvature, astigmatism was difficult to suppress. A better design could probably be achieved with high-index glasses.

Before a focal reducer is used for serious astrophotography, especially with a Schmidt-Cassegrain system, the design should be optimized for the refocused telescope.

15.3 Remarks on Achromatic Combinations

In chapters 13, 14, and 15, we have seen a variety of two-lens combinations with positive or negative power. These include focal correctors, distant field flatteners, focal extenders, and focal reducers. To be achromatized these lenses must consist of a positive and a negative element.

When the combination has a positive power (for instance, the focal reducer), the positive lens is made from a crown glass, while the negative is a flint. For negative-power two-lens combinations, the situation reverses, and the positive element must be flint while the negative one is crown.

In every system we analyzed from the literature, the powers of the positive and negative element of the combinations were found to follow the thin-lens achromatization formula quite closely. This formula, derived in section 21.13.3 (the section on refractor design), states that the focal lengths of the positive and negative elements of an achromatic doublet are inversely proportional to the Abbe numbers of the glasses. In order to facilitate the correction of spherical aberration and spherochromatism, glasses are chosen so that the flint has a somewhat higher refractive index than the crown glass; this is true for both positive and negative lens combinations. The difference between the Abbe numbers of both glasses should not be too small or the design will require strongly curved surfaces, which hampers the correction of monochromatic aberrations.

The choice of glass remains a matter of trial and error. We found many designs in the literature with common flint and crown glasses. When it is important to reduce aberrations to a minimum, a special category of glasses should be considered: the lanthanum glasses. They combine a relatively high refractive index with a relatively low dispersion (see fig. 4.16), allowing shallower curved surfaces and facilitating the correction of aberrations. These glasses are not normally used in production telescopes because of their high cost. For the amateur telescope maker, however, the situation can be different. Because the lenses under consideration are often rather small, and because only one element made of lanthanum glass is sufficient, this glass is a viable alternative. The lanthanum glasses lie in the upper part of the glass chart.

When a two-lens combination is designed for use with a particular telescope, after the designer has determined the powers of the positive and the negative elements, the lenses should be bent to minimize the axial and off-axis aberrations. We have found that the best bending for minimal on-axis aberrations does not always coincide with the best bending for off-axis aberrations. In order to achieve the best off-axis correction, a slight amount of axial spherical aberration must be allowed. When the lenses are separated (i.e., not cemented), the inner radii of curvature may be different, giving the designer an extra degree of freedom to correct aberrations. Separating the elements is particularly recommended when coma from the telescope must be corrected.

Chapter 16

Eyepieces for Telescopes

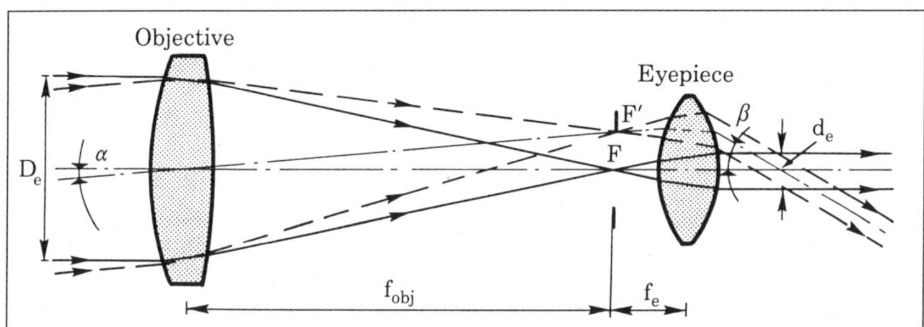

Fig. 16.1 *Image Formation in a Telescope.*

16.1 Introduction

The eyepiece is a vital part of every telescope used visually. The eyepiece is, in fact, the part which makes it possible to examine the image formed by the objective on the focal surface more closely than is possible with the unaided eye. Despite its importance, however, this component receives relatively little attention in the amateur-oriented literature. In this chapter we examine the optical characteristics and performance of a number of telescope eyepieces. We do not discuss how to design eyepieces because that rather specialized aspect of telescope design goes beyond the scope of this book.

A most important aspect of the eyepiece (self-evident but not always recognized by telescope users) is that the focal ratio of the light cone entering the eyepiece, f_e/d_e, is the same as the focal ratio of the objective system, or f_o/D_e, when there are no other optical elements between the objective and the eyepiece. This equality of focal ratio has important consequences, particularly in the off-axis performance of eyepieces. This is especially evident when eyepieces are used in combination with fast objectives, i.e., those with focal ratios under *f*/5. This is discussed fully in section 16.3.

For a thorough understanding of the ray paths through an eyepiece, we direct the reader to chapter 3, and to fig. 3.10 in particular. In that chapter, we defined the focal length of an eyepiece, its exit pupil, and the eyepiece diaphragm.

The angle β in fig. 16.1 between the axis-parallel and oblique exit bundles corresponds with half of the apparent field of this particular eyepiece. Eyepieces with a total apparent field, 2β, of 60° or more are called wide-field eyepieces. The

Scale: Focal Length	Name of Inventor	Apparent Field in Degrees	Year of Prototype	Remarks
	Huygens	50	1703	The first compound two lens eyepiece. With this type lateral color could be eliminated
	Ramsden	40	1783	
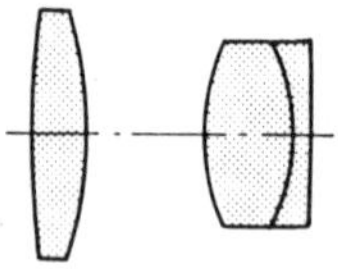	Kellner	50	1849	Incorrectly called an achromatised Ramsden
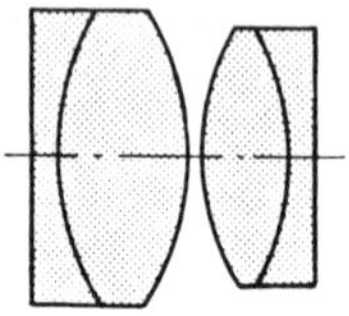	Plössl	50	1860	Also called symmetrical
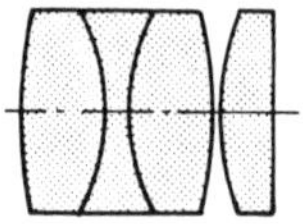	Abbe	50	1880	Also called orthoscopic because of its low degree of distortion
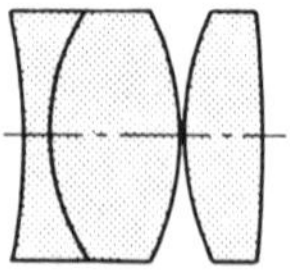	König	50	1915	
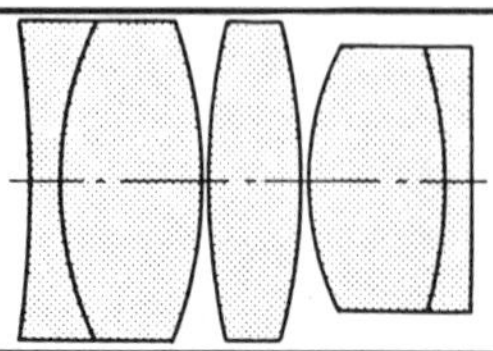	Erfle	70	1917	The first genuine wide angle eyepiece developed for military applications

Fig. 16.2 *Milestones in the Development of Telescope Eyepieces.*

Table 16.1

ε	Apparent Field
0.51	30°
0.69	40°
0.86	50°
1.03	60°
1.20	70°

size of the apparent field can be found with an accuracy of a few degrees, provided the eyepiece has only moderate distortion, from the ratio:

$$\varepsilon = \frac{\text{eyepiece diaphragm diameter}}{\text{eyepiece focal length}} \quad \textbf{(16.1.1)}$$

using table 16.1. Precise determination of the apparent field of a particular eyepiece requires an optical ray trace.

16.2 Eyepiece Types

The telescope eyepiece has had a long history of development. Between the very first compound two-lens design, the Huygenian, and the modern multi-element eyepiece, some 280 years have passed. Without making any claims for completeness, fig. 16.2 shows the important milestones in this development. The various designs are shown at the same scale.

Eyepieces generally bear the name of the original designer. Most of the types shown are not the original designs, but are modern modifications. Using modern high index glass, the axial image quality exceeds that of the prototype, and, in many cases, a larger apparent field has been realized with the same or improved sharpness at the edge of the field. The Huygenian and Ramsden have more or less gone out of use except in cheap telescopes. Long focal length Huygenian eyepieces, however, are still employed in combination with slow objective systems, for instance, in Schiefspieglers, and also in professional observatories. The reason for this will be explained later on.

Before World War II two-lens eyepieces were appreciated because of their low internal reflections. After the war anti-reflection coating techniques were improved to such an extent that multi-lens eyepieces with improved optical performance are now used to advantage.

A single-lens eyepiece suffers badly from the following aberrations: lateral color, astigmatism, curvature of field, and distortion. In addition, the field of view is very narrow. Huygens showed that lateral color, the most disturbing aberration, is corrected by choosing two lenses of the same glass having focal lengths f_1 and f_2, spaced a distance d apart:

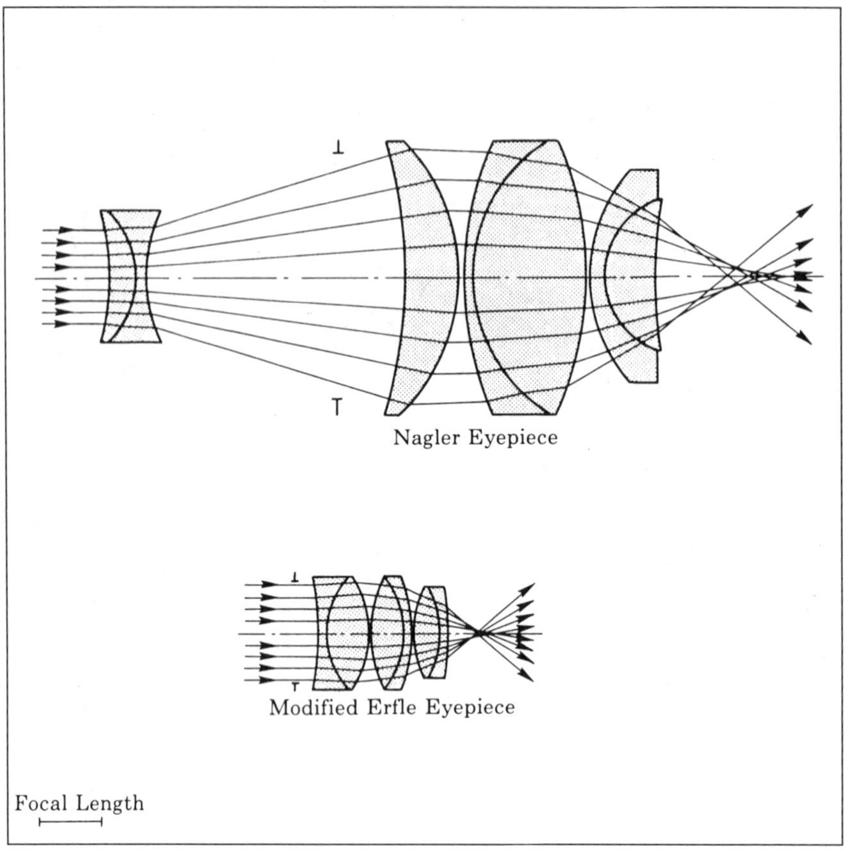

Fig. 16.3 *Cross Sections of Two Ultra-Wide-Field Eyepieces.*

$$d = \frac{f_1 + f_2}{2} . \tag{16.2.1}$$

Today, modified Huygenian eyepieces are made in such a way that the eye lens has half the focal length of the other lens. The lens closer to the objective serves as a field lens: as a result, the angular field of view is larger than in the single lens eyepiece. The most important aberration remaining is curvature of field.

Ramsden also chose two lenses of the same glass for his design, but set $f_1 = f_2 = d$. With this spacing, however, the exit pupil coincides with the eye lens, and the intermediate image with the field lens. To avoid these effects, the lenses are set closer together, causing lateral color to exceed that of the Huygenian. However, the intermediate image of the modified Ramsden lies in front of the field lens, and the exit pupil lies outside the eyelens.

With simple two-lens eyepieces, all available degrees of freedom must be employed to control spherical and chromatic aberration and coma. When designers substituted two-element (crown and flint) lenses, as did Kellner, they gained

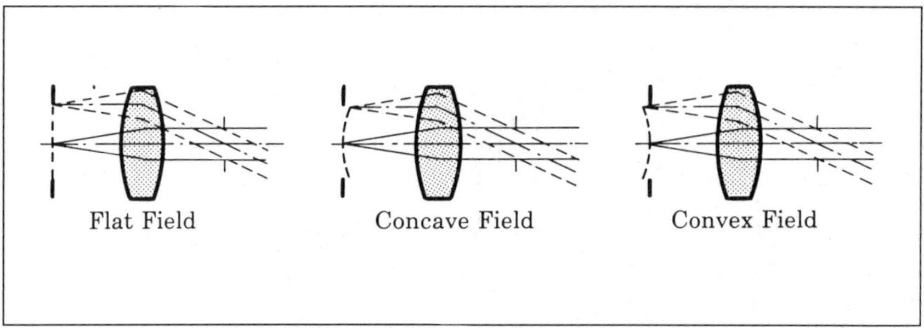

Fig. 16.4 *Flat and Curved Focal Surfaces in Eyepieces.*

the freedom to correct other aberrations and improve eye relief. The Huygenian and Ramsden suffer from eye relief only 0.2 to 0.3 times the focal length. Modern achromatic eyepiece designs offer eye relief between 0.4 and 0.9 times the focal length of the eyepiece, and sometimes even more.

We especially wish to direct the reader's attention to the König eyepiece and Plössl eyepiece designs because many modifications of these types are available, and they have become popular among amateurs.

The Abbe eyepiece is known for its low distortion, and therefore is referred to as an "orthoscopic" eyepiece. The first genuine wide-field eyepiece was the Erfle. In its original form, it consisted of two achromatic doublets with a biconvex lens in between. Since its invention in 1912, numerous modifications have appeared. Although a field larger than 50° is not strictly necessary for astronomical observation, the Erfle and other wide-field eyepieces are popular among amateurs.

We have included in our analysis two eyepieces having very large (i.e., 82°) apparent fields. The first is a modified six-element Erfle eyepiece; the second is a new development, the Nagler eyepiece. The Nagler was developed for the amateur market in 1980. Cross sectional drawings of these new designs are given in fig. 16.3. Note that the scales of figs. 16.2 and 16.3 are different.

16.3 Aberrations and Other Eyepiece Characteristics

The major difference between an objective and an eyepiece, as far as the designer is concerned, is the image angle it must cover. In contrast to an objective, which seldom exceeds 2°, the image angle of an eyepiece may exceed 80°. As a consequence, the designer must pay careful attention to the correction of off-axis aberrations.

Axial color and spherical aberration are normally well corrected in eyepieces. However, for use with objectives faster than *f*/5, spherical aberration in the eyepiece has often been insufficiently suppressed. Correction of coma seems to be easy because in every eyepiece the authors examined by means of ray tracing, coma was either small or absent.

Astigmatism and curvature of field are by far the most difficult to correct.

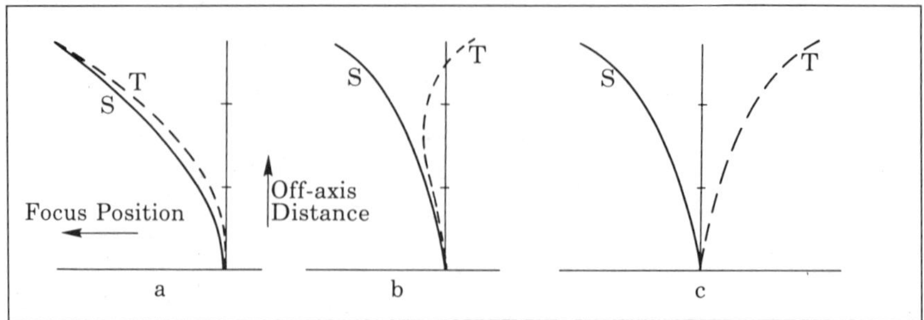

Fig. 16.5 *Tangential and Sagittal Focal Surfaces in Telescope Eyepieces.*

These aberrations must be treated in combination because correction of curvature of field often leads to an increase in astigmatism, and vice versa. The field of an eyepiece is usually curved: fig. 16.4 shows flat, concave, and convex focal surfaces. In conventional eyepieces, when the eyepiece is well corrected for astigmatism, the image surface is concave. Within limits, though, field curvature can be compensated by the accommodation power of the eye, i.e., its ability to focus in order to accommodate for viewing distance.

When an eyepiece suffers from field curvature, the accommodation required of the eye for a simultaneously sharp image at both the center and edge increases in inverse proportion to the focal length of the eyepiece. This means that for eyepieces with field curvature (for instance, the Huygenian), those of relatively long focal length are more satisfactory than an eyepiece with same design but having a short focal length.

If we assume that the focal surface of the objective is flat and the field of the eyepiece is concave, the observer should first focus on the edge of the field with an unaccommodated eye, then accommodate as much as necessary to view the center. Over a certain level—approximately three diopters for young observers and one diopter for older observers—accommodation becomes fatiguing to the eye. Therefore an observer will normally observe with an unaccommodated eye and focus on the center of the field. This means the edge of the field will appear unsharp.

To avoid this situation, designers generally prefer a flat focal surface, and will accept some astigmatism to get it. As we saw in fig. 4.9, when astigmatism is present, two focal surfaces exist: the tangential and sagittal focal surfaces. Fig. 16.5 shows three typical cases of the positions of the tangential and sagittal focal surfaces in eyepieces. Note that in the presentation of such curves, we always assume the eyepiece lies to the left of the curves and, according to a convention to be explained in section 16.4, light travels from the eye to the focal surface, from left to right.

Fig. 16.5(a) shows a situation in which a strong inward-curving focal surface (concave toward the eyepiece) occurs. This is typical of the Huygenian eyepiece. Although astigmatism is well corrected, this situation is undesirable, particularly for short focal length eyepieces, because the eye cannot easily compensate the field curvature by accommodation.

In fig. 16.5(b) we show an intermediate case, one with more astigmatism but a less strongly curved field.

In fig. 16.5(c), the tangential and sagittal focal surfaces are symmetrically placed with respect to a plane perpendicular to the optical axis, thus avoiding field curvature altogether. As noted above, many designers prefer this correction; such curves are typical of modern eyepieces.

As we saw in fig. 4.9, the image of a star on the tangential focal surface appears as a horizontal focal line, and on the sagittal focal surface as a vertical focal line. Midway between these surfaces the image of a star appears as a circular blur of light. The diameter of this blur is inversely proportional to the focal ratio of the light cones entering the eyepiece. If the focal ratio is *f*/5, the diameter of the blur will be three times as large as that of an *f*/15 light cone. This is the reason why the edge unsharpness of a particular eyepiece used in combination with an *f*/5 telescope will be much more detrimental than in combination with an *f*/15 telescope. Users of fast Newtonians often attribute image softness at the edge of their eyepiece field to the coma of the mirror. In reality, however, eyepiece astigmatism usually dominates coma because the eyepiece is not corrected for such a fast light cone.

A particularly annoying aberration is lateral color, in which the magnification of the image depends on the color of the light. This manifests itself by more or less strongly colored fringes around objects off the optical axis.

Another aberration is distortion. Distortion becomes especially important in wide-field eyepieces. In discussing distortion, however, one must clearly distinguish between rectilinear distortion and angular magnification distortion. For terrestrial telescopes, it is often required that straight lines in the focal surface look straight in the eyepiece. For zero rectilinear distortion the following relationship should apply:

$$y = f \cdot \tan\beta \qquad \textbf{(16.3.1)}$$

where y is the off-axis distance in the focal plane, β the image angle from the optical axis, and f the focal length of the eyepiece.

For astronomical observation, however, it is important that the angular magnification remain constant over the field. For instance the angular distance between double stars should be the same and a round object (a planet) should retain its shape whether viewed in the center of the field or at the edge of the field. In this case, the following relationship should apply:

$$y = f \cdot \beta \qquad \textbf{(16.3.2)}$$

where β is expressed in radians.

With zero angular magnification distortion, straight lines on a focal plane appear curved in a pincushion fashion, with the curvature becoming greater the farther they lie from the center. It is impossible to correct an eyepiece simultaneously

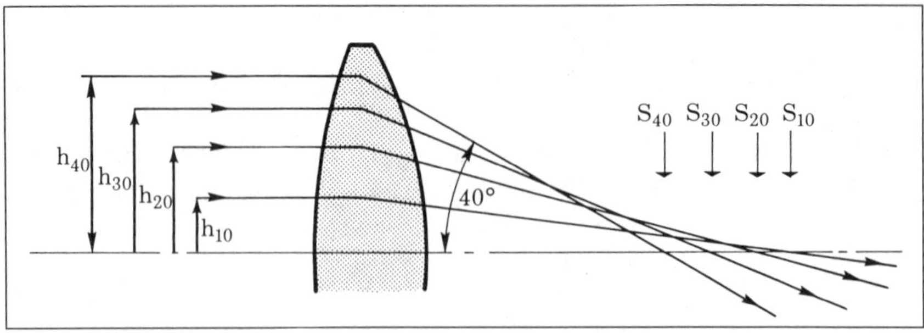

Fig. 16.6 *Intersection Points for Axis-Parallel Principal Rays.*

for rectilinear distortion and angular magnification distortion.

Yet another phenomenon of eyepieces is spherical aberration of the exit pupil. Although not an aberration in the normal sense, because it does not influence the image sharpness, in eyepieces with long focal lengths, wide fields, or both, it can be annoying. In fig. 16.6., we show an eyepiece as a single lens. Four principal rays (i.e., central rays of bundles) enter it parallel to the optical axis, at different distances from the optical axis. Popular treatments of telescope optics state that all exit beams leaving the last eyepiece lens will intersect in the exit pupil. This means that the principal rays of the beams intersect in the same point on the optical axis.

In reality, this is only an approximation. Generally, the rays with large exit angles will intersect the optical axis nearer the last eyepiece lens than rays with moderate or small exit angles. For eyepieces with moderate apparent fields, 50° for example, and short focal lengths, this is not detrimental. For ultrawide eyepieces and eyepieces with long focal lengths, the observer may need to move his eye toward the eyepiece in order to see the outer parts of the field. Parts of the field lying between the center and the edge may then become invisible. This manifests itself as fleeting bean-shaped shadows moving around the field, a phenomenon referred to as the kidney-bean effect. In the treatment of the wide-field eyepieces in section 16.5, this effect will be explored more in detail.

Shadows resulting from the kidney-bean effect should not be confused with another effect which occurs when a telescope is used for daytime terrestrial observation. In full daylight, the pupil of the eye is 2 mm to 3 mm. If the exit pupil of the telescope is small too, then it is difficult for many observers to hold their heads steady enough to keep the pupil of the eye in the exit pupil of the telescope. Small side-to-side shifts of the head may cause the image to disappear partly or totally, or a black shadow to move across the field.

The same phenomenon can occur when the observer tries to observe at the field edge of a wide-field eyepiece. Because the eye pivots in its socket with the center of rotation behind the pupil, the pupil shifts sideways. The eye can lose the image unless the observer simultaneously shifts his head to the side. These effects are much more important during daytime observation than during the night because the pupil is then so much larger.

Users of Newtonian and Cassegrain-like telescopes often find their instruments difficult to use at low magnification during the daytime because of a shadow in the center of the field. This results when the small pupil of the eye lies in the image of the secondary mirror, which is typically ⅓ the diameter of the exit pupil. The eye receives no light when the observer looks straight into the eyepiece, but when the eye pivots, the edge of the field can be seen.

To astronomers, it is important that the pupil of the eye be larger than the exit pupil of the telescope so the exit pupil determines the diameter of the pencil of light entering the eye. In daytime observation, for instance with a binocular, an entirely different situation may occur.

Consider the case of the 7 by 50 binocular. Such a binocular has an exit pupil of $^{50}/_{7}$ = 7.1 mm. Binocular objectives normally have a focal length of 200 mm, or a speed of *f*/4. If the full 7.1 mm exit pupil were used during daytime observations, the simple Kellner eyepieces would exhibit intolerable astigmatism because this type of eyepiece is insufficiently corrected to handle an *f*/4 light cone. However, in the day, the eye pupil is typically only 2.5 mm, so the exit pupil is vignetted to this diameter, and the effective focal ratio of the objective is *f*/11.4—much more favorable for the eyepiece. Of course, only a small part of the 50 mm entrance pupil is used; in this case, the effective entrance pupil is only 17 mm!

During the night the eye pupil expands to approximately 7 mm, so the full *f*/4 light cone can pass through the instrument. Of course the eyepiece will exhibit large off-axis aberrations, but at this low light intensity, visual acuity drops by a factor 10 to 20, so the off-axis unsharpness of the binocular is not disturbing to the eye.

This explains why the 7 by 50 binocular can be used during daytime and during the night, and give acceptable performance, despite the fact that a fast objective and simple eyepieces are used.

The aberrations we have discussed above will be examined by means of spot diagrams in section 16.4. We emphasize that these spot diagrams exhibit no aberrations due to the objective; all aberrations shown originate in the eyepiece itself. Generally the off-axis image aberrations of the objective, including its field curvature, are relatively unimportant in comparison to the eyepiece aberrations, as we show in section 16.7. Consequently, the off-axis ray-trace results of an eyepiece give, in most cases, an accurate picture of the off-axis performance of the telescope as a whole, when it is used visually.

16.4 Ray-Tracing Eyepieces

A thorough evaluation of the optical performance of an eyepiece requires considerably more work than does evaluation of an objective system. Let us see why this is so. In chapter 4 we explained how to ray trace reflecting or refracting objectives. In principle the same methods can also be applied to eyepieces, but there are significant differences and several complications. In ray-tracing an objective, rays travel from a star and come to focus. For tracing eyepieces, we reverse the direc-

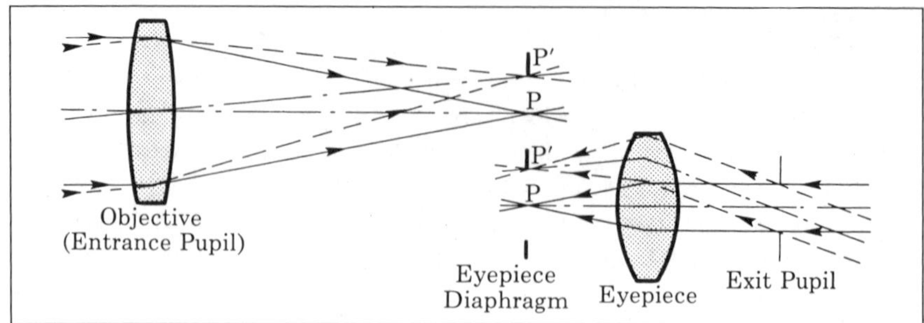

Fig. 16.7 *Ray Tracing Objective and Eyepiece.*

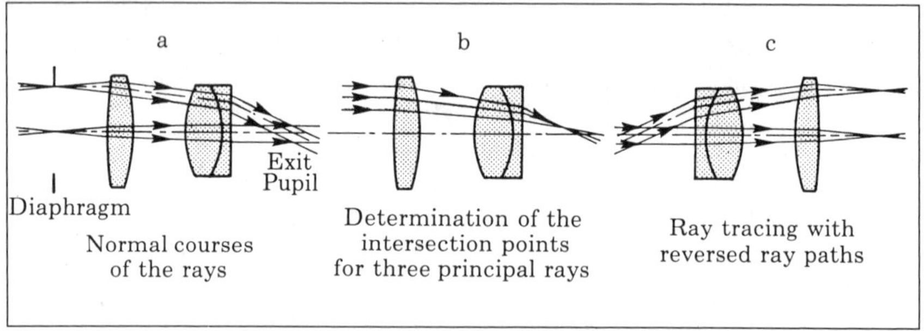

Fig. 16.8 *The Ray-Tracing Procedure for a Kellner Eyepiece.*

tion of travel: parallel beams of light travel from the eye to the eyepiece and come to focus at the focal plane (see fig. 16.7). This calculation leads us to the same conclusions as would tracing real rays through the eye, but is easier than tracing the true course of the light rays.

Before discussing the calculation scheme, several points deserve mention. Construction data of an eyepiece, such as the axial thicknesses of lenses, distances between lenses, and radii of curvature, are generally given for a focal length of 100 mm, and in the normal sequence of the lenses, with light traveling from left to right (see table 16.2). On this basis the sizes of the blurs produced by the eyepiece are calculated. The image sharpness is determined by the apparent diameter of the spot diagrams. For a focal length of 100 mm, one minute of arc corresponds to 0.0291 mm in the focal plane. We then compare the angular sizes of the calculated spot diagrams with a maximum allowable spread criterion. Near the center of the field the blur should not exceed one minute of arc, the maximum resolving power of the eye for bright objects. At the edge of the field we allow a spread of five minutes of arc.

What problems do we encounter in tracing an eyepiece that we do not encounter when tracing an objective? Recall that the first mirror or lens surface encountered is normally the entrance pupil of an objective. However, when we trace an eyepiece, the "entrance" pupil lies "in front of" the eye lens, outside the system, so the size of the entering pencil is limited by a fictitious diaphragm at the position of the exit pupil.

Another problem is that the exact position of the fictitious entrance pupil is not known at the outset because of the spherical aberration of the exit pupil. Before we can begin the ray trace, we must determine the exact positions of the intersecting points, *S*, for a set of angles, as shown in fig. 16.6. The calculation, shown graphically in fig. 16.8, proceeds as follows:

Decide the off-axis angles for which spot diagrams should be calculated. For a 50° field, you might use 0, 10, 20, and 25° off-axis; for a 60° field, 0, 10, 20, and 30° off-axis; for an 80° field, 0, 10, 20, 30, and 40° off-axis.

Next, determine the intersecting points *S*. (This process is shown in fig. 16.6 for a single lens eyepiece, and in fig. 16.8(b) for a Kellner eyepiece.) Begin by finding the off-axis distances of the horizontally entering rays, *h*, shown in fig. 16.6. These rays are the central (principal) rays of entering light cones which exit the eyepiece at 10, 20, etc., degree angles from the optical axis.

Unfortunately, no direct method of doing this exists, so a trial and error method is necessary. Table 16.1 may be useful in this respect. Search for *h* by moving these rays up and down until the exit angles are exactly 10°, 20°, etc. From the ray heights, calculate the intersection points, *S*, along the optical axis.

Now reverse the eyepiece. This is the equivalent of changing the direction of travel of the light with respect to that of the foregoing calculation.

Establish the exact position of the focal surface by tracing a paraxial bundle of green light (546 nm); then trace the other parallel bundles of light for the various off-axis image angles. The central rays of these pencils pass through the intersection points, *S*, as calculated in the first step.

Calculate the initial set of spot diagrams for a flat focal surface. These should be done in three colors (blue, green, and red) and for a range of focal ratios, for example, *f*/5, *f*/10, and *f*/15. The diameters of the entering beams for an eyepiece with a focal length of 100 mm will be 20 mm for an *f*/5 objective, 10 mm for an *f*/10 objective, and 6.67 mm for an *f*/15 objective.

The focal surface may be curved rather than flat, which further complicates the analysis. If the off-axis spot diagrams seem excessively large, shift the focus for them with respect to the paraxial focal plane to see if they can be made smaller. If they can be, then the sizes of the spot diagrams must be calculated for the curved focal surface and then analyzed to determine whether the eye is able to compensate for the field curvature by accommodation.

This scheme of eyepiece analysis clearly involves more complications than analyzing an objective system does. While the analysis of an objective is largely straightforward, a thorough analysis of an eyepiece requires many trial calculations.

16.5 Ray-Trace Results for Eyepieces

In making a general study of the performance of eyepieces, the designer soon discovers that every eyepiece type has numerous variants and modifications. Moreover, the development of certain types (the Plössl, for example) is still going on.

Table 16.2 Part 1
Construction Data for Nine Telescope Eyepieces
(Focal Length of 100 mm)

	Huygenian	Ramsden	Kellner	Plössl	Abbe
Dist[a]	–49.53	21.24	62.47	59.65	56.38
R_1[b]	91.6	∞	404.5	∞	184.07
T_1[c]	16.8	12.2	21.86	10	31.31
M_1[d]	607567	517642	620603	667330	607494
R_2	∞	–80.14	–163.43	101.8	–77.53
T_2	101	86.8	45.76	50	6.25
M_2	Air	Air	Air	564604	762265
R_3	45.8	64.11	86.82	–101.8	77.53
T_3	7.2	9.6	34.42	5	34.31
M_3	607567	517642	603606	Air	607595
R_4	∞	∞	–79.47	93.3	–156.06
T_4			5.72	35	0.6
M_4			728284	564608	Air
R_5			–1190	–97.1	101.04
T_5				10	21.91
M_5				667330	728284
R_6				∞	∞
T_6					
M_6					
R_7					
T_7					
M_7					
R_8					
T_8					
M_8					
R_9					
T_9					
M_9					
R_{10}					
T_{10}					
M_{10}					
R_{11}					
Eye relief[e]	10/20.18	10/33.34	10/47.70	10/68.99	10/81.00
	20/17.35	20/16.46	20/43.93	20/63.32	20/80.88
	25/15.08	25/–	25/40.83	25/58.60	30/82.34

a. The distance from the focal plane to the first lens can be negative when this focal plane lies between the lenses.

b. Radius of curvature

c. Axial thickness or distance

d. Medium, glass or air, six digit code

e. The distance between the exit pupil and the eye lens for various exit angles (angle degree/distance mm)

Table 16.2 Part 2
Construction Data for Nine Telescope Eyepieces
(Focal length of 100 mm)

	König	Erfle 5-lens	Erfle 6-lens	Nagler 7-lens
Dist[a]	51.155	39.615	23.357	–155.60
R_1[b]	–225	–465.1	–466	–376
T_1[c]	10	9.8	9.1	38
M_1[d]	755276	648338	648338	717295
R_2	83.6	136.13	131.5	–140
T_2	50	55.01	61.4	15
M_2	623569	552635	517642	620603
R_3	–102	–179.54	–156	235
T_3	0.7	0.7	0.9	405
M_3	Air	Air	Air	Air
R_4	110	246.05	194.5	–756
T_4	33.3	39.31	52.2	84
M_4	607567	487704	557587	620603
R_5	–458	–246.05	–141	–299
T_5		0.7	9	5
M_5		Air	728284	Air
R_6		107.02	–247	529
T_6		52.31	0.9	15
M_6		552635	Air	717295
R_7		–176.04	143	252
T_7		9.8	38.7	168
M_7		728284	557587	620603
R_8		–3020.62	–168	–529
T_8			9.1	5
M_8			728284	Air
R_9			–451	252
T_9				15
M_9				717295
R_{10}				127
T_{10}				84
M_{10}				620603
R_{11}				756
Eye relief[e]	10/97.42	10/57.53	10/63.47	10/185.26
	20/91.61	20/55.73	20/59.96	20/173.82
	30/88.21	30/49.93	30/54.31	30/152.72
		35/46.75	40/47.04	40/121.31

a. The distance from the focal plane to the first lens can be negative when this focal plane lies between the lenses.
b. Radius of curvature
c. Axial thickness or distance
d. Medium, glass or air, six digit code
e. The distance between the exit pupil and the eye lens for various exit angles (angle degree/distance mm)

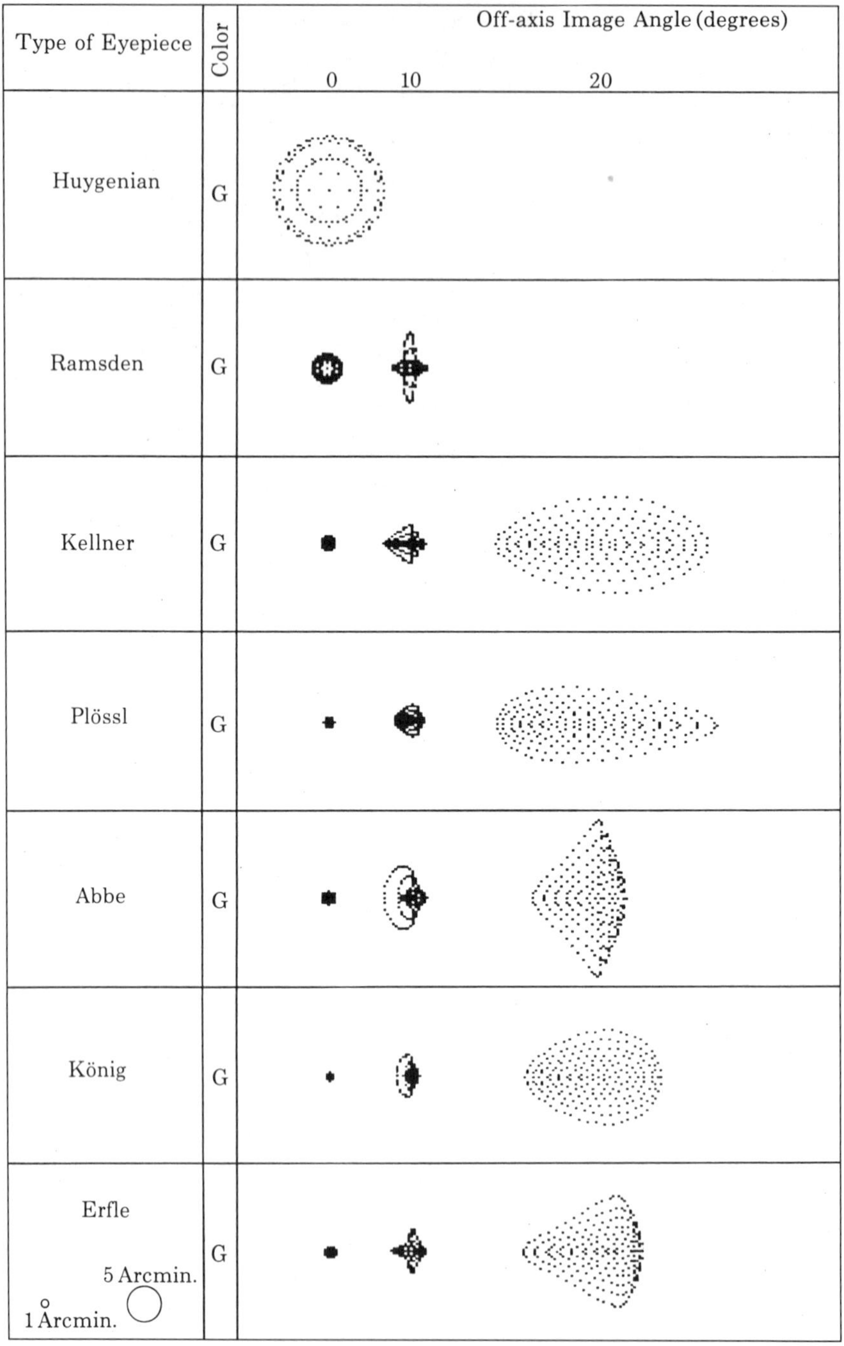

Type of Eyepiece	Color	Off-axis Image Angle (degrees) 0 10 20
Huygenian	G	
Ramsden	G	
Kellner	G	
Plössl	G	
Abbe	G	
König	G	
Erfle 5 Arcmin. 1 Arcmin.	G	

Fig. 16.9 *Optical Performance of Seven Eyepieces for an f/5 Light Cone and Flat Field.*

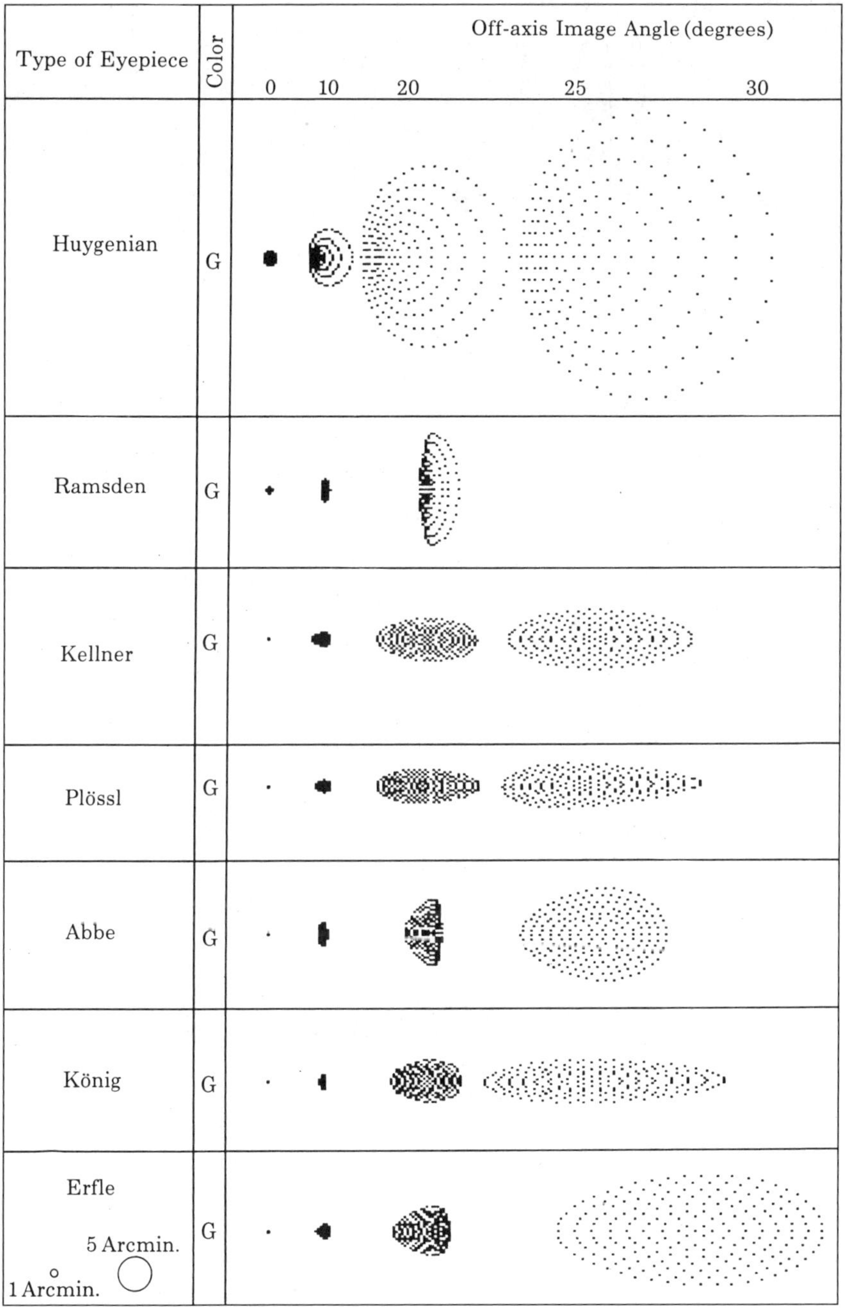

Type of Eyepiece	Color	Off-axis Image Angle (degrees) 0	10	20	25	30
Huygenian	G					
Ramsden	G					
Kellner	G					
Plössl	G					
Abbe	G					
König	G					
Erfle 5 Arcmin. 1 Arcmin.	G					

Fig. 16.10 *Optical Performance of Seven Eyepieces for an f/10 Light Cone and Flat Field.*

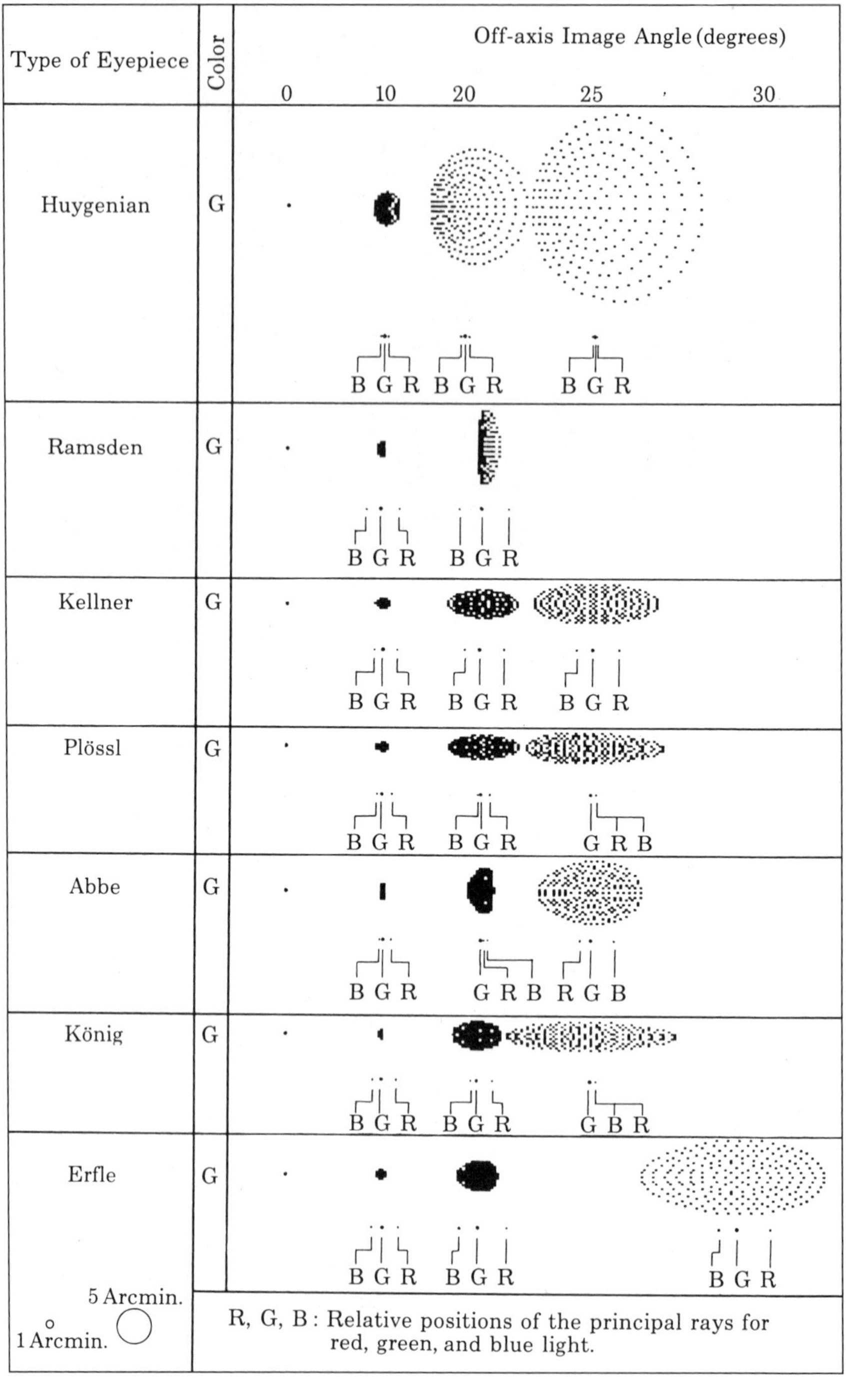

Type of Eyepiece	Color	Off-axis Image Angle (degrees) 0	10	20	25	30
Huygenian	G		B G R	B G R	B G R	
Ramsden	G		B G R	B G R		
Kellner	G		B G R	B G R	B G R	
Plössl	G		B G R	B G R	G R B	
Abbe	G		B G R	G R B	R G B	
König	G		B G R	B G R	G B R	
Erfle	G		B G R	B G R		B G R
5 Arcmin. 1 Arcmin.		R, G, B : Relative positions of the principal rays for red, green, and blue light.				

Fig. 16.11 *Optical Performance of Seven Eyepieces for an f/15 Light Cone and Flat Field.*

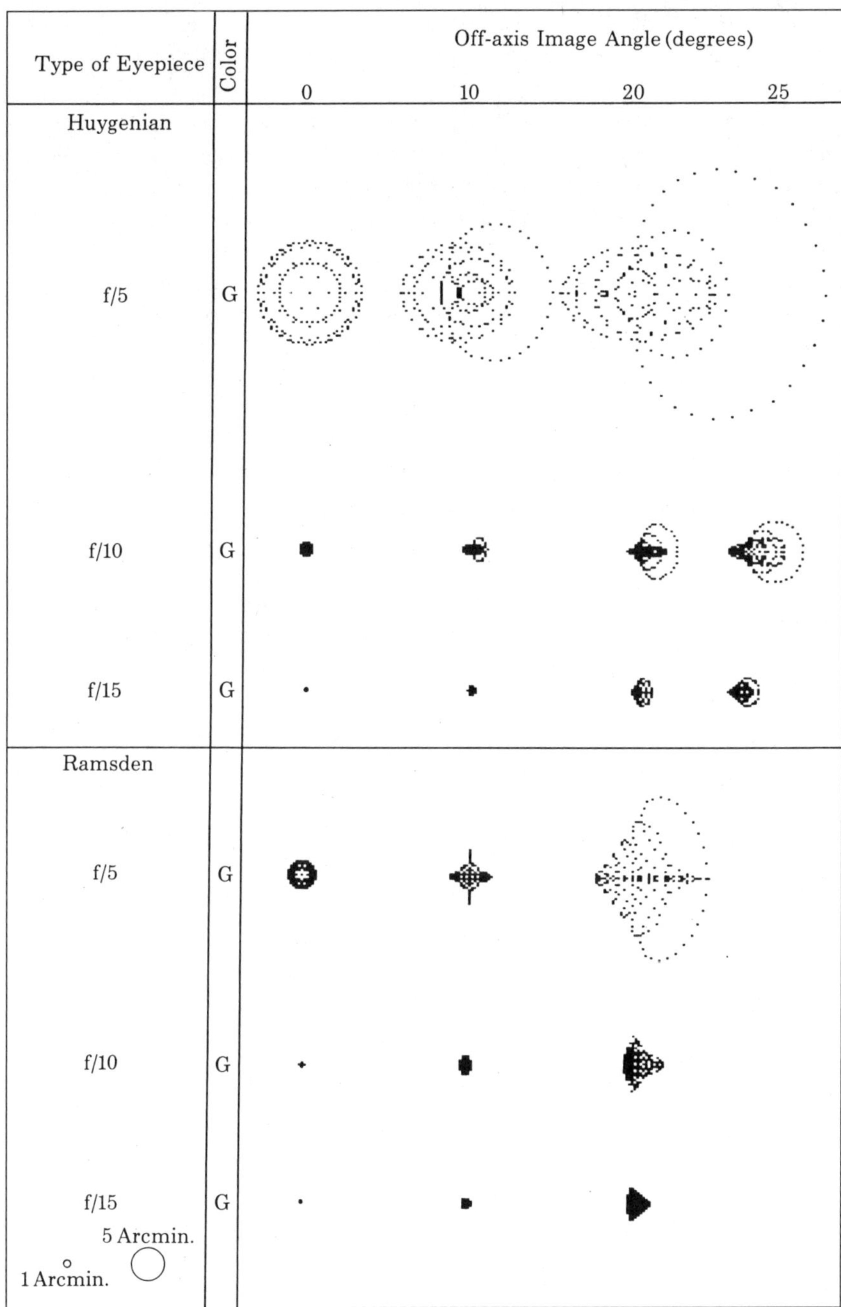

Fig. 16.12 *Optical Performance of Two Eyepieces for Various Light Cones and Curved Field.*

Luckily, though there are many designs, we need not contend with an endless proliferation of names. As long as the configuration of lenses remains the same, a design retains the name of its original designer.

Eyepiece manufacturers seldom publish construction data for their products. The construction data given in table 16.2 were obtained from many sources. In studying eyepiece characteristics, one is entirely dependent upon data published in the literature. It is hardly possible, based on a literature study alone, to draw specific conclusions about the performance of a type in comparison with others.

In the end, the best and most effective method to judge a particular eyepiece remains testing and comparing it by actual observation. Nevertheless, it is possible to determine the broad characteristics of various designs. Analyses were done for all the eyepieces shown in figs. 16.2 and 16.3, but the results for the two groups are presented in a somewhat different way.

Figs. 16.9, 16.10, and 16.11 show the results for the eyepiece types shown in fig. 16.2, the Huygens, Ramsden, Kellner, Plössl, Abbe, König, and Erfle. These spot diagrams are for green light, for a flat focal surface, for appropriate off-axis image angles. Fig. 16.9 shows spot diagrams for *f*/5; fig. 16.10, for *f*/10; and fig. 16.11, for an *f*/15 focal ratio.

In order to obtain an impression of lateral color (color aberration outside the optical axis), compare the relative position of the intersecting points of the principal rays in the focal plane for blue and red with respect to green as shown in fig. 16.11. We have omitted here the complete spot diagram for each color because of their similarity in shape and size to the green spot diagrams. Moreover spot diagrams larger than 40 minutes of arc are not shown in these figures because of their large size. Of the eyepieces shown, the Huygenian and the Ramsden suffer most obviously from curvature of field. In fig. 16.12, we show spot diagrams for the best focal surface. It appears that the image sharpness for the curved focal surface is much better than for a flat focal surface.

From figs. 16.9 through 16.12, we draw the following conclusions:

1. Image sharpness depends strongly on the focal ratio of the objective. None of the eyepieces is sufficiently well corrected to give excellent performance over its entire field when combined with an objective having a focal ratio of *f*/5, mainly as a result of astigmatism.

2. Even with an *f*/15 objective, the image blur at the edge of the field considerably exceeds five minutes of arc.

3. Lateral color is moderate except in the Ramsden, Kellner and Erfle. The Huygenian exhibits particularly low lateral color.

4. The Huygenian exhibits a large amount of spherical aberration at *f*/5. It cannot even form an acceptable axial image at that focal ratio.

5. The Huygenian and Ramsden exhibit better sharpness for the curved field than for the flat field. Critical observation of off-axis objects

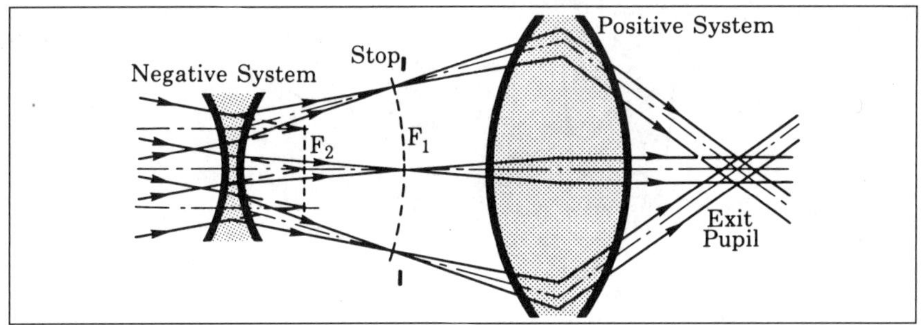

Fig. 16.13 *Fundamentals of the Nagler Eyepiece.*

requires refocusing by moving these eyepieces forward. Then, of course, the center of the field will become unsharp to the extent the eye is unable to accommodate.

To use a Huygenian with a 50° field for an image that is simultaneously sharp at both center and edge requires an accommodation of one diopter for an eyepiece having a focal length of 110 mm. Recall that the amount of accommodation is inversely proportional to the focal length. This, and conclusions (4) and (5) explain why Huygenian eyepieces with focal lengths up to 150 mm are still used in professional observatories, in combination with large refractors.

In the foregoing we have seen that astigmatism and curvature of field strongly influence the image sharpness outside the optical axis, particularly when used with a fast (i.e., *f*/5) objective system. Although eyepiece designers had already tried for many years to eliminate these aberrations, it was not until 1980 that a major breakthrough, the Nagler eyepiece, occurred. Because of this, we give this design a more detailed treatment than other eyepiece designs.

We now examine the optical performance and characteristics of the Nagler eyepiece compared to a six-lens modified Erfle. Both eyepieces have a 13 mm focal length and an apparent field of 82°. Fig. 16.3 shows both at the same scale. The Nagler is much larger, and contains more elements (7) than the Erfle. Unlike most modern eyepieces, the field diaphragm of the Nagler is between the lenses. One drawback is that, as we will show, the Nagler in this particular form is feasible only in short focal lengths.

In order to understand the principle of the Nagler eyepiece, imagine the system reduced to its simplest form, as shown in fig. 16.13. To the left of the stop, on the side nearer the objective, is a negative lens, and to the right of the stop is a positive lens.

As mentioned above, it is possible to reduce astigmatism in a positive eyepiece system at the expense of introducing field curvature. By placing a negative lens between the objective and the positive eyepiece, however, we can compensate this field curvature. The reason is that the negative lens introduces curvature in the intermediate focal surface, F_1. If this is chosen to match that of the positive lens, the resulting combination of lenses can be free of both astigmatism and field

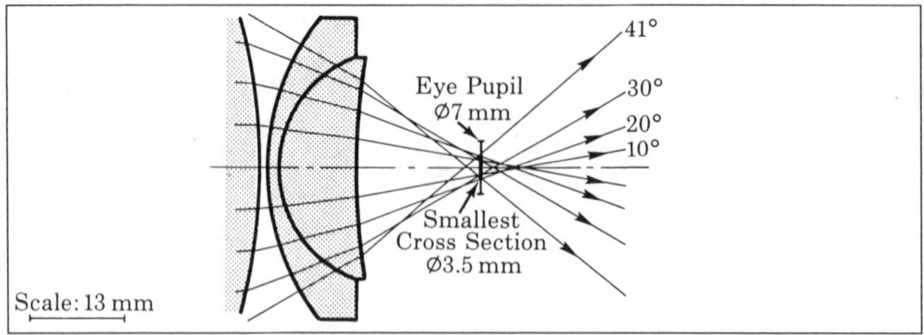

Fig. 16.14 *Spherical Aberration of the Exit Pupil in the Nagler Eyepiece.*

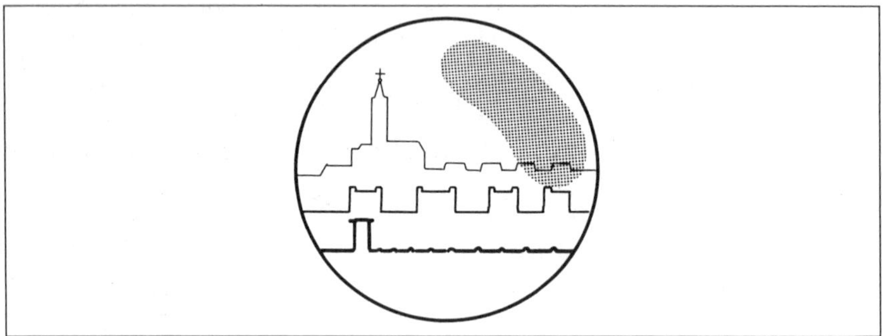

Fig. 16.15 *The Kidney-Bean Effect.*

curvature in the imaginary focal surface, F_2.

The design principle of the Nagler is not totally new. The combination was proposed long ago by Smyth; hence, the negative lens is called a Smyth lens (see ref. 16.1). In a more recent publication (ref. 16.4), this possibility is again pointed out. Despite the excellent optical correction which can be obtained with such a combination, this design is rarely seen because the strong divergence of rays in the negative lens section means the positive lens section must be large, resulting in relatively high manufacturing costs and weight.

This is the case with the Nagler eyepiece: for a focal length of 13 mm, the eyepiece is 62 mm in diameter and weighs 24 ounces. A 25 mm eyepiece of the same design would be 120 mm in diameter and weigh 10 pounds! For this reason alone, the Nagler eyepiece would be used only for short focal lengths.

However, these eyepieces also suffer from spherical aberration of the exit pupil, and the resultant kidney-bean effect. Fig. 16.14 shows a scale drawing of the eye lens of a 13 mm Nagler eyepiece with its exit rays. In order not to complicate the drawing, the principal rays of each parallel exit bundle, but not the bundles themselves, are drawn. (In fig. 16.13, we show the bundles.) The shift of the intersection points with the optical axis at increasing exit angles is readily visible. We list the value of the shift in table 16.2.

It should be emphasized that this "spherical aberration" must not be confused with longitudinal spherical aberration (LA); it does not lead to unsharpness. However, in order to capture all the exit rays, the pupil of the eye must be as large as the smallest cross section of the bundles. The smallest cross section of the principal rays is 3.5 mm, and the bundles would be 3 mm diameter for an *f*/5 objective. The smallest cross-section of all exit beams together would then be 6.5 mm.

This means that the 13 mm Nagler use is restricted to observation when the pupil of the eye is wide open. When the eyepiece is used with a pupil of 2 or 3 mm, some of the exit beams will never reach the retina. Depending on the position of the eye with respect to the eyepiece, parts of the image will not be visible, resulting in the kidney-bean effect, shown in fig. 16.15. For Nagler eyepieces with focal lengths shorter than 13 mm, the situation is more favorable. They can be used for daytime observation and by older individuals whose maximum dark adapted pupil is smaller than 7 mm.

It is interesting to note that three years after the introduction of the original Nagler, the eyepiece was redesigned to eliminate spherical aberration of the exit pupil. The improved design is called the Nagler 2, and is available in longer focal lengths than the original Nagler design.

One especially favorable characteristic of the original Nagler eyepiece is the large eye relief, roughly 1.2 times the focal length. Large eye relief is realized by the combination of two effects. First, the positive part of the eyepiece has a focal length which is 1.7 times that of the system, so the eye relief is that of a longer focus system. Second, since the negative element diverges the rays, the exit rays converge less steeply. The eye relief of the 13 mm Erfle is only 0.6 times the focal length.

Based on the data given in table 16.2, we analyzed the optical performance of the Nagler and Erfle eyepieces at 0, 10, 20, 30 and 40° off-axis, for red, green, and blue light, at focal ratios of *f*/5, *f*/10 and *f*/15. Figs. 16.16 and 16.17 show the resulting spot diagrams.

Let us look first at the Erfle eyepiece. The specimen we traced is a relatively well-corrected representative of the many modifications of Erfle designs we checked. Nonetheless, fig. 16.16 shows how swollen the spot diagrams are, especially for fast focal ratio (*f*/5) and large off-axis angles. At the edge of the field, the blur circle has an angular diameter of some 60 minutes of arc—twice the angular diameter of the full moon!

Another striking point concerns the lateral color. This aberration is determined by the shift of the spot diagrams for red and blue with respect to each other. This shift varies between 5 and 9 minutes of arc for the Erfle. Correction of lateral color seems to be difficult in Erfle type eyepieces because all Erfle designs we checked suffer from this aberration.

Spot diagrams for the Nagler are shown in fig. 16.17. It is impressive to see how well astigmatism has been suppressed in this design. For a 20° field angle, it is a factor of 6 better than the Erfle; for a large field angle (40°) it is a factor of 10 better than the Erfle.

Lateral color is also under control. It lies between 1.5 and 3 arc minutes, a

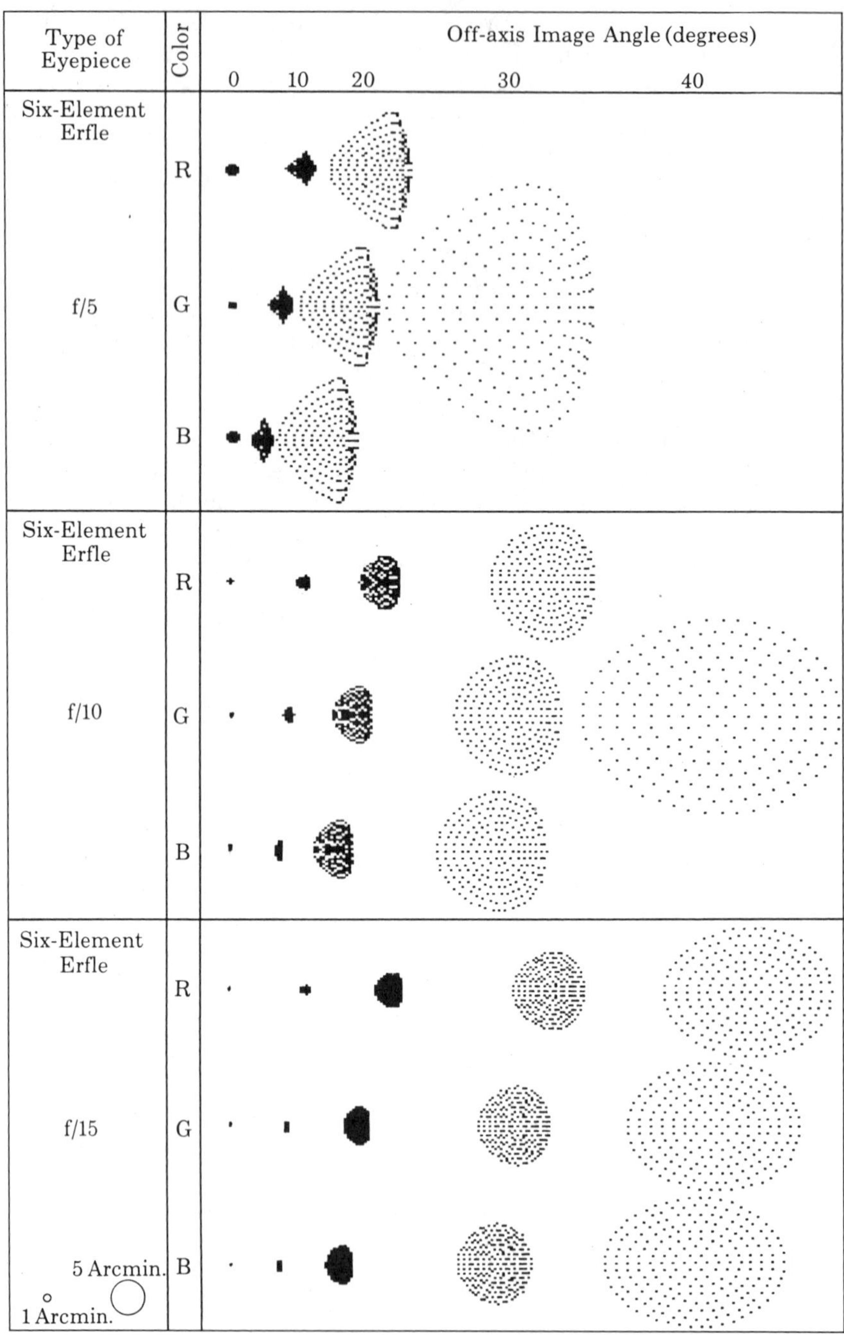

Fig. 16.16 *Optical Performance of a Modified Erfle Eyepiece.*

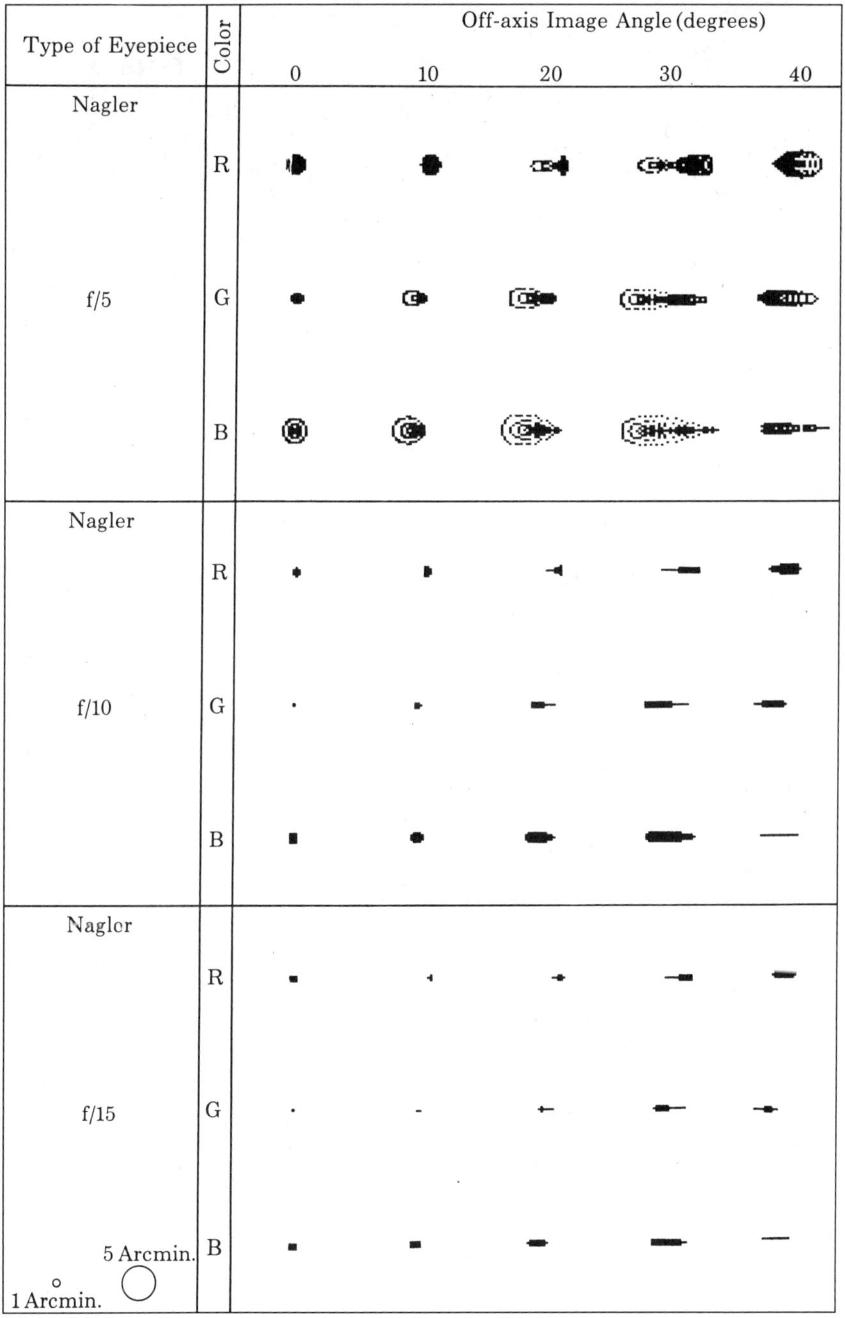

Type of Eyepiece	Color	Off-axis Image Angle (degrees) 0	10	20	30	40
Nagler	R					
f/5	G					
	B					
Nagler	R					
f/10	G					
	B					
Nagler	R					
f/15	G					
5 Arcmin. 1 Arcmin.	B					

Fig. 16.17 *Optical Performance of the Nagler Eyepiece.*

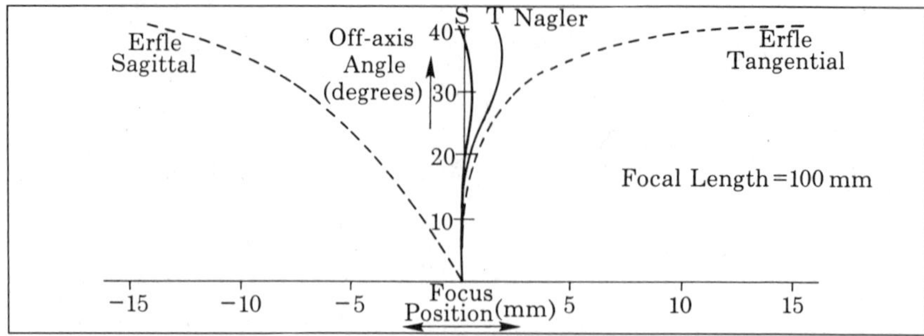

Fig. 16.18 *The Tangential and Sagittal Focal Surfaces for the Erfle and Nagler Eyepieces.*

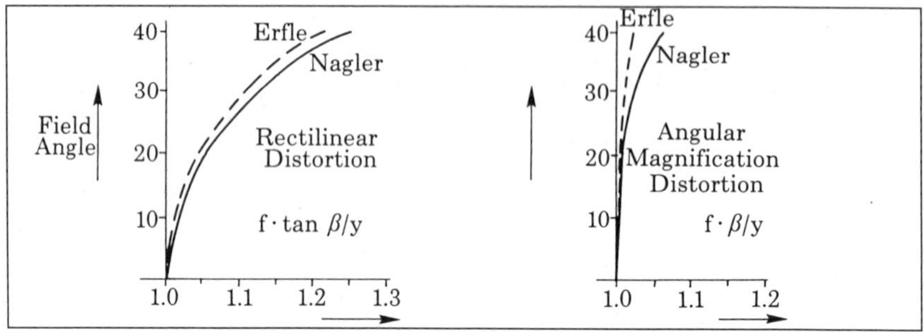

Fig. 16.19 *Distortion in the Erfle and Nagler Eyepieces.*

factor of 3 better than the Erfle. The Nagler has a slightly outward curving field; the spot diagrams were calculated for the flat focal plane. In this particular case, though, the eye can easily accommodate since an accommodation of less than one diopter is necessary. For the curved focal surface, the off-axis spot diagrams are smaller and rounder than those in fig. 16.17.

Fig. 16.18 shows the positions of the tangential and sagittal focal surfaces for a focal length of 100 mm. In this case, as in fig. 16.8c, light travels from left to right, from the eye to the focal surface. The graph clearly shows how far astigmatism in the Nagler has been suppressed relative to the Erfle.

We also investigated the degree of distortion. As we noted above, it is impossible to correct simultaneously for both rectilinear distortion and angular magnification distortion. Fig. 16.19 gives the results of the calculations. Rectilinear distortion is considerable in both eyepieces, meaning that straight lines in the outer field are pincushions. The angular magnification distortion, which is more relevant in astronomical observation, is low and probably invisible.

16.6 Eyepieces Used for Projection

We have compared various eyepieces for visual use, with parallel exit beams. When eyepieces are used for projection, typically for lunar or planetary photography, the exit beams do not remain parallel, and the imaging characteristics will be changed (fig. 16.20). The image angle will generally be smaller than it is for visual

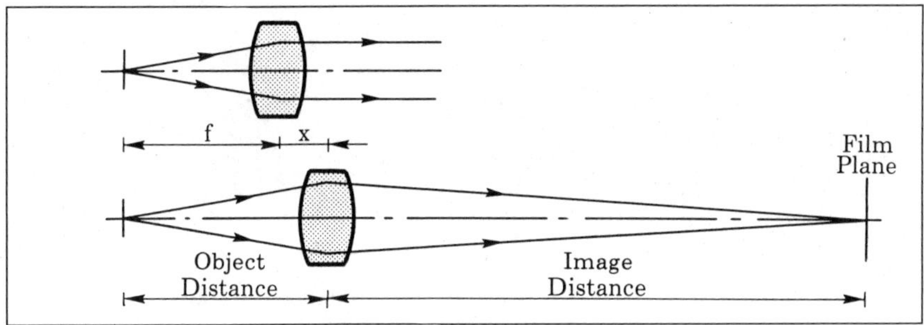

Fig. 16.20 *Eyepieces with Parallel and Converging Exit Beams.*

applications, typically 5° or less off-axis.

For a given projection ratio, gamma, $\gamma = f_{combination}/f_{objective}$, the eyepiece movement, x, can be calculated from:

$$x = \frac{f}{\gamma}. \tag{16.6.1}$$

Because of the large variety of telescope focal ratios, eyepieces, and projection ratios possible, spot diagrams are not given. However, since the eyepiece forms a real image in projection, calculation of spot diagrams for a particular system is straightforward.

16.7 The Performance of Objective-Eyepiece Combinations

16.7.1 Introduction

In the foregoing we have examined the optical performance of telescope objectives and eyepieces as separate entities. The visual observer, however, is far more interested in their combined performance than in the performance of either taken alone. It is often difficult to predict, based on spot diagrams of each alone whether aberrations will be diminished or reinforced when they are combined. Both situations can occur.

Recall that in section 13.4 we analyzed the Jones-Bird telescope objective. The image appeared to be free of both spherical aberration and coma, but suffered from strong astigmatism and field curvature—indeed, field curvature is some 16 times as strong as for a Newtonian reflector of the same aperture and focal ratio. Despite this, various users have reported (see refs. 16.5 and 16.6) that these aberrations are hardly visible in actual use.

Bird and Bowen, the instrument's designers, explained this effect as due to partial compensation by opposing eyepiece aberrations. It was this that inspired us to investigate various objective-eyepiece combinations.

We examine, by means of ray tracing, several objective-plus-eyepiece combinations with a view to gaining insight into how to obtain a final image of high quality. In particular, we are interested in discovering to what extent the aberration

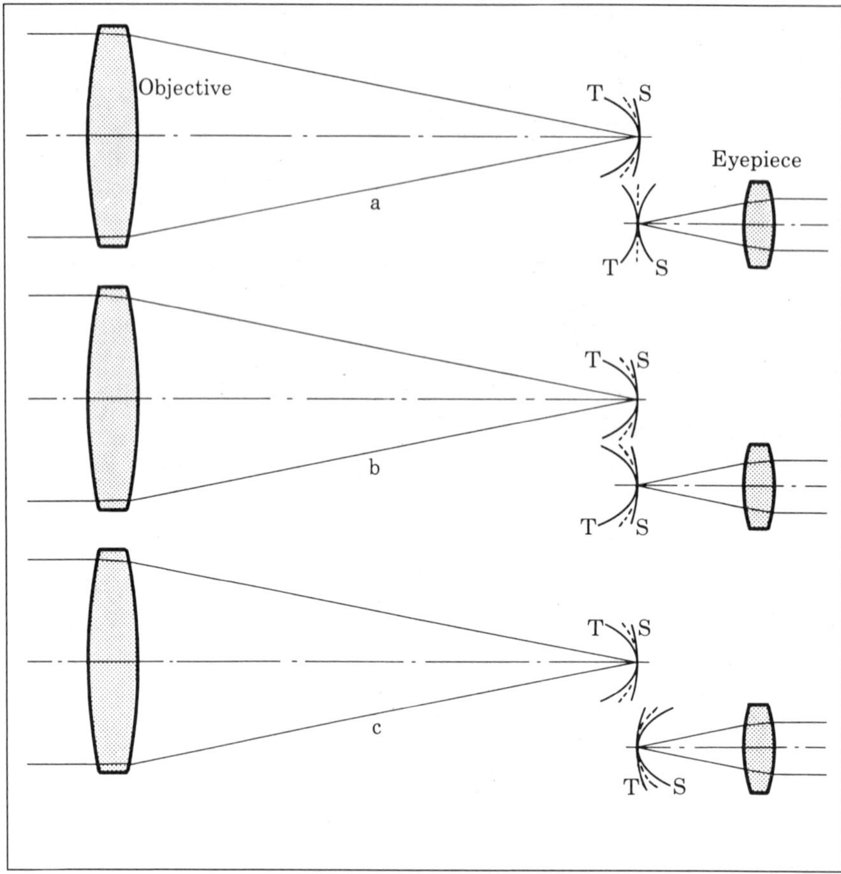

Fig. 16.21 *The Relationship Between Sagittal and Tangential Field Curvature for Eyepieces and Objectives.*

of an objective may be compensated by opposite aberrations in the eyepiece. If, by means of suitable design, the compensation of aberrations could be fully realized, we could make virtually aberration-free telescopes without the need of designing both objectives and eyepieces that are aberration-free.

16.7.2 Astigmatism and Field Curvature

As we have seen in previous chapters, most objectives suffer from both astigmatism and field curvature. In most cases we observe undercorrected (or positive) astigmatism in which the tangential focal surface is located nearer the objective than is the sagittal focal surface. Between the tangential and sagittal focal surfaces lies the average curved field, a somewhat imaginary construct. For most objectives—refractors, Newtonians, and Cassegrain-like systems—this surface is inward curving (fig. 16.21).

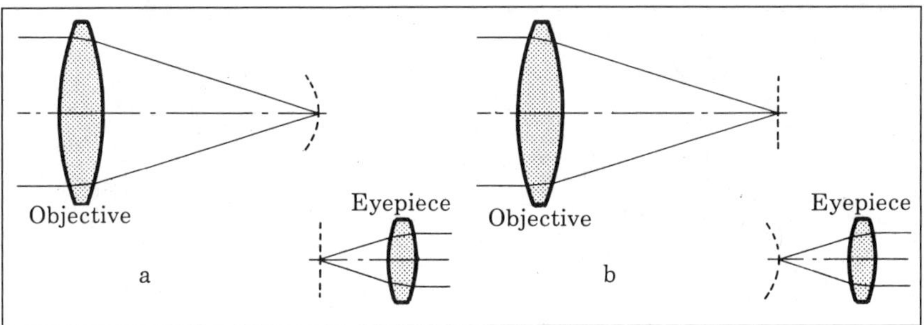

Fig. 16.22 *Matching Field Curvature in Eyepieces and Objectives.*

In eyepieces, the situation is generally reversed: the tangential focal surface lies farther from the eyepiece than the sagittal surface, giving rise to overcorrected (or negative) astigmatism. The case that most often occurs is shown in fig. 16.21a. Note that the tangential surfaces and sagittal surfaces are positioned on the same side of the average focal surfaces. Here astigmatism is partially compensated but not eliminated, and some field curvature remains.

When the corresponding focal surfaces coincide exactly (i.e., if both the distances between the tangential and sagittal focal surfaces and their radii of curvature are the same), we have a system without noticeable astigmatism or field curvature. This is because the aberration generated by one system, the objective, is just compensated by the other, the eyepiece. This ideal situation is shown in fig. 16.21b.

In addition to the two cases described, there is a third case in which astigmatism is eliminated but field curvature remains, shown in fig. 16.21c. This case occurs when the distances between the tangential and sagittal focal surfaces are the same, but their curvatures are not. In this case, it is impossible to obtain a uniform focus over the whole field; the eyepiece must be moved forward and back or the observer's eye must accommodate when observing different parts of the field.

While designers are free to pursue the ideal case in telescopes with fixed or permanently attached eyepieces, most observers want astronomical telescopes having interchangeable eyepieces. Most of the time then, the telescope functions somewhere between the cases shown in fig. 16.21a and 16.21b: astigmatism and field curvature are reduced but not eliminated.

16.7.3 Accommodation of the Eye

The eye tends to automatically seek optimum image sharpness by means of accommodation, or changing its focus. This enables the observer to compensate, at least in part, for incompletely-corrected field curvature. The relaxed eye is focused for distant viewing; accommodation involves contracting the muscles around the lens so it focuses closer. The eye will not focus farther than its relaxed focus.

We distinguish two cases where the eye can correct field curvature by ac-

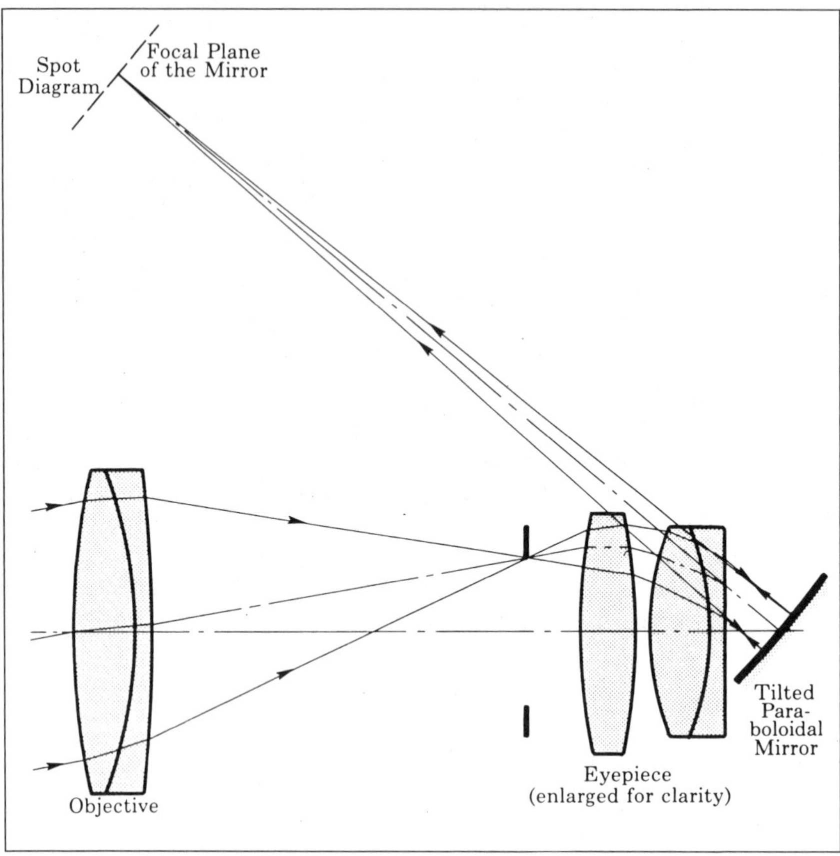

Fig. 16.23 *Analyzing the Performance of an Objective-Eyepiece Combination.*

commodation. These are shown in fig. 16.22. In fig. 16.22a, the field of the objective curves inward while the field of the eyepiece is flat. In order to be able to accommodate, the observer must focus on the edge of the field with the unaccommodated eye. The eye can then accommodate for the center of the field.

In fig. 16.22b, the field of the objective is flat, while the field of the eyepiece curves outward. In this case, the observer must focus on the center of the field using the unaccommodated eye. The eye can then accommodate as necessary to focus the edge of the field.

In the analyses that follow, we show the accommodation, expressed in diopters, required for each field angle in each objective-eyepiece combination examined.

16.7.4 Analyzing Objective-Eyepiece Combinations

Telescopes are afocal, that is, the objective-eyepiece combination does not form a real image. In principle, both the entering and exiting bundles are parallel beams.

To simulate how the eye images the exit bundle on its retina, we must place an image-forming system in the exit pupil, a system with little or no inherent aberration. We have chosen a paraboloidal mirror with such a long focal ratio that aberrations caused by the mirror itself are entirely negligible. The mirror is placed in the exit perpendicular to the exit bundle under examination, as shown in fig. 16.23.

To position the mirror properly, we first determine the location of the axis of the bundle (i.e., the principal ray) with respect to the optical axis of the system itself. This is where the eye would be positioned during actual observation through a telescope. This intersection is not fixed, but generally shifted toward the eyepiece as the exit angle increases due to spherical aberration of the exit pupil. The computer program developed for this task automatically positions the mirror and determines its tilt angle for every exit bundle examined.

If the bundle emerging from the eyepiece consists of parallel rays, the mirror focuses them to a single point in its focal plane. If the rays are non-parallel, the image formed by the mirror will be spread, and may be represented in a spot diagram. The size of the blur shown by the spot diagram can be expressed by the angle it subtends from the mirror, corresponding to the angular blur that would be seen by the eye. Furthermore, by calculating the position of best focus, we were able to calculate the degree of accommodation needed to give the sharpest image.

16.7.5 Combinations Examined

As examples of the performance of objective-eyepiece combinations, we chose four objectives in combination with two eyepieces. We deliberately chose relatively fast objectives because the aberrations would be large. The objectives are a 200 mm *f*/6 Newtonian, a 200 mm *f*/6 Jones-Bird telescope, a 100 mm *f*/8 two-element fluorite refractor (a scaled-down version of the design given in section 6.3), and a 100 mm *f*/5.5 four-lens objective with reduced field curvature and astigmatism relative to a two-element objective (ref. 16.7). Of these objectives only the Newtonian has any coma; the others are aplanatic.

The eyepieces are the same as those investigated in section 16.5, a 13 mm modified Erfle with six elements and the 13 mm Nagler design with seven elements. Both eyepieces have fields in excess of 80° and a field diameter of approximately 17 mm. Note that the tangential and sagittal surface curves shown in fig. 16.18 are reversed relative to those shown in fig. 16.21, because of the reversed position of the eyepiece.

16.7.6 Results of Ray Tracing

Figs. 16.24, 16.25, and 16.26 show the results of tracing the eight combinations of four objectives and two eyepieces. For each combination three rows of five spot diagrams are shown. Each row shows star images 0, 10, 20, 30, and 40° off-axis; they represent the star images as seen by the eye in the exit pupil of the eyepiece, and their sizes are expressed in arcminutes. Near the field center, the spread of the

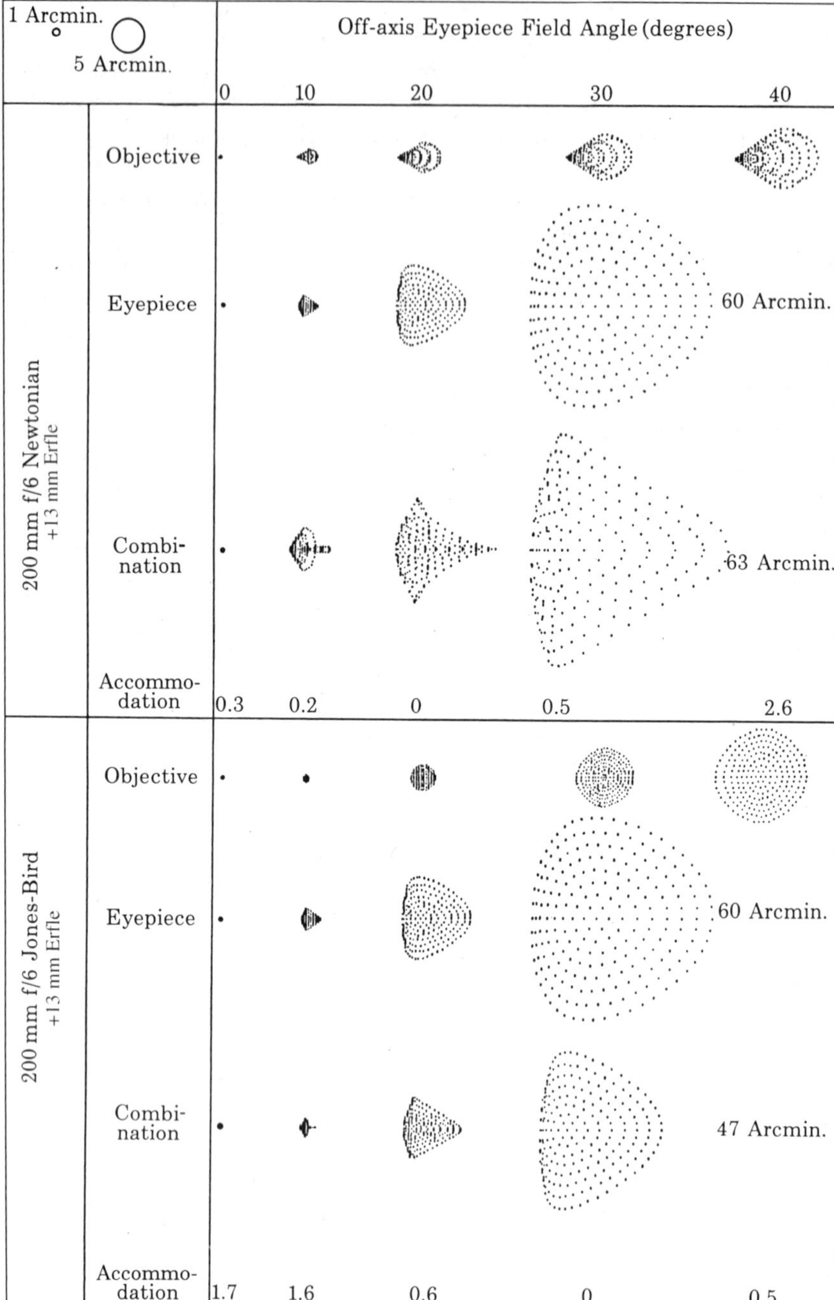

Fig. 16.24 *Optical Performance of Two Erfle-Telescope Combinations.*

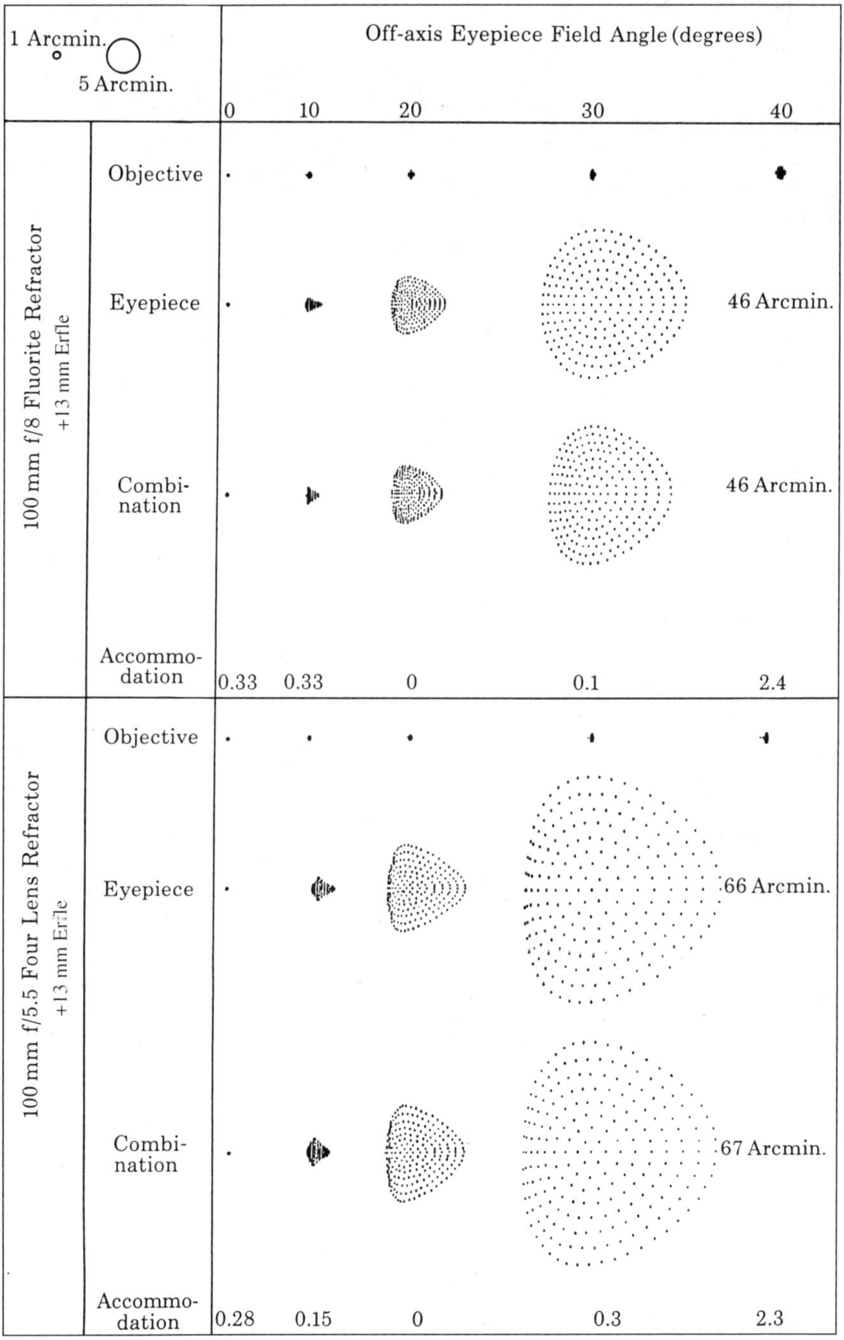

1 Arcmin. 5 Arcmin.

Off-axis Eyepiece Field Angle (degrees)

		0	10	20	30	40
100 mm f/8 Fluorite Refractor +13 mm Erfle	Objective					
	Eyepiece					46 Arcmin.
	Combination					46 Arcmin.
	Accommodation	0.33	0.33	0	0.1	2.4
100 mm f/5.5 Four Lens Refractor +13 mm Erfle	Objective					
	Eyepiece					66 Arcmin.
	Combination					67 Arcmin.
	Accommodation	0.28	0.15	0	0.3	2.3

Fig. 16.25 *Optical Performance of Two Erfle-Telescope Combinations.*

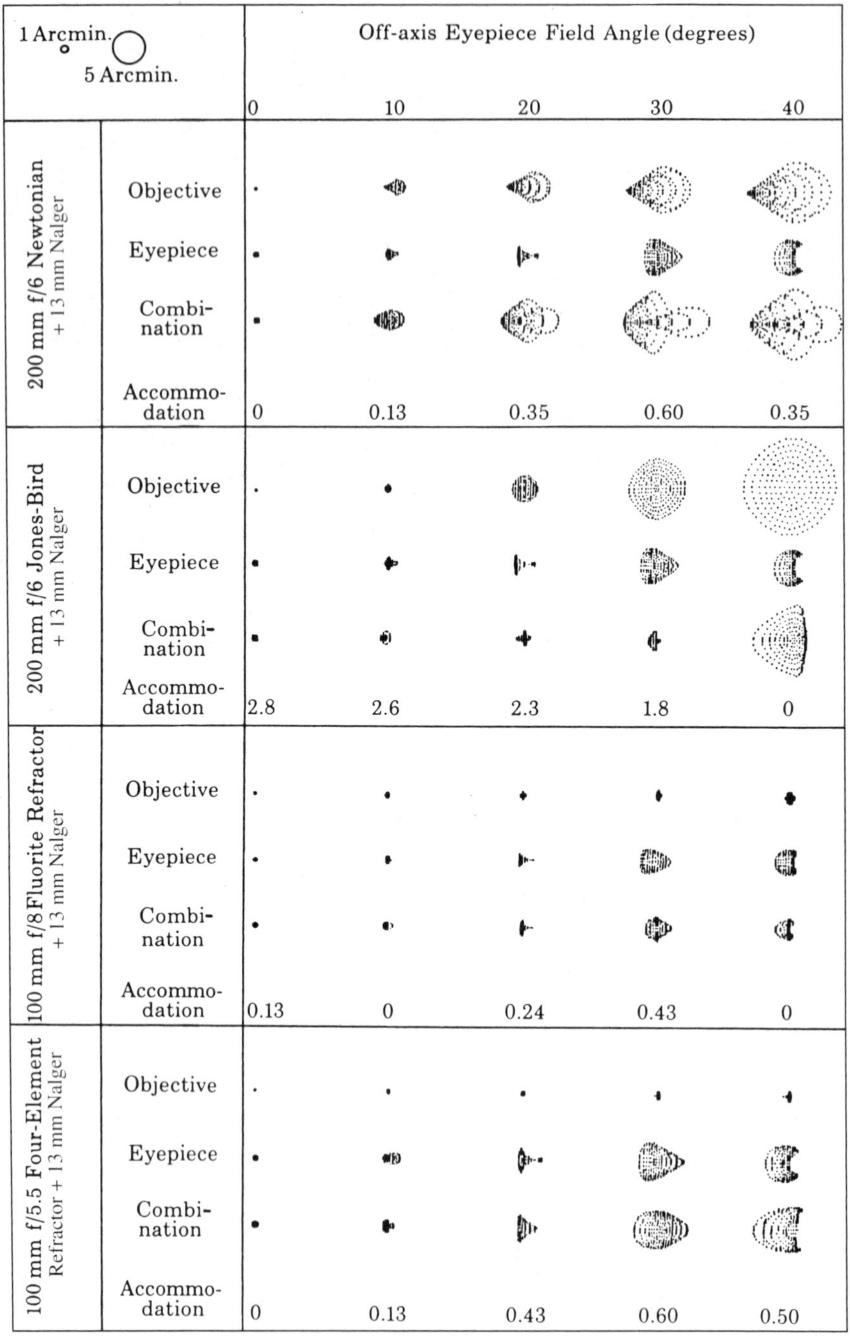

1 Arcmin. ○ 5 Arcmin.		Off-axis Eyepiece Field Angle (degrees)				
		0	10	20	30	40
200 mm f/6 Newtonian + 13 mm Nalger	Objective					
	Eyepiece					
	Combination					
	Accommodation	0	0.13	0.35	0.60	0.35
200 mm f/6 Jones-Bird + 13 mm Nalger	Objective					
	Eyepiece					
	Combination					
	Accommodation	2.8	2.6	2.3	1.8	0
100 mm f/8 Fluorite Refractor + 13 mm Nalger	Objective					
	Eyepiece					
	Combination					
	Accommodation	0.13	0	0.24	0.43	0
100 mm f/5.5 Four-Element Refractor + 13 mm Nalger	Objective					
	Eyepiece					
	Combination					
	Accommodation	0	0.13	0.43	0.60	0.50

Fig. 16.26 *Optical Performance of four Nagler-Telescope Combinations.*

spot diagram should not exceed one minute of arc. At the edge of the field a spread of five minutes of arc is allowable. Spot diagrams are given for the optimally curved field.

The three rows are labeled as follows:

Objective: Spot diagrams for the objective alone. The spot diagram represents the angular size of the blur that would be seen if it were observed with an aberration-free eyepiece of 13 mm focal length.

Eyepiece: Spot diagrams for the eyepiece alone. The spot diagram shows the aberration that would result from the eyepiece used with an aberration-free objective. (Since these are computed for a curved focal surface, they differ from those shown in figs. 16.16 and 16.17, for the flat focal surface.)

Combination: Spot diagrams for the objective-eyepiece combination.

A final row of figures labeled "Accommodation" gives the accommodation required of the eye, expressed in diopters. Note that the point of zero accommodation need not lie at the field center or edge, but may lie between. All accommodations are positive because a normal eye cannot accommodate beyond infinity.

We began by discussing possible compensation of the astigmatism of an objective by compensating astigmatism in the eyepiece, but in most cases it appears that just the opposite has happened: astigmatism of the eyepiece is partially compensated by astigmatism in the objective.

Full compensation will occur only when the astigmatisms are equal and have opposite signs. Unfortunately, for conventional eyepieces such as the Erfle, the astigmatism of the eyepiece is much greater than the astigmatism of the objective. Only in the new generation of low astigmatism eyepieces, such as the Nagler, does astigmatism approach the levels found in most objectives.

Spot diagrams presented in figs. 16.24 and 16.25 show the off-axis performance encountered when different objectives are combined with the 13 mm Erfle eyepiece. Note that accommodations up to three diopters are sometimes required. In several cases, we see spot diagrams exceeding 40 minutes of arc. Where this has occurred, we have substituted a number giving the diameter in minutes of arc.

Two findings are especially significant:

First, eyepiece astigmatism strongly dominates the net astigmatism of the combination, and astigmatism from the objective plays only a small part. In most cases, a compensation effect, though it may be present, will hardly be noticeable in practical observation when conventional eyepieces are used.

Second, eyepiece astigmatism strongly dominates the coma seen in the Newtonian reflector, particularly in the outer parts of the field. This confirms the statement made in section 16.3. Owners of fast Newtonians often ascribe the unsharpness at the edges of the field to the coma

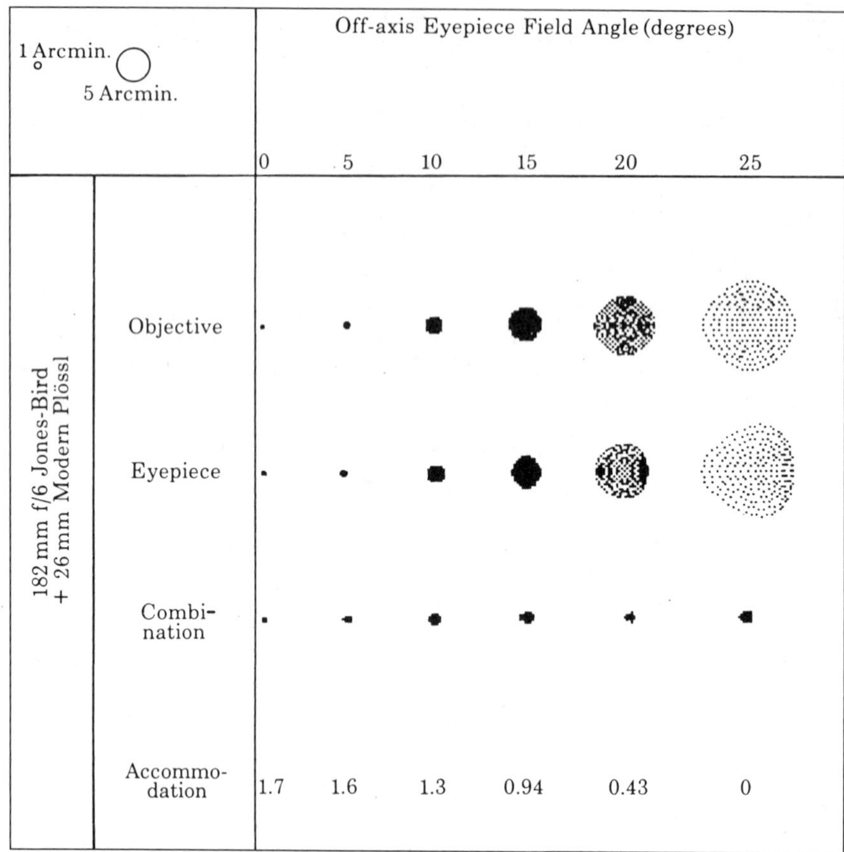

Fig. 16.27 *An Optimized Objective-Eyepiece Combination.*

of the paraboloidal mirror. In reality, it is the astigmatism of the eyepiece that causes the blurring.

Fig. 16.26 displays spot diagrams for the 13 mm Nagler eyepiece. Note that these blurs are much smaller, the result of greatly reduced astigmatism in the Nagler design. In such combinations, noticeable compensation is far more likely.

Note that the comatic blur in the Newtonian is not reduced. Coma compensation generally does not occur because most eyepieces are designed in such a way that they have no inherent coma. It is possible, however, to design coma-correcting eyepieces (see ref. 16.9).

Among the combinations examined, that of the 200 mm Jones-Bird objective and 13 mm Nagler eyepiece (see fig. 16.26, second series of rows) best illustrates the "compensation effect" we originally sought (see table 16.3).

Fig. 16.27 illustrates the performance of an optimized objective-eyepiece combination. The objective is a 182 mm *f/6* downscaled version of the 200 mm Jones-Bird telescope, while the eyepiece is a recent Plössl design with a 26 mm

Table 16.3
200 mm *f*/6 Jones-Bird + 13 mm Nagler Eyepiece

Field Angle, (°)	0	10	20	30	40
Spot Size					
Objective, arcminutes	0.3	1	4	9	15
Eyepiece, arcminutes	1	2	3.2	5.5	5.5
Combination, arcminutes	1	2	2.3	3	11
Compensation, arcminutes	0	0	+1.7	+6	+4

focal length (ref. 16.8). Because they compensate, astigmatism at the edge of the field is greatly reduced. A 224 mm *f*/6 Jones-Bird with a 32 mm Plössl would be another compensating combination.

16.7.7 Discovering Favorable Objective-Eyepiece Combinations

From the foregoing, we see that for some combinations it is possible to obtain a great deal of mutual compensation for astigmatism and field curvature. But how can we predict which will be the favorable cases? Computer ray tracing is helpful, of course, but not particularly feasible. The more like the situation shown in fig. 16.21b, the better the chance of compensation. To predict the off-axis performance requires a detailed knowledge of the tangential and sagittal focal surfaces of the objective and the eyepiece. Although aberration curves are sometimes available for objectives, most manufacturers do not supply the aberration curves of their eyepieces. Coma of the objective generally disturbs the possibility of mutual compensation because, as we have seen, most eyepieces are coma-free. However, a coma-correcting eyepiece has been designed (see ref. 16.9).

In the absence of adequate eyepiece aberration data, the most expedient way to find good objective-eyepiece combinations remains empirical testing of many different combinations. However, from the foregoing it should be clear that the new generation of low-astigmatism eyepieces, and the introduction of coma-correcting eyepieces, will offer the telescope maker the best chance to achieve this goal.

Chapter 17

Deviations, Misalignments, and Tolerances

17.1 Introduction

The performance of a telescope is degraded not only by image aberrations but also by surface inaccuracies, deviations in assembly, and misalignments. In this chapter we examine these problems. However, because it would be impractical to treat them in combination, we have restricted ourselves to discussing each of these factors in turn.

A single surface may have errors only in its accuracy and in its radius of curvature. Several surfaces in combination may also deviate from the design ideal in their positions relative to one another. We divide these errors into axial shifts, transverse shifts, and tilt errors. Of course, all kinds of combinations of these deviations can and do occur.

17.2 Surface Accuracy

We place high demands on the surface accuracy of optical components. These are, however, considerably different for reflecting and refracting surfaces. When a ray is reflected, the effect of a surface error is doubled. For refracting surfaces the error depends on the refractive index and is proportional to $(n-1)$. While this means that higher refractive index materials have smaller surface tolerances, it also means that the surface accuracy required in refracting surfaces is generally lower than for reflecting surfaces. Moreover, since refractive elements have two surfaces, their errors may either neutralize or reinforce each other.

The extent to which surface inaccuracies will influence optical performance is difficult to predict. Because optical surfaces shape the wavefront, the more optical surfaces used, the greater the requirement for accuracy in each surface in the system.

This picture, however, fails to take into account the complexity of the wavefront. If an error of given wavefront deviation is restricted to a small part of the wavefront, the performance of the system will be better than it would be if an error of the same magnitude were smoothly distributed over the entire surface. However, the deleterious effect of numerous small errors may well exceed that of a single smoothly distributed error even though the wavefront deviation is the same in both

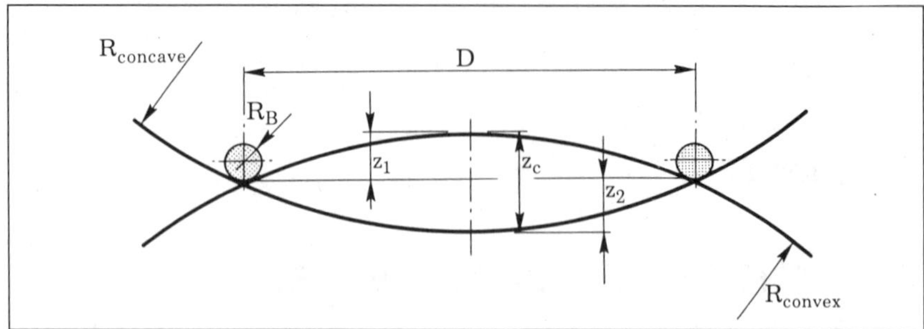

Fig. 17.1 *Measuring Surface Curvature with a Ball Spherometer.*

cases.

This means that surface errors are very difficult to model mathematically. We can easily enough ray-trace the effects of a mathematical deviation from the correct spherical or aspheric figure by altering the appropriate variables in the ray-trace program. Real-world surface errors are more difficult to quantify. Suffice it to say that such defects should be kept smaller than the largest tolerable figuring errors.

Optical surfaces generated by optical surfacing machines often have better average accuracies than surfaces made by hand. However, most amateur telescope makers work by hand. Hand-finishing optical surfaces allows local retouching, which is particularly important for aspheric surfaces. In the case of all-spherical telescope designs, hand-working also permits the elimination of residual spherical aberration by aspherizing.

Amateurs usually check the radius of curvature of an optical surface by means of a spherometer. With a spherometer, the sagitta of a section of the spherical surface is measured. The sagitta may be measured relative to a ring or a configuration of balls. Both the ring spherometer and the ball spherometer are used by ATMs.

Of the two, the ball spherometer is preferable. A ring spherometer must have an exceedingly accurately made ring. When the ring has a finite edge thickness, its diameter is different for convex and concave surfaces. Furthermore, the precious optical surface is easily damaged by the sharp edges of this kind of spherometer.

The ball spherometer (Fig. 17.1) consists of three small balls mounted on a flat plate on a circle, with a micrometer spindle in its center. Obviously, the ring has been replaced by three balls. In large ball spherometers, the instrument consists of a bar with one ball at one end and the remaining two balls mounted close together at the other end.

We find the radius of the surface from:

$$R_{\text{concave}} = \frac{\frac{D^2}{4} + Z^2}{2Z} + R_B \qquad (17.2.1)$$

$$R_{\text{convex}} = \frac{\frac{D^2}{4} + Z^2}{2Z} - R_B \qquad (17.2.2)$$

where:

R_{concave} = radius of curvature of a concave surface
R_{convex} = radius of curvature of a convex surface
D = diameter of the circle the balls are mounted on
Z = measured value of sagitta
R_B = radius of balls (all balls same diameter).

From these formulae it follows that the value of Z must be measured very accurately, and D must be well-known. This is particularly the case when the diameter of the spherometer is small in comparison with the radius of the surface, because the sagitta will be very small then. The accuracy of the calculated radius of curvature is approximately M/Z, where M is the value of the smallest increment on the micrometer's scale and Z the measured value of the sagitta. Suppose, for example, that the smallest increment of the measuring instrument is 0.001 mm and the smallest acceptable accuracy of the radius is 1%. The minimum sagitta that must be measured to attain this accuracy is 0.1 mm, since 1% = 0.01, and Z = 0.001/0.01 = 0.1.

The user should apply a drum micrometer rather than a dial micrometer. Dial micrometers are handy devices but often suffer quite large deviations over their range, so they should be used only to monitor changes, or for rough measurements. The diameter of the spherometer, D, must be known with an accuracy of 0.01 mm or better. For accurate measurements, the diameter of the spherometer should be as large as possible, and a flat reference plate should be available to precisely "zero" the spherometer.

17.3 Deviations and Misalignment

Misalignment degrades telescope performance. It may stem from errors in assembly, result from the flexibility of the telescope tube, or occur because of temperature changes. In general, the deviations that produce misalignment are caused by the manufacturing process, the assembly of the optical components, and the use of the instrument.

To some certain degree, manufacturing deviations cannot be eliminated, but they must remain within some specified limit. The maximum allowable deviations are called tolerances. In the case of a lens, the tolerances may refer to the radii of curvature, center thicknesses, and variation in edge thickness.

The smallest measurable deviation from the nominal radius of curvature de-

Table 17.1
Optical Component Tolerance Guidelines

Diameter	50	150	250	350	Comment
Radius of curvature			1%		reasonable (by hand)
			0.25%		very accurate (by hand)[a]
Diameter	0.5	0.75	1	1.5	rough
	0.2	0.4	0.6	0.8	accurate
	0.05	0.1	0.15	0.2	very accurate
Center thickness			0.3		accurate
			0.1		very accurate
Edge thickness difference	0.05	0.08	0.12	0.15	accurate
	0.01	0.01	0.015	0.02	very accurate

a. When commercial radiused tools are available, it is possible to generate surfaces to nominal value.

pends on the spherometer used. Often it is difficult to make a lens to a certain center thickness. This is true particularly when both radii of curvature of a lens are sufficiently accurate already, but the center thickness should still be brought within the desired tolerance.

Advanced amateur optical workers can often meet close tolerances. The amateur can spend more time on the optics than a professional optical workshop because the manufacturing time need not be held within limits. One particular problem for the amateur is wedge error (figs. 17.2 and 17.3). For eliminating wedge error, the professional optical workshop has a centering machine. This is seldom the case for the amateur telescope maker, so correcting wedge error can be a time-consuming job.

It is difficult to give generally valid tolerances for optical components. However, for diameters between 50 mm and 350 mm, the guidelines in table 17.1 may be used (all dimensions are in millimeters). A general rule is that halving the tolerance doubles the manufacturing time to meet that tolerance. This means that a component specified for a tolerance of 0.2% typically takes eight times longer than the same component "spec'ed" with a 1.5% tolerance.

Assembly deviations are the differences between the positions of the optical components after they have been mounted and the design ideal. Here the actual distance between components is important. When components are close together, close tolerances are not difficult to meet. For example, in an airspaced doublet, where the edge distance may be several tens to a hundred microns, thin foils can be used as precise spacers. When the distance is larger, spacing rings can be used. Generally, these require larger tolerances than foil.

Another factor is the play that lenses must have in their mounts. This play may be small, but it cannot be reduced to nothing. In order to center a lens accurately, centering cones may be ground into the edge of the lens. This technique is usually not available to amateur telescope makers, however.

The situation is entirely different when optical components have large dis-

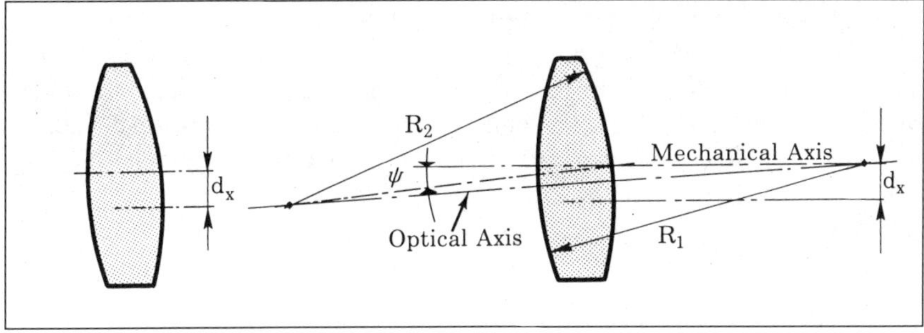

Fig. 17.2 *Parameters of a Biconvex Lens with a Tilted Surface.*

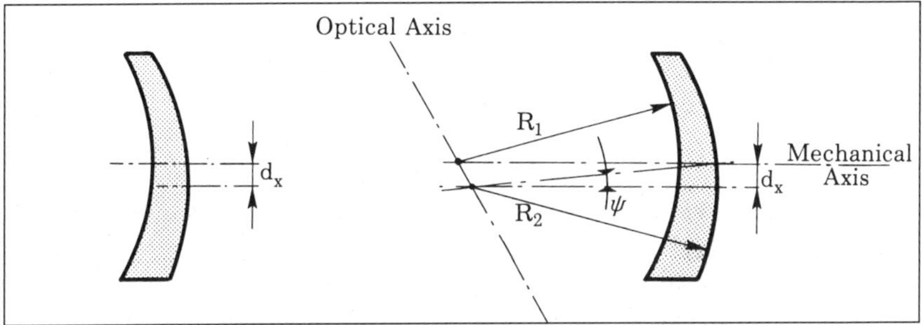

Fig. 17.3 *Parameters of a Meniscus Lens with a Tilted Surface.*

tances with respect to each other, as is the case in reflecting telescopes and catadioptrics. Spacers are normally not used. Instead every component has its own mount, which is attached to the tube or to a spider. Because it is difficult to find reference points to position the components correctly, these elements are normally provided with adjustment devices. With these devices it is possible to align the elements in an indirect way, for example, by studying the reflections of elements in one another, inspecting the Airy disk, or measuring the focus position. In most cases, such alignment procedures are rather time-consuming.

Further deviations from ideal alignment occur when a telescope is used. These may result from thermal expansion and deformations of the tube. Thermal expansion causes changes in the length of the tube and also in the distance between the optical components. When the tube is made of a synthetic material, the thermal expansion is much larger than is the case for metal tubes. Over a 30° Celsius (54°F) temperature difference, a tube 1.5 meters long of aluminum will expand 1.1 mm; of polyvinyl chloride, 3.6 mm; of polyethylene, 10 mm. When the distances are critical, for instance the distance between the mirror and the film holder in a Schmidt camera, compensating elements or ultra-low-expansion Invar bars must be built in.

When the telescope tube deforms or bends under its own weight, misaligned and tilted elements occur. Long tubes may bend differing amounts depending on the tilt angle of the telescope. This is especially a problem when a synthetic material is not reinforced. Glass-reinforced resin ("fiberglass") works quite well.

As with lens cells, mirror cells in large telescopes must allow some play. These cells may deform under the weight of a large element, causing the element to be misaligned. Large spider-mounted secondary mirrors are especially at risk, since the cell must be supported in the light path by thin spider legs, and must often weigh as little as possible.

Some telescopes have moveable optical components. In some catadioptric Cassegrainians, for example, the primary mirror is moved for focusing. When there is looseness or play in the mirror support system, misalignment of this mirror with respect to the other components may occur.

17.4 Influence of Deviations and Misalignments

We illustrate the three possible deviations of an optical surface in fig. 17.4, using the secondary mirror of a Cassegrainian telescope as our example. In this case, a small axial shift of an optical surface does not usually lead to serious aberrations. The most striking effect caused by an axial error is the shift of the focal plane. The magnitude of this shift depends on the power of the surface. The stronger the power is, the larger this shift. The aberration caused by an axial shift of the surface is mainly spherical aberration and to a lesser extent also coma and astigmatism. Transverse shifts and tilt angles are much more serious, causing coma and astigmatism in the axial image. A transverse shift of the image also occurs, which may be especially noticeable if the system is used for astrophotography.

In a single spherical lens, a transverse error of one surface with respect to the other manifests itself as a tilt error. The lens then works like a prism, and axial lateral color occurs. This holds true not only for glass lenses, but also for air lenses such as the air space in a doublet. There is no difference between transverse and tilt errors in lenses with spherical surfaces. This is illustrated in figs. 17.2 and 17.3 for two types of lenses. A transverse error of d_x has the same effect as a tilt angle of arcsin d_x/R.

Knowing that the optical axis is the connecting line of both centers of curvature, it is clear from fig. 17.2 and 17.3 that a wedged meniscus is much more detrimental than a wedged biconvex lens.

The tilt angle corresponds with a transverse error of:

$$R \cdot \sin\phi = d_x . \qquad \textbf{(17.4.1)}$$

Conic surfaces are quite similar in this respect to spherical surfaces, except that a transverse error cannot be replaced by a tilt error. However, a different situation occurs for certain nearly flat aspheric surfaces. A Schmidt corrector's surface is rather insensitive to tilt error. Because the lens has little optical power and has mainly a corrective action, coma introduced by a tilt will be low. Some astigmatism is introduced.

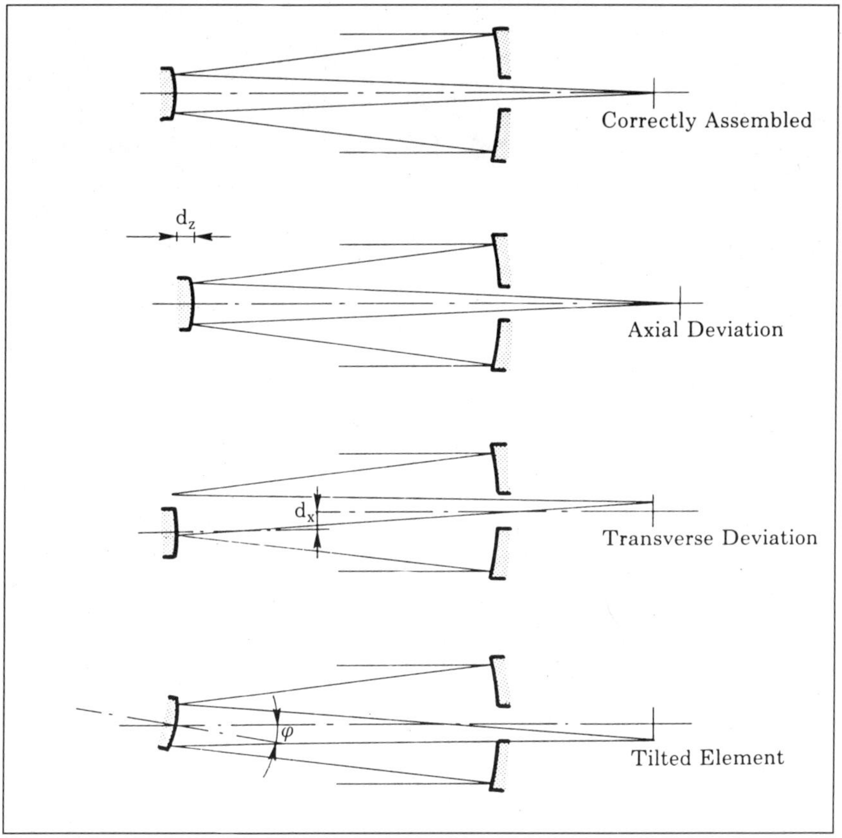

Fig. 17.4 *Assembling Errors in a Cassegrain Telescope (Deviations Strongly Exaggerated for Clarity).*

17.5 Tolerance Analysis

Every telescope designer is faced with the problem of finding what deviations are allowable before intolerable image aberrations will occur. Naturally, it is important to know what tolerances the telescope maker can reasonably achieve. Following this, the designer must determine the allowable deviations; for this, a tolerance analysis is required. Here, the designer tries to determine, for each optical component in the system, what deviations and misalignment errors are allowed before intolerable image aberrations occur. For components that will be mounted in adjustable supports, it is not necessary to make a misalignment analysis.

For visual systems, the allowable aberrations must be compared to the Airy disk. For photographic systems, the photographic criterion should be applied. Calculations should be carried out both on-axis and off-axis, and, when refractive elements are used, for more than one color.

Sometimes the number of required calculations can be reduced considerably. When spherical surfaces are involved, transverse errors can be directly related to

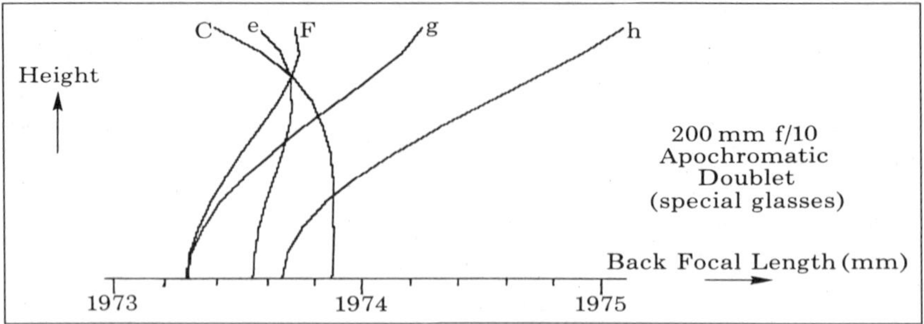

Fig. 17.5 *LA-Curves for an Apochromatic Doublet.*

tilt errors, as we have shown. In the case of Cassegrainian telescopes, the transverse and tilt errors are interchangeable for both mirrors.

When refractive elements are involved, the tolerance of the refractive index must also be investigated. Normally optical systems with refractive elements are designed on the basis of the refractive indices given in the glass catalog. The glass manufacturer will supply the glass with the actually measured refractive indices. If these deviate from the values given in the glass catalog, the system should be re-analyzed and possibly optimized for the actual glass.

The amateur telescope maker considering a new optical system should guard against "peak designs." These are optical designs which may offer very high optical performance, but have tolerances which are so tight that minuscule deviations produce intolerable aberrations. Tolerance analysis can be carried out rather quickly with our ray-tracing computer program.

17.6 Correcting Manufacturing Deviations

Manufacturing deviations from a specified optical design can often be corrected during further manufacturing or assembly. While a complete discussion lies outside the scope of this book, we include a simple example to illustrate this technique. We have chosen a refracting optical system with four surfaces, a highly-corrected 200 mm *f*/10 objective.

The design employs two special glasses, both quite expensive, and achieves very good correction for color. (For the design procedures employed, see section 21.13.) The nominal design appears in the tables on the following page.

Because of its close tolerances, this system may be considered a "peak design." Fig. 17.5 shows the spherical aberration curves for five colors. Note that the LA curves for C, e, and F light intersect at the 82% zone, so that this two-element objective is apochromatic when the refractive indices are in exact accord with the values given above.

The nominal design has been chosen so that when the OSC value for the edge zone is zero, the intersection points of the edge ray and paraxial ray coincide in green light (compare this with section 21.13.5). This will allow us to evaluate the

Table 17.2
Two-Element Apochromat
(Dimensions in mm)

R_1	1143.8432	
T_1	28.5577	FK 52
R_2	–463.4415	
T_2	2.000	Air
R_3	–471.807	
T_3	12.000	KzFSN2
R_4	–2685.629	
T_4	1973.238	Air
F_{eff}	2000.000	

Table 17.3
Refractive Indices

	FK 52	KzFSN2[a]
n_e	1.487471	1.560818
n_C	1.484238	1.555208
n_F	1.490179	1.565518
n_r	1.483204	1.553394
n_d	1.486052	1.558361
n_g	1.493375	1.571111
n_h	1.496008	1.575776

a. Since we designed this system, KzFSN2 has been replaced by KzFS2

effects of manufacturing deviations on LA, OSC, and color correction quite easily.

Each surface and thickness has its own tolerance. This can be calculated with the aid of a tolerance analysis. However, for the practical lens maker, it would be too cumbersome to finish every surface and thickness to its individual tolerance. This is especially true for a peak design, since the tolerances will be so close. (Of course, the nominal system specifications should remain the lensmaker's goal.)

Rather, it is more practical to compensate for the unavoidable deviations that occur in the beginning stages of manufacture by taking compensatory steps later in the manufacturing process. First, however, it is necessary to know which of the surfaces are most critical and which are the least critical. To do this, we carry out a sensitivity analysis (which we emphatically point out should NOT be confused with a tolerance analysis!).

Work as follows: Select a reasonable deviation in the sagitta measurement. A deviation that can be detected easily with most spherometers would be 0.01 mm. For each of the surfaces, calculate the new radius of curvature corresponding to this hypothetical manufacturing deviation. Ray-trace the lens, varying the radi-

Table 17.4
Sensitivity Analysis
(Deviation on the spherometer is 0.01 mm)
(Spherometer diameter is 200 mm)
(Dimensions in mm)

Surface	Radius	Z	Z+0.01	New Radius	ΔR
1	1143.8432	4.380	4.390	1141.147	–2.696
2	–463.4415	10.917	10.927	–463.046	0.396
3	–471.807	10.719	10.729	–471.391	0.416
4	–2685.629	1.862	1.872	–2671.876	13.753

Table 17.5
Resulting Aberrations

	Deviation (mm)	LA (mm)	OSC (–)
Nominal	0	0	0
R_1	2.696	–0.141	$1.62 \cdot 10^{-5}$
R_2	0.396	–0.502	$22.7 \cdot 10^{-5}$
R_3	0.416	–0.537	$24.3 \cdot 10^{-5}$
R_4	13.753	–0.055	$1.71 \cdot 10^{-5}$
T_1	0.1	+0.001	0
T_2	0.1	+0.149	$-6.08 \cdot 10^{-5}$
T_3	0.1	+0.000	0

us of one curve at a time; then tabulate the deviations in the LA and OSC values and the color correction. Repeat this procedure for the lens thicknesses. A deviation of 0.1 mm is a reasonable manufacturing tolerance.

The influence of these deviations on the color correction is negligible in this particular case, so no color data are given here.

The values of R_2 and R_3 are the most sensitive to manufacturing deviations. R_1 and T_2 are considerably less fussy, while R_4, T_1, and T_3 are quite insensitive. A variety of methods can be used to compensate the aberrations caused by these deviations, but the best (and, unfortunately, most time-consuming) method is called the matching principle.

The Matching Principle: We begin by making the most sensitive surface first. When this is finished, if the radius appears to deviate from design specifications, we change other parameters to compensate. In effect, we must redesign the system, using the new radius of curvature of the most sensitive surface as a fixed value.

The second-most sensitive surface is made, and again, if there is any detectable deviation, a new design based on two fixed parameters is made. This process continues until the last surface is completed. Thickness variations must be taken into account, of course. This method, which produces a system closely matching the performance of the nominal design, is best suited to making a single critical

Table 17.6
Correction Strategies Compared

Method	Action	LA	OSC
Compensation	$R_3 = R_3 + 0.389$	0	0
Aspherize R_1	$SC_1 = -0.067$	0	2.37 · 10–5
Aspherize R_4	$SC_4 = -0.862$	0	2.64 · 10–5
Air Space	$T_2 = T_2 -0.337$	0	–2.12 · 10–5

system. A computer design program should be available for best results.

Simple Compensation: A somewhat simpler alternative to the matching principle also begins with the most sensitive surface. However, the deviation in the first surface is corrected, as much as possible, by altering the second most sensitive surface. Although the final system will be less well corrected than a system built by the matching principle, in most cases it will be acceptable.

In our example, we would alter the radius of curvature of R_3. By applying a correction to the radius proportional to the sensitivity of this surface, we determine a new radius of curvature. Assuming an error in R_2 of 0.396, note that we must compensate for LA variation by a factor of (–0.502/–0.537) times the deviation for R_3. R_3 then becomes R_3 +0.389. Note that R_3 must be made very accurately.

Asphetizing: We can correct LA by aspherizing one or more of the optical surfaces. This method can become quite time-consuming. It is desirable to apply this correction to the surface which requires the least aspheric correction.

Changing the Air Space: The simplest and quickest compensation is changing the air space between the lenses. We compare the relative sensitivities of R_2 and T_2, then multiply T_2 times this difference: –0.502/0.149 · 0.1 = –0.337. The new T_2 becomes T_2 –0.337. Obviously this method will not work if the air space becomes negative or when edge contact occurs.

The optical errors produced by these methods are shown in table 17.6.

In this example, the compensation method eliminates both LA and OSC, providing R_3, which is nearly as sensitive as R_2, can be made precisely enough. Of the two aspherizing alternatives, R_1 is preferable since it requires relatively little correction. Respacing this objective is by far the simplest method, and results in a new T_2 of 1.663 mm.

Chapter 18

Resolution, Contrast, and Optimum Magnification

18.1 Introduction

Every telescope is limited in its ability to resolve or separate details in the object observed. This property is called resolving power. In this chapter, we discuss the most important factors influencing resolving power.

Let us begin by distinguishing clearly between resolving power for point sources such as double stars, and resolution as it applies to extended objects such as planetary surface details. Most treatments for amateur astronomers touch only on resolving power for double stars. Much more interesting—but also more complex—is the meaning of resolving power for extended objects. This subject, particularly the role of contrast transfer, forms the basis for most of this chapter.

Specifically, the definition of resolving power used in this chapter, expressed in line pairs per millimeter (the so-called linear resolving power) should not be confused with the angular resolving power, expressed in arcseconds. The (theoretical) linear resolving power is connected with the focal ratio of the telescope and for a given focal ratio is independent of the aperture. For a given focal ratio *f/D* and a wavelength λ, in millimeters, the linear resolving power is:

$$SL = \frac{D}{f \cdot \lambda} \text{ lp/mm.} \quad \textbf{(18.1.1)}$$

For example, if *f/D* = 10 and for green light (λ = 555 nm):

$$SL = \frac{1}{10 \cdot 5.55 \cdot 10^{-4}} = 180 \text{ lp/mm.} \quad \textbf{(18.1.2)}$$

The angular resolving power depends on the aperture, and for a given aperture is independent of the focal ratio. For an aperture *D*, the angular resolving power is:

$$SA = \frac{\lambda}{D} \text{radians} = \frac{206265 \cdot \lambda}{D} \text{ arcseconds.} \quad \textbf{(18.1.3)}$$

For an aperture of 100 mm:

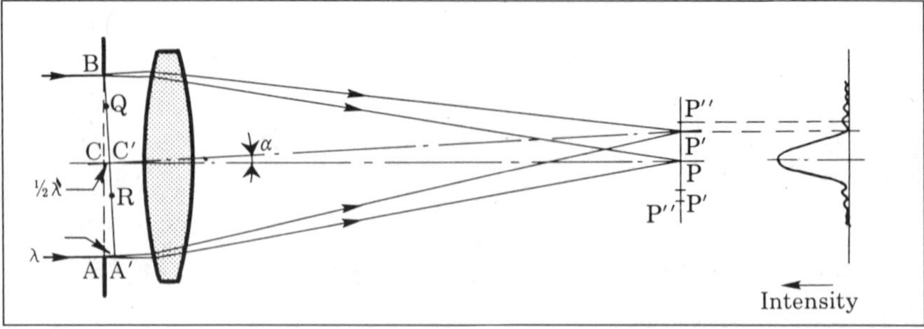

Fig. 18.1 *Diffraction Image of an Ideal Lens.*

$$SA = \frac{206265 \cdot 5.55 \cdot 10^{-4}}{100} = 1.14 \text{ arcseconds.} \tag{18.1.4}$$

The linear resolving power *SL* and the angular resolving power *SA* are related through the focal length, *f*, of the optical system:

$$SL = \frac{206265}{SA \cdot f}. \tag{18.1.5}$$

From the example above, in which $SA = 1.14$ arcseconds and $f = 1000$ mm,

$$SL = \frac{206265}{1.14 \cdot 1000} = 180 \text{ lp/mm} \tag{18.1.6}$$

18.2 Resolving Point Sources

The resolving power of a telescope depends on the size of the diffraction pattern and the distribution of intensity in it. Every telescopic image is made of overlapping images of the optical system's response to a point source of light, the point spread function.

Diffraction for a lens with a narrow slit is shown in fig. 18.1. Monochromatic light from a distant point source enters an aberration-free lens through a slit of width *AB*. For a flat cross section, all the light has the same phase. Because all light waves from AB retain the same phase at point *P*, these waves reinforce each other, so that at *P* a high intensity exists. According to the principle of Huygens-Fresnel, every point that is hit by a light wave acts like a new center of waves. Point *C*, for example, sends light toward both *P* and *P´*.

Suppose we choose a line *BA´* in such a way that the distance *AA′* is equal to the wavelength, λ, of the light. This means $CC' = \lambda/2$. Light arriving at *B* differs a half wavelength from light from *C′*, and light rays from *B* and from *C′* arrive at *P′* exactly out of phase. They destructively interfere, so *P′* is dark. The same argument holds for the points *Q* and *R*, which differ by a half wavelength when $BQ = C'R$. On the line *BA′* one can find for every point *Q* another point R at which the light will interfere destructively so that *P′* is dark.

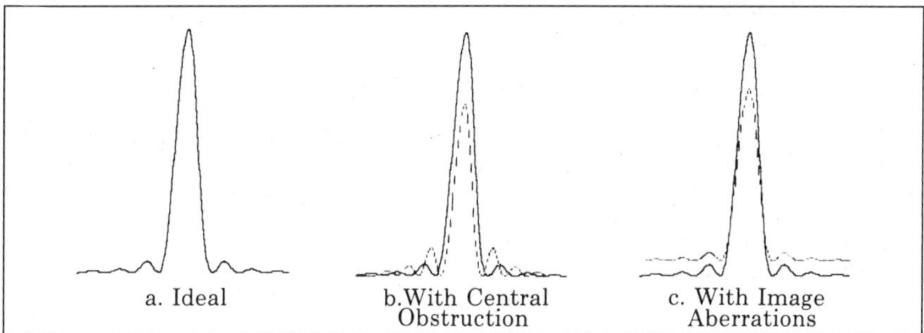

Fig. 18.2 *Light Intensity Distributions in the Airy disk.*

Between P and P' the light arrives only partially out of phase. These points will always be darker than P. We say that a diffraction maximum exists at point P. Mathematical analysis shows that at certain distances outside P', for instance P'', intensity maxima with a much lower intensity than P will occur, and that the intensity of these points decreases the farther they are from P.

From fig. 18.1, it can easily be understood that:

$$\sin\alpha = \frac{P'P}{CP'} = \frac{\lambda}{A'B}. \tag{18.2.1}$$

When the angle α is sufficiently small that $\sin\alpha = \alpha$, then angle PCP' becomes:

$$\alpha = \frac{\lambda}{AB} \text{ radians.} \tag{18.2.2}$$

This says that the width of the diffraction maximum is proportional to the wavelength and inversely proportional with the width of the slit.

For a circular aperture, usually the case in telescopes, the situation is more complicated. The analysis must be carried out by integrating over the whole entrance pupil. In 1835, Airy succeeded in making the mathematical analysis of this problem and found exact values for the diameters and intensity distribution in the diffraction pattern formed by a circular entrance pupil. In his honor, we call the diffraction pattern the Airy pattern and the central light spot the Airy disk.

The intensity distribution in the Airy pattern is shown in fig. 18.2a. At the center is an intensity maximum. This is separated by a dark ring from the first diffraction ring, which again is separated from the second diffraction ring, and so on.

Airy calculated that for an ideal system with a circular aperture, the central disk contains 84% of the total light from a point source. The first diffraction ring contains 7%, the second 3%, and so on. The diameter of the first dark ring is $2.44 \cdot \lambda/D$ radians, where λ is the wavelength of the light and D is the aperture. (For more detail on computing the Airy disk, see ref. 18.13)

Because one radian is 206,265 arcseconds, for green light (555 nm) the angular diameter of the first dark ring is $280/D$ arcseconds when D is in millimeters.

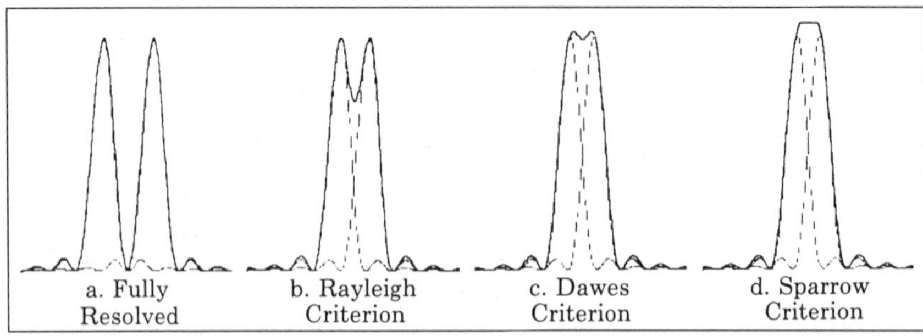

Fig. 18.3 *Resolution for Equally Bright Double Stars.*

For red light (656 nm) the angular diameter is 330/*D* arcseconds, and for violet light (404 nm), 203/*D* arcseconds.

It should be clear that the image of a star in white light is a composite of numerous overlapping diffraction patterns in various colors. (Under laboratory conditions, this can be seen in a color-free instrument as rainbow rings around the Airy disk.) Because the eye has its greatest sensitivity at 555 nm, we usually take 280/*D* arcseconds as the angular diameter of the first dark ring.

From the size of the Airy disk, we might expect that a telescope could easily distinguish two Airy disks separated by their own angular diameter (see fig. 18.3a). Rayleigh showed that stars with intensity maxima separated by roughly half that distance can be distinguished visually. Fig. 18.3b shows diffraction patterns separated by 140/*D* arcseconds.

The image seen by the eye is the sum of the light patterns. For stars of equal brightness, the intensity in the dip between the stars' maxima is 74% of the maximum. Although Rayleigh defined this angular distance as the resolving power of a telescope, he realized that even closer stars could be distinguished. For small instruments Dawes empirically determined the resolving power is 117/*D* arcseconds, which implies a dip of only 3.2% between the intensity maxima (see fig. 18.3c).

To the values cited above, however, the following should be added:

1. The Dawes criterion is strictly valid only for white double stars consisting of two sixth magnitude components, viewed with a 150 mm telescope.
2. The diameter of the Airy disk, thus the resolving power, depends on the wavelength of the light. The Dawes criterion, for instance, does not apply to red double stars.
3. For stars of unequal brightness, the dip in the combined Airy pattern will be less favorable, so distinguishing the star images will be more difficult. Lewis found a 3 times worse resolving power for a double star pair with magnitudes 6.2 and 9.5, and 8 times worse for a pair with magnitudes 4.7 and 10.

4. The Dawes criterion is only valid when the diffraction pattern has the ideal intensity distribution. Optical aberrations affect this distribution and decrease the resolving power of the telescope.
5. Air currents smear the combined Airy pattern of the double star. Only after careful examination under good seeing conditions can definitive conclusions with respect to the optical performance of telescopes be drawn.

18.3 Resolving Power and Contrast for Extended Objects

Amateurs most often test their telescopes on double stars. But the ability to resolve close doubles does not always measure the capacity to resolve details on lunar and planetary surfaces, or "extended objects." For the observation of this kind of detail in extended images, the transfer of contrast by the optical system is of great importance.

An image of an extended object is far more complex than an image of a point source. The image consists of a multitude of details having different size, shape, color, brightness, and contrast—a virtually infinite number of bright and less bright point sources. Each of these contributes a diffraction pattern to the focal plane, so the final image is the composite of the overlapping diffraction patterns.

Large uniformly illuminated surfaces are uniformly illuminated in the image as well. No unsharpness is visible. Noticeable diffraction effects are present only at the borders of surfaces with different brightness. In the case of a bright surface and an adjacent dark surface, diffracted light encroaches into the dark border, causing blurring and unsharpness of the border line.

A thin dark line on a bright background is "greyed," while a bright line on a dark background is widened. These effects are visible particularly when these lines have an angular width comparable with or smaller than the diffraction pattern. Depending on the shape, size, brightness, contrast, and color of the object observed, the influence of diffraction on the final image will be different. Since the image of an extended detail can be very complicated, it is difficult to find a representative and reproducible method to define the resolution of an optical system for this kind of image.

However, in 1946, P. M. Duffieux developed the concept of contrast transfer for optical systems (ref. 18.1). This approach yields considerable insight into what happens in the image forming process. For details to be visible, they must have sufficient contrast. If the image contrast lies below the eye's visibility threshold, then the detail will be invisible.

Image contrast depends not only on the inherent contrast in the object (e.g., the contrast of faint markings on Saturn), but also on how much contrast the optical system transfers from the object to the image plane. Contrast transfer is the key for understanding, for example, why a planetary detail may be visible in one telescope, but not in another of the same aperture.

Resolving power and contrast transfer are both quality criteria for every tele-

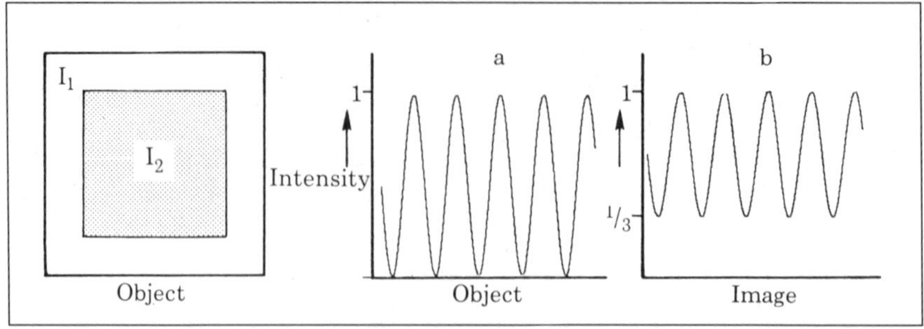

Fig. 18.4 *The Definition of Contrast.*

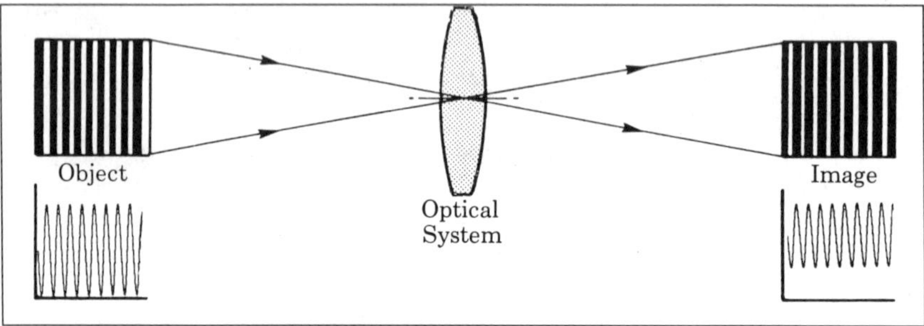

Fig. 18.5 *Contrast Transfer by an Optical System.*

scope. Today it is possible to measure the contrast transfer of an optical system with special equipment, and the relation between image contrast and resolution can be determined for every point in the image plane.

18.4 Contrast Transfer in a Perfect Optical System

An optical system may be characterized as perfect if, in the absence of diffraction and providing no obstructions such as a secondary mirror or spider are present, it would produce a point image of a point source. When a telescope meets these criteria, then the Airy diffraction pattern will have the ideal intensity distribution.

Before discussing the contrast transfer of this perfect system, let's define contrast and contrast transfer (see fig. 18.4). The contrast, C, between two adjacent surfaces with the intensities I_1 and I_2 is defined as:

$$C = \frac{I_1 - I_2}{I_1 + I_2} \qquad \textbf{(18.4.1)}$$

when $I_1 > I_2$. Using this definition, the contrast always lies between 0 and 1 (or 0% and 100%).

Suppose we examine two sinusoidal intensity distributions in a grating of parallel lines. If the intensity varies between 0 and 1 (as shown in fig. 18.4a), the contrast between the highest and lowest intensity is:

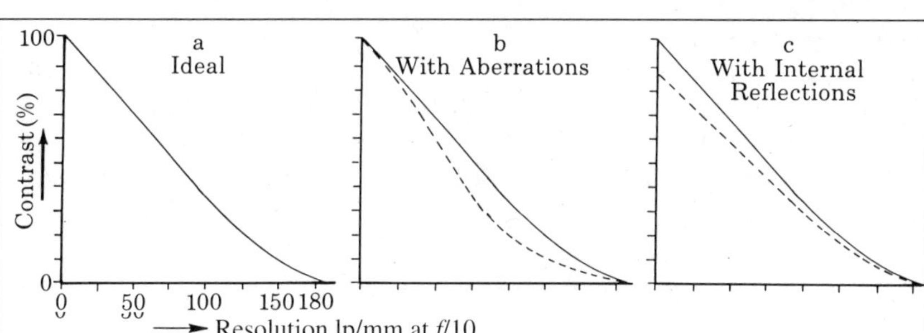

Fig. 18.6 *Contrast Transfer Curves for Ideal and Imperfect Optical Systems.*

$$C = \frac{1-0}{1+0} = 1 \ (= 100\%). \tag{18.4.2}$$

But if the intensity varies from 0.33 and 1 (fig. 18.4b), the contrast is:

$$C = \frac{1-0.33}{1+0.33} = 0.5 \ (= 50\%). \tag{18.4.3}$$

In principle, contrast transfer is measured by placing a grating having a sinusoidal intensity distribution as an object in front of the optical system, then measuring the contrast of the resultant image (see fig. 18.5). The ratio between image contrast and object contrast is called the contrast transfer coefficient, *CT*.

$$CT = \frac{C_{\text{image}}}{C_{\text{object}}}. \tag{18.4.4}$$

The contrast transfer coefficient for the example in fig. 18.5 would be:

$$CT = \frac{0.5}{1} = 0.5 \ (= 50\%). \tag{18.4.5}$$

Each combination of a bright line and a darker line is called a line pair (lp). A coarse target has a small number of line pairs per millimeter in the target grating, while a fine target has a high count of line pairs per millimeter (lp/mm). To evaluate an optical system, we vary the spacing of line pairs in the grating, and measure the contrast in the image.

As the number of the line pairs increases, the optical system renders them with lower and lower contrast because every point in the object is represented by a diffraction pattern in the image. This diffraction pattern scatters light around every image point so that the dark places in the image are illuminated by diffracted light. This effect becomes more important as the distance between elements in the image approaches the size of the diffraction pattern.

At some value the image contrast is reduced to zero. The image of the grating will then be uniformly bright and without any structure. This is the highest resolving power the system can attain.

A graph showing the relation between the contrast transfer coefficient and the number of line pairs in the image is called the contrast transfer function (CTF), the modulation transfer function (MTF), or the optical transfer function (OTF). Duffieux found that the contrast transfer function for a perfect system is a smoothly decreasing monotonic curve, as shown in fig. 18.6a. The curve shown in the figure is for a perfect telescope with a focal ratio *f*/10, imaging light with a wavelength of 555 nm (green).

18.5 Contrast Transfer for Imperfect Optical Systems

CTF curves are extremely useful because we can compare the performance of imperfect optical systems with the curve for a perfect system. Since the CTF of real systems is the accumulation of both diffraction effects and various aberrations, we gain information about the magnitude of image aberrations.

The curves for imperfect systems generally lie below the ideal CTF curve. This means that for the same resolution, the image contrast of the imperfect system is lower than that of a perfect system. Spherical aberration, coma, astigmatism, and chromatic aberration are important, but the CTF curves also reveal unsmoothness in the optical surfaces and the influence of the spider and secondary mirror.

Image aberrations usually lower the CTF curve more at large numbers of line pairs per millimeter than at low numbers because the diffraction rings are brightened at the cost of the Airy disk (Figs. 18.2b and 18.2c). Only very severe aberrations illuminate the image plane at large distances from the Airy disk. The influence of aberrations on the CTF is shown in fig. 18.6b.

Other contrast-diminishing effects in an optical system are internal reflections and stray light. They reduce the contrast by the same percentage for all resolution values (fig. 18.6c).

18.6 Central Obstructions

One subject of ongoing, and seemingly never-ending, discussion among amateurs, and therefore probably interesting for the reader, is the reduction in contrast caused by the central obstruction in a mirror telescope. This happens in Newtonians, Cassegrains, and catadioptric telescopes.

First, how are the Airy disk and diffraction pattern affected? This is shown in fig. 18.2b. The diameter of the Airy disk for an obstructed system is actually slightly reduced with respect to an unobstructed system, but the rings are brightened at the expense of the central intensity of the Airy disk. Fig. 18.7 shows CTF curves for systems with the linear obstruction ratio, η, equal to 0% (i.e., no obstruction), 25%, 50%, and 75% (ref. 18.4). Table 18.1 gives the influence of η on the distribution of light in the Airy pattern for typical refractors (0%), Newtonian

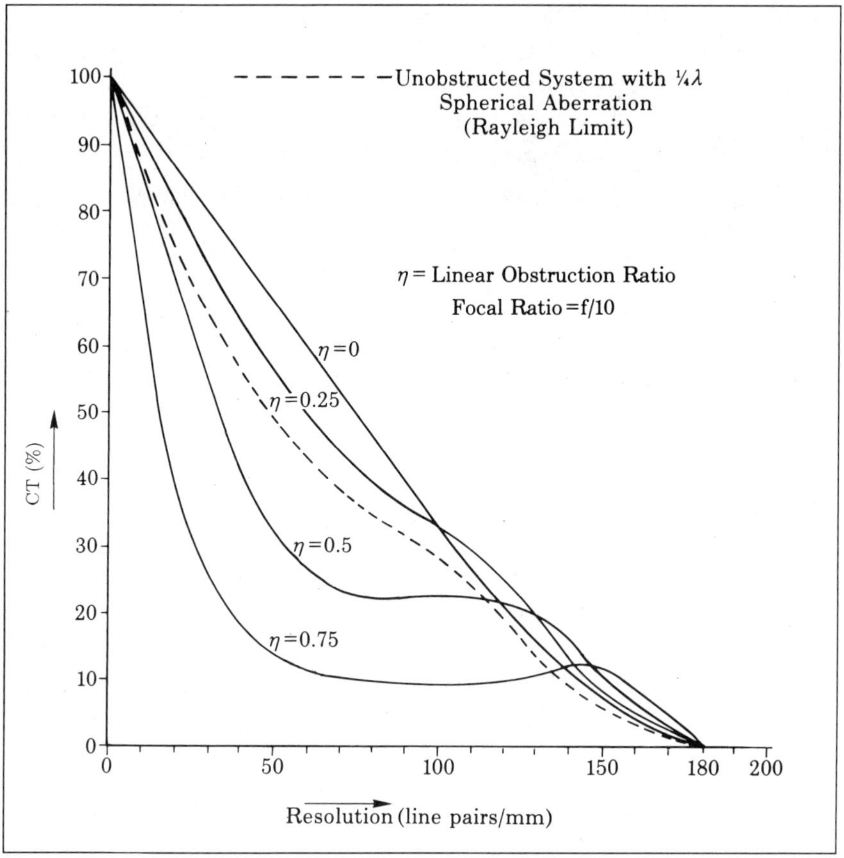

Fig. 18.7 *Contrast Transfer for Obstructed Systems.*

reflectors (25%), and Cassegrain instruments (50%).

Table 18.1
Distribution of Light in the Diffraction Pattern

Obstruction (%)	0	25	50
Airy disk (% of total)	84	73	48
First ring (% of total)	7	18	35
All other rings (% of total)	9	9	17
Airy disk diameter[a]	280	262	230

a. In arcsecond-millimeters (of aperture) for green light.

There is clearly a loss of contrast in obstructed systems. But surprisingly, for higher resolutions, contrast appears somewhat enhanced in obstructed systems. The reason for this is that the Airy disk diameter is slightly reduced when a large central obstruction is introduced.

Fig. 18.7 also shows, as a dotted line, the CTF for an unobstructed system with ¼-wave of spherical aberration, corresponding with the Rayleigh limit on wavefront errors. We draw the following conclusions:

1. For a 25% obstruction, the loss of contrast is not severe; only 15% with respect to the unobstructed system at 60 lp/mm.
2. An obstruction of 50% causes considerable loss of contrast, some 55% at 70 lp/mm.
3. Contrast in a system having a 75% obstruction, which would hardly ever be used in practice, would be very poor.
4. The CTF curve for a system having 1/4-wave of spherical aberration corresponds roughly with an obstruction of 30%. In other words, a 30% obstruction is approximately as bad for image quality as a 1/4-wave error in the optical system.

18.7 Obstructed Telescopes for Visual Use

Let us now compare a typical obstructed system, one with a 40% obstruction (typical of a Cassegrain), with an unobstructed system, in terms of their CTF curves. We will assume that both systems have the same aperture and focal ratio, *f*/10 in our example shown in fig. 18.8. Although the comparison looks rather complicated at first glance, we'll go through it step by step and see what parameters matter when extended details are observed visually in a telescope.

Theoretically, the maximum resolving power for both systems is the same. Both resolve 180 lp/mm in the focal plane. However, the effective resolution of the system in combination with the eye depends on the minimum contrast the eye can detect. The eye, however, is not a simple detector. Its contrast threshold depends on the brightness of the image and the angular distance between the lines. Moreover, it is different for every individual.

For brightly illuminated objects, for instance the moon, we'll assume a minimum required contrast for fine detail of 5%, and for coarser patterns, a somewhat lower minimum contrast. This is shown in curve 5 in fig. 18.8. (Strictly speaking, this curve is not the same for both telescopes because the obstruction blocks 16% of the light, but for the purposes of this discussion we will neglect this minor difference.) In order to distinguish details for dimly illuminated objects, the eye needs higher image contrast; this is shown in curve 6.

The maximum resolution for the eye/telescope combination occurs at the intersection of the CTF curve and the minimum contrast curve for the eye. Here, however, another very important factor, the intrinsic object contrast, enters. Curves 1 and 2 show the contrast transfer from an object having an intrinsic contrast of 100%. Many astronomical objects, particularly planetary surfaces, have a much lower intrinsic contrast. Curves 3 and 4 show the image contrast as seen in the telescope for objects with 10% intrinsic contrast.

Regarding the resolving power of the combination of a telescope and the eye,

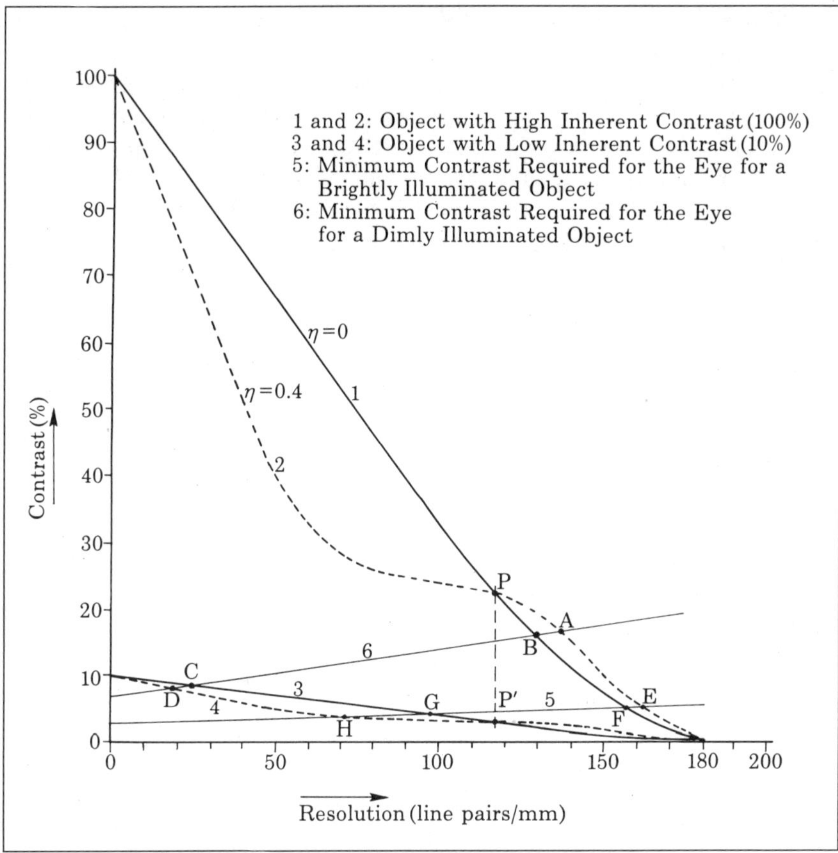

Fig. 18.8 *Comparison of Two Telescopes for Visual Use.*

we can now draw a number of conclusions:

1. The resolving power is higher for brightly illuminated objects than for dimly illuminated objects. Note and compare points F and B.
2. The resolving power for high-contrast objects is higher than for low contrast objects. Note and compare points F and G.
3. The resolving power for high-contrast objects such as double stars can be slightly higher for an obstructed system than for an unobstructed system used in very fine seeing. Compare points *E* and *F*.
4. The resolving power of an obstructed system for low-contrast objects such as planetary detail is less than that of an unobstructed telescope. Compare points *H* and *G*.

This last situation occurs whenever the intrinsic contrast of the object is low enough that the crossover point (*P′*) falls below the eye's contrast threshold curve. This conclusion is confirmed by the experiences of Pickering and Dall: "For the observation of low-contrast extended details, an unobstructed telescope is superior to a telescope having a central obstruction." (ref. 18.5.)

What is the largest obstruction that will not noticeably degrade a telescope's resolving power? Unfortunately no consensus exists in the literature on this subject. Anton Kutter, a well-known advocate of unobstructed telescopes, wrote in 1975: "In my opinion the limit is a diameter of the obstruction of 10% of the aperture." (ref. 18.12.) Horace Dall found that the influence becomes noticeable when the diameter of the central obstruction exceeds 20% (ref. 18.6), while investigators Steel (ref. 18.7) and Bouwers (ref. 18.10) put the limit at 25% to 30% of the entrance pupil. These last values correspond, at least approximately, to the Rayleigh limit.

However, this does not always imply that a telescope with a 30% central obstruction will perform in an essentially perfect manner. If it suffers from any additional image aberrations or surface errors, the effects will accumulate in the contrast transfer diagram. The final system may then no longer be diffraction limited.

This means that centrally obstructed systems must be considerably more accurate than unobstructed systems to attain the same net performance. In other words, Rayleigh's ¼-wave criterion, while perhaps adequate for an unobstructed system, is insufficient for telescopes with a central obstruction!

18.8 Residual Aberrations

We have seen that a central obstruction causes a loss of contrast which decreases the resolving power of the telescope, and that this effect is particularly important for objects with low intrinsic contrast. The same reasoning applies to telescopes with residual aberrations. The problems mostly likely to affect the axial performance of an amateur telescope are spherical aberration and surface roughness.

It can be shown that the residual aberrations result in relatively little loss of resolving power for high contrast objects. However, for objects with a low intrinsic contrast, a relatively large drop in resolving power will take place. This corresponds with the experience of Conrady (ref. 18.11), who found that residual aberrations and slightly imperfect worked optical surfaces do not much influence the resolving power of the telescope for high contrast objects, and that the effects are much more noticeable for objects with low intrinsic contrast.

18.9 The Value of the Contrast Transfer Function

The CTF curve gives a better overall picture of the telescope's optical quality, and certainly yields far more information, than testing on double stars possibly can. It takes into account not only the accumulation of diffraction effects but also the imperfections in the optical system, and in this realm, not only errors of fabrication but also of design. The net capability of the telescope finds its expression in the position of the contrast transfer curve with respect to the idealized curve.

For visual observation of low contrast details on extended objects—plane-

Table 18.2
Resolving Power of the Eye

contrast (%)	89	70	49	38	27	16	10	6	2.5
resolving power (arc-sec)	74	76	80	82	84	89	95	105	127

tary detail—it is difficult to define a meaningful resolving power for a telescope. Parameters such as brightness of the image, intrinsic contrast, obstruction ratio, image aberrations, and magnification, as well as the contrast sensitivity and visual acuity of the eye must be taken into account. Because of this, any definition of resolving power is always subject to strict conditions.

Finally, because the resolving power for high contrast objects is not sensitive to optical errors, it is obvious that the common practice of testing telescopes (and camera lenses) with charts consisting of black and white bars is a poor test of optical quality. Conclusions drawn on the basis of such charts do little to predict the performance of a telescope on objects with a low intrinsic contrast. Test charts with dark grey and light grey lines are more suitable for testing the performance of a telescope.

18.10 Optimum Magnification

Telescopic magnification for a given detail is optimum when the visibility of that detail decreases at both lower or higher magnification.

We often see the following stated: The resolving power of the unaided eye amounts to 60 seconds of arc, and the telescope's resolving power is $120/D$ arc-seconds. The optimum magnification is $60/(120/D)$, or $0.5 \cdot D$, where D is in millimeters.

In this reasoning several important factors have been overlooked. The resolving power of 60 seconds of arc refers to a brightly illuminated test chart with high contrast, and is not valid for everyone. Danjon (ref. 18.8) gives the values for the resolving power of the unaided eye for well illuminated test charts shown in table 18.2.

For lunar and planetary objects the contrast is generally much lower than 100%, so that a higher magnification is necessary to resolve the detail. Coleman (ref. 18.9) studied the optimum magnification of a telescope. He observed test charts with black/white and grey/white grids with a variety of intrinsic contrasts. His results are given in fig. 18.9.

Depending on the intrinsic contrast of the object observed (5% to 100%), Coleman found that the optimum magnification lies between $1.1 \cdot D$ and $2.7 \cdot D$, where D is the objective diameter in millimeters. This is considerably higher than the $0.5 \cdot D$ derived by the faulty logic above. In judging Coleman's results, however, bear in mind that the experiments were carried out under laboratory conditions, using brightly illuminated test charts and optimal seeing conditions. For less

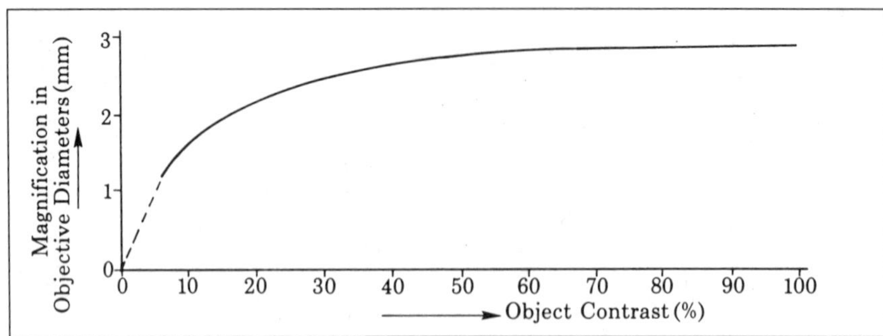

Fig. 18.9 *Optimum Magnification of a Telescope for Various Object Contrasts.*

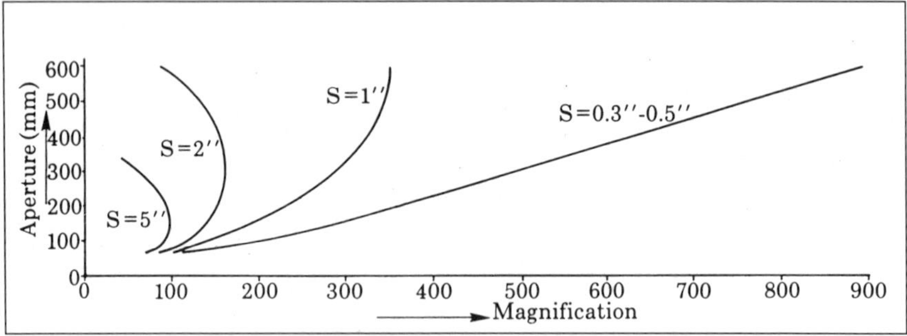

Fig. 18.10 *Optimum Magnification in the Presence of Scintillation.*

brightly lit objects, the optimal magnification will be lower. This is also true when air currents are present.

Moreover, from the data underlying fig. 18.9, it follows that the gain in resolving power above $1.0 \cdot D$ magnification is no larger than 5%. In other words, at $1.0 \cdot D$ magnification the observer already enjoys 95% of the maximum resolving power the telescope can show him. Even at $0.5 \cdot D$ magnification, roughly 80% of the telescope's full resolving power is attained.

According to Texereau, an experienced planetary observer, the optimum diameter of the exit pupil for observing planetary details is 0.8 mm. This implies a magnification of $1.25 \cdot$ D. This is in accord with Coleman's results for objects with low (10%) contrast.

Treatments of optimum magnification are usually based on perfect atmospheric conditions. In reality, seeing conditions must always be taken into account. Under conditions of poor seeing, large instruments are relatively disadvantageous because the influence of air currents increases with the square of the entrance pupil diameter.

In fig. 18.10, we show how the optimum magnification and the objective diameter for observing lunar and planetary details depends on scintillation. Scintillation, expressed in arcseconds, is classed as: 0.3 arcsec, exceptionally good; 1 arcsec, good; and 5 arcsec, severe turbulence.

From the diagram, we can see that for every degree of scintillation, there is an optimum objective diameter. For severe turbulence, it is 150 mm; and for good seeing, 500 mm to 600 mm. For exceptional seeing, the aperture is off the scale of the chart. Note that for observation of planetary detail, instruments larger than 300 mm are only useful under good seeing conditions.

Chapter 19

Opaquing and Vignetting

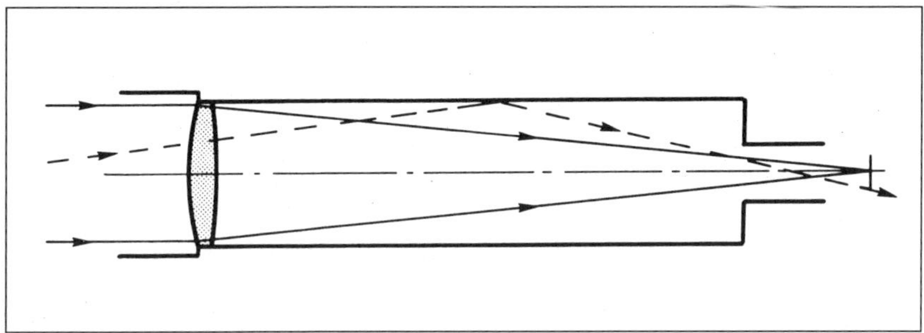

Fig. 19.1 *Reflections in an Unbaffled Telescope Tube.*

19.1 Introduction

Few telescope makers pay enough attention to opaquing, or blocking stray light in their telescopes. Good opaquing is of the utmost importance for daytime observation, and it is vital for the best night sky viewing and photography. Eliminating stray light produces images with the highest contrast that can be obtained from a given optical system.

The authors have evaluated a number of commercially available telescopes with respect to their opaquing. Most manufacturers pay far too little attention to this subject. Glossy baffle tubes, shiny telescope walls, and glaring internal stops are the rule rather than the exception. Yet these evils can be avoided at relatively low cost.

Consider the internal reflection in a refractor in which no opaquing measures have been taken (fig. 19.1). Off-axis light simply reflects from the tube wall and reaches the focal plane. In theory, at least, reflections from such surfaces are diffuse reflections since the light is supposed to scatter in all directions. However, for small angles of incidence, these surfaces tend to become mirror-like (see fig. 19.2), producing specular reflections. Since light usually enters telescopes at small angles to the tube axis, this phenomenon occurs frequently.

The authors have measured the reflectance for flat black paint, flock paper, and black velvet. For normal incidence, the reflectance of black paint is low, and that of flock paper and black velvet is near zero.

However, at grazing angles of incidence, the reflectance, especially of black

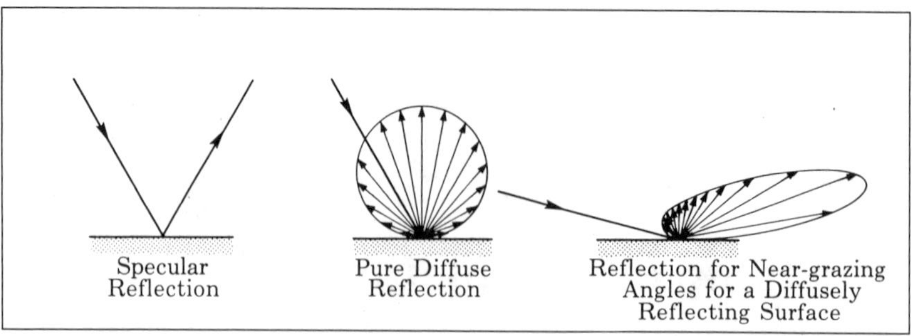

Fig. 19.2 *Reflection from Various Surfaces.*

paint, increases rapidly. At grazing angles of incidence, flock paper and especially black velvet represent a considerable improvement over black paint. The reflectance of flock paper is about twice that of black velvet. Black paint reflects about a hundred times as much.

Therefore, if there is no space for mounting annular baffles, it is often worthwhile to cover the inside of a telescope with black velvet or flock paper rather than black paint. When using black velvet the direction of the pile should be taken into account.

The best, but most expensive, method of preventing reflections from entering the focal plane is to install annular baffles. For a maximum effect, such baffles must have carefully designed diameters and spacings. Baffles require a larger tube diameter, but the improved performance usually justifies the expense.

Baffling and shielding Cassegrain-like systems is notoriously difficult. Reflections occur along the inside walls at grazing angles of incidence, and, in most cases, there is not enough space to install annular baffles. The inner surfaces of the baffles and shields used should be scored, roughened, or threaded, then painted matte black to break up the internal reflections and reduce glare. Covering some surfaces with velvet or flock paper can be useful as well.

19.2 Baffles for Refractors and Newtonians

Fig. 19.3 shows how annular baffles are to be placed in a refractor. The positions must be chosen so that no place on the wall illuminated by the entrance pupil can be seen from the focal plane. Examples are given for different tube diameters. The designer starts by drawing the auxiliary lines *A–A*, *B–B* and *C–A*. The graphic construction of the baffles is then quite easy. When the instrument has a lens shade, the auxiliary lines can start at the edge of this shade.

Note that the shaded areas cannot be seen from any point of the focal plane. The unshaded areas can be seen, but are illuminated at near-normal incidence by strongly reduced scattered light. Fig. 19.3 shows clearly the price of reducing the tube diameter: the number of baffles increases rapidly.

The position of the baffles in a Newtonian can be found in a similar way (fig.

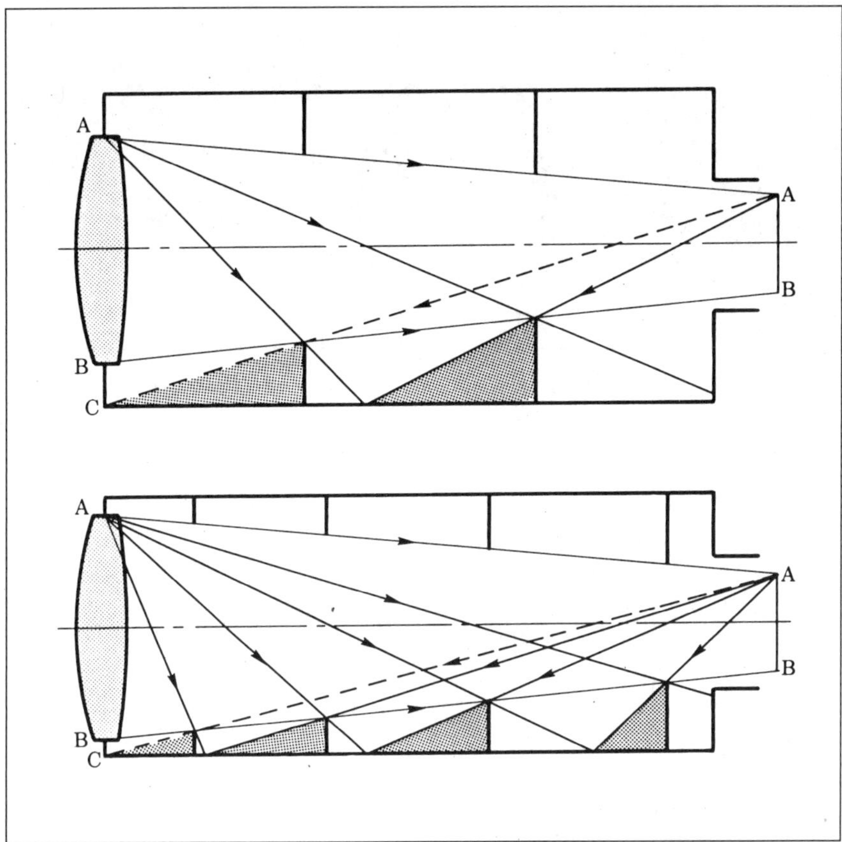

Fig. 19.3 *Baffle Positions in a Refractor.*

19.4). The fundamental difference is that in a refractor the wall is illuminated directly only from the entrance pupil. In a Newtonian, illumination occurs not only from the open end of the tube but also from the mirror. The designer starts by determining diameter D_2, chosen to be larger than D_1 so as to avoid vignetting oblique bundles. The principle of the design is that no part of the tube wall *A–J* should be visible from point *Q* (which corresponds to point *Q′*). The first auxiliary line is *QA*, after which the construction proceeds to *B*, *C*, *et cetera*. From the section of tube wall *J–K*, you will see that no light can reach the focal plane *P′Q′* via the secondary mirror.

However, if the secondary mirror is relatively small, light reflected by the wall opposite the eyepiece and between *J* and *K* reaches the eyepiece directly. It is often useful to place additional baffles or black velvet at the front side of the tube opposite the eyepiece.

As is the case with the refractor, the number of baffles increases rapidly as the tube diameter decreases. To be effective, the baffles must be matte black,

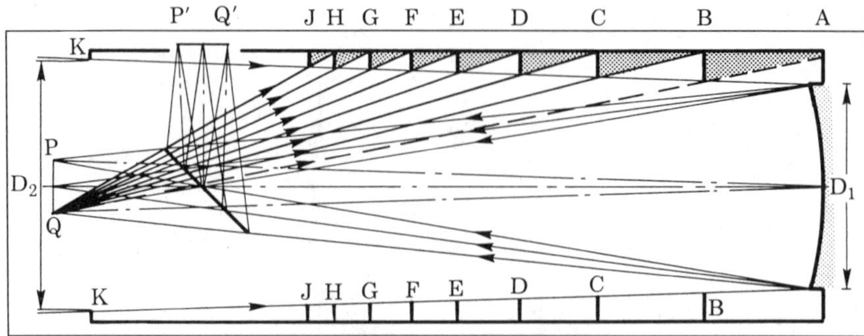

Fig. 19.4 *Baffle Positions in a Newtonian.*

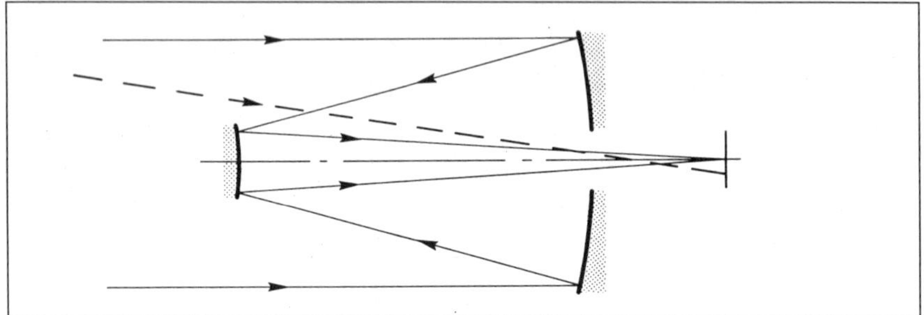

Fig. 19.5 *Stray Light in an Unbaffled Cassegrain.*

rough-surfaced, and sharp-edged. When constructing baffles, it is useful to make a large scale drawing (preferably 1:1 or even larger) of the instrument. This is also true for designing the baffle tube system in a Cassegrain, as described in the following section. Of course, baffle design can be carried out analytically, and only simple mathematics is necessary.

19.3 Baffling for Cassegrain Telescopes

In Cassegrain optical systems, light entering the tube can reach the focal plane directly without being reflected by the mirrors (fig. 19.5). For daytime observation, serious loss of contrast can take place. In order to prevent this, Cassegrain telescopes must be provided with baffle tubes. The shape and size of the baffles depends strongly on the configuration of the mirrors and the size of the field that must be free of vignetting. When a large unvignetted field is desired, the diameters of the baffle tubes and the central obstruction become quite large. Usually a certain off-axis light loss must be accepted to reduce the diameters of the baffle tubes.

Basically the baffle system must be designed in such a way that on the optical axis no vignetting occurs. So, for a beam of light that is parallel to the optical axis, all light available must reach the focal plane. Before discussing the method for designing a baffle system for a Cassegrain telescope, it should be remarked

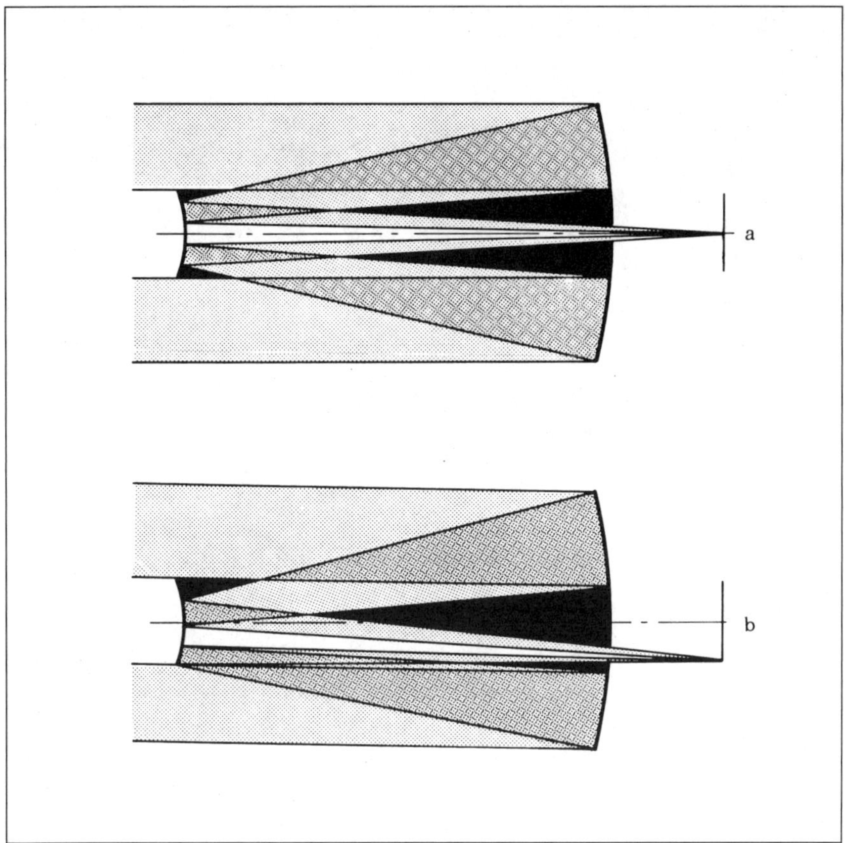

Fig. 19.6 *Axial and Off-axis Rays in a Cassegrain.*

that this is not a straightforward process, but an iterative process of successive approximations that goes on until an acceptable compromise between various conflicting demands has been attained.

We will first discuss the case in which a minimal light loss at the edge of the field is allowed. This implies that no constraints are to be applied with respect to the maximum allowable central obstruction. The designer starts by making a large scale drawing showing the axial light and off-axis bundle illuminating the edge of the desired field. (This is shown as two separate drawings in fig. 19.6. In a single drawing, it will immediately become clear how baffles can be placed.) This determines the size of the secondary mirror. Next, for both cases determine which parts of the mirrors do not participate in the image-forming process; then draw the regions through which no image-forming light passes. These are shown as black regions in fig. 19.6.

For the bundle parallel to the optical axis, two black zones occur. For axial light shielding, we could place in these zones two cylindrical baffles, one, called

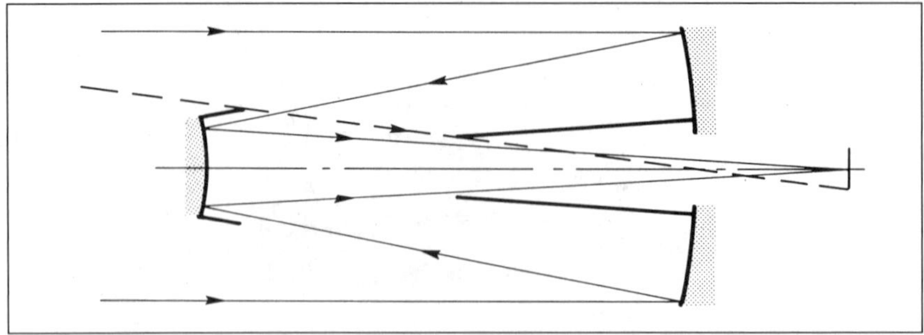

Fig. 19.7 *Baffles for a Cassegrain Telescope.*

the rear baffle tube, projecting from the primary mirror toward the secondary mirror; the other, the front baffle tube, extending from the secondary toward the primary.

However, if the oblique bundle must pass through these tubes without vignetting, the rear baffle tube must be shorter and have a somewhat different position. We also note that if the front baffle must be a cylinder, it should be omitted completely to avoid vignetting of the oblique bundle. If this is allowed, an oblique entering ray could reach the focal plane directly.

To prevent the oblique ray from reaching the focal plane directly, make the front baffle a diverging cone around the secondary mirror, as shown in fig. 19.7. The degree of divergence is adjusted to fit the oblique bundle. It should be clear that the conical front baffle increases the diameter of the central obstruction. This is the price that has to be paid to avoid vignetting at the edge of the field of view.

We will now consider the case which is most often encountered, in which a small central obstruction is desired and a considerable vignetting of the field is allowed. In the literature, a 30% to 40% light drop-off at the edge of the field is considered normal for photographic use, and produces negligible deterioration in image quality. For visual use, even larger values of light drop-off may be acceptable in order to obtain the smallest possible central obstruction, and only a small field of view is necessary.

Usually the front tube is assumed to be cylindrical in order to prevent enlarging the central obstruction, but the rear baffle tube may be made conical, because the diameter at the front end of this tube should be as small as possible in order to block direct oblique rays but not block the center of the bundle reflected from the primary mirror, while the back end should be large to prevent additional and unnecessary obstructions in the off-axis bundle. This important point will be elucidated further on. Although the baffle design must be carried out in such a way that no vignetting on the optical axis occurs, it is inevitable that oblique bundles will be vignetted to some degree.

Although all Cassegrain-like systems are similar, the designer should be aware that the size and placement of baffle tubes also depend on the position of the entrance pupil, the limiting stop of the system, and the optical power of refrac-

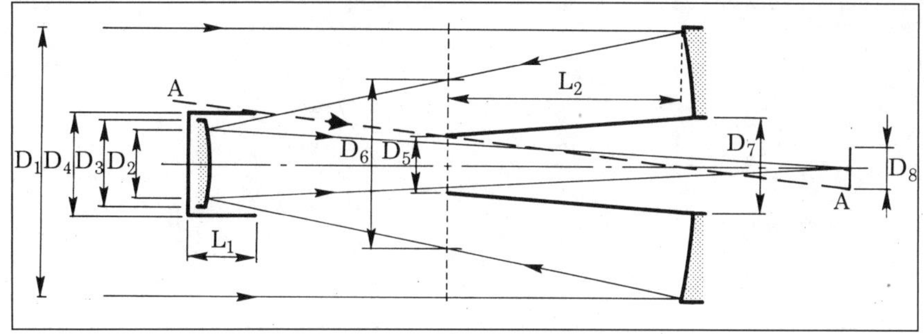

Fig. 19.8 *Parameters for Cassegrain Baffles.*

tive elements in catadioptric systems. For a two-mirror Cassegrain, the entrance pupil usually coincides with the primary mirror. In the Schmidt-Cassegrain design, the entrance pupil is located at the Schmidt corrector. This makes the position of oblique bundles in a Schmidt-Cassegrain somewhat different from those in a Cassegrain, and this influences the layout of the baffle tubes.

In a Maksutov-Cassegrain, the entering light bundle will be widened and displaced by the power of the meniscus corrector lens. This influences the baffle system more than the near-zero power Schmidt corrector does in the Schmidt-Cassegrain.

The design of a baffle tube system for Cassegrain telescopes consists of series of successive approximations. Every step should be accompanied with a complete ray tracing for straight and oblique bundles. Because the process is difficult, we give an example in the next section.

The following guidelines for the design of a baffle tube system in a Cassegrain telescope may be helpful. Refer to fig. 19.8.

- D_1 is the entrance pupil, i.e., the aperture.
- D_2 is the secondary mirror illuminated by the axial bundle.
- D_4 is the largest allowable central obstruction.
- D_3 is the secondary mirror illuminated by the most oblique bundle, unless this diameter is larger than D_4.
- D_8 is the desired field of view.
- D_7 should not be larger than D_4.
- The ratio D_5/D_6 should not be larger than the ratio D_4/D_1. If it is, the front end of the rear baffle tube forms an artificial central obstruction in the conic bundle between the primary and secondary mirror. (This constraint is easily overlooked by beginning amateur designers.) Note also

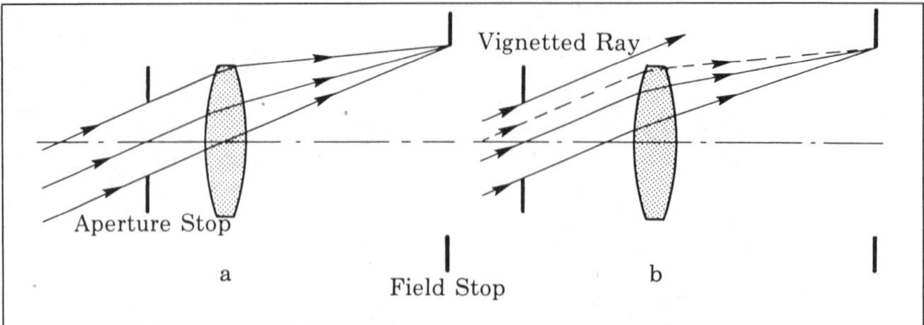

Fig. 19.9 *Aperture Stops, Field Stops, and Vignetting.*

that this condition changes when the telescope is a catadioptric Cassegrain.

In designing the baffle tubes, don't forget that the rear baffle tube can be stepped if this is easier to make than a conical baffle. The front baffle length, L_1, should be just enough to nearly touch the axial beam, or slightly shorter. The length of the rear baffle, L_2, must be such that the tip of the rear baffle lies on line *A–A*, which runs from the tip of the front baffle to the edge of the desired field of view.

It is difficult to design a baffle tube system for a 200 mm Cassegrain-like telescope having a central obstruction smaller than 30% of the entrance pupil. When a smaller obstruction is pursued in a single-minded fashion, the baffle system that results causes such excessive vignetting that the useful field is considerably reduced.

Sometimes the designer of a baffle tube system carries out a "stop analysis." In this analysis, the designer discovers how the various stops "look" from the focal plane. Vignetting and stray light sources are readily visible for both axial and oblique beams. Readers interested in this subject should consult reference 19.2.

The literature contains a variety of articles on the design of Cassegrain baffles (refs. 19.3–19.5). In the most advanced of these (ref. 19.5), Alan W. Greynolds gives a computer optimization method which designs the baffle tubes as well as additional annular baffles to reduce stray light to an absolute minimum.

For visual use, we shouid point out that it is possible to use a telescope entirely without baffle tubes. To intercept stray light, a small diaphragm can be placed behind the eyepiece, in the plane of the exit pupil. This method can only be used for visual observing, and is of no value for direct photography through the telescope.

19.4 Stops and Vignetting

In the foregoing sections, we have mentioned stops, vignetting, and light drop-off several times. In this section we examine these subjects more fully. In fig. 19.9a, we see a single lens with a diaphragm in front of it. This diaphragm forms the limiting stop, for both the axial and oblique beams. It, and not the lens, determines the

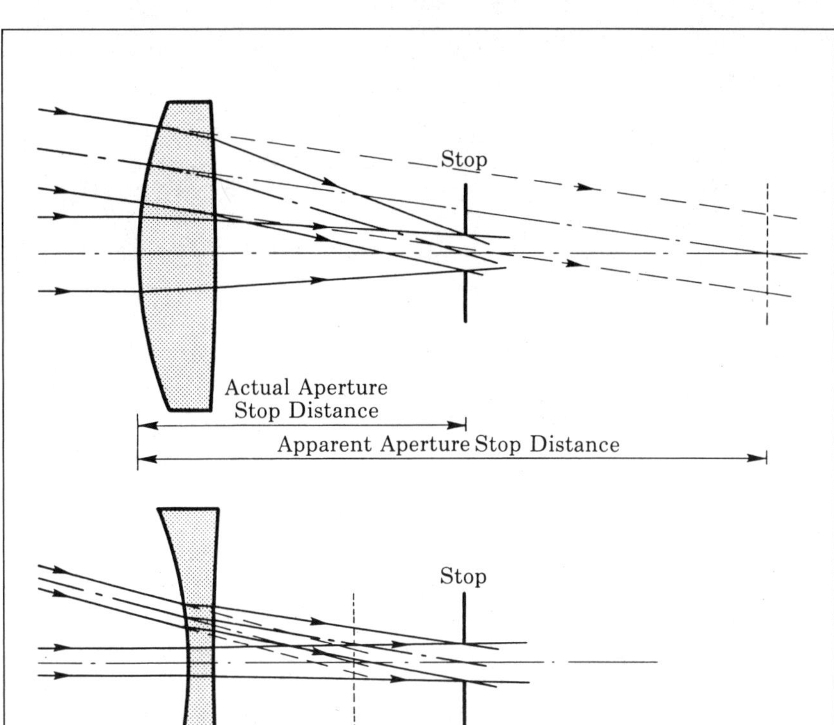

Fig. 19.10 *The Effect of Preceding Optical Elements on Apparent Stop Size.*

aperture and thus the light-gathering power of the system. Such a diaphragm is called an aperture stop.

An aperture stop also determines the position of oblique beams and the place where they pass through the lens. This particular point is important because the position of the stop often influences the off-axis image aberrations.

Each objective system also contains another type of stop—a field stop. Field stops limit the size of the image field and determine the most oblique beam that reaches the focal plane. They may be square or rectangular, as they are in a camera, or circular, as they are in telescopes used in combination with an eyepiece. The situation shown in fig. 19.9a is a simple case in which the size and the position of the oblique beam are determined entirely by the aperture stop and the field stop.

In fig. 19.9b, however, quite another situation occurs. The diameters of the aperture stop and the field stop are the same as in fig. 19.9a, but now part of the oblique beam is cut off by the edge of the lens. This situation often occurs in amateur telescopes, and is called edge vignetting. Rays more oblique than a certain

Table 19.1
Edge Vignetting

Type	Aperture stop	Edge vignetting by
Newtonian	Primary mirror	Secondary mirror
Refractor	Objective	Focuser tube
Cassegrain	Primary mirror	Secondary mirror, Baffle tubes
Schmidt/Mak Camera	Corrector	Primary mirror
Schmidt/Mak Cassegrain	Corrector	Primary mirror, Secondary mirror, Baffle tubes
Schmidt-Cass Camera	Separate stop or Corrector	Primary mirror, Secondary mirror

angle get cut off, resulting in light loss towards the edge of the field. In the chapters on the design of various optical systems, this point has come up several times.

The aperture stops in fig. 19.9 were external to the optical system. When the stop is inside the system, the situation changes. In a typical camera lens, fig. 19.10, the aperture stop is an adjustable diaphragm between the lens elements. When the front element has positive power, the ray bundle is narrowed, so its original diameter is larger than the stop. The reverse occurs when the first element is negative.

Apart from the change in size, the entering oblique bundle "sees" the aperture stop shifted to a new position. This means that the tilt point of the oblique bundles does not occur at the actual position of the stop, but at its apparent position. (In this case, the entrance pupil coincides with the apparent aperture stop.) Before a skew-ray-trace analysis can be carried out, both the diameter of the entering bundles and the position of the apparent stop must be calculated.

In an amateur telescope, vignetting is caused by other lens or mirror rims, or, particularly in Cassegrain-like telescopes, by the baffle tube system. Table 19.1 gives the positions of the aperture stop and the place where edge vignetting usually occurs. In a number of cases central vignetting will also occur as a result of a central obstruction.

Locating vignetting and determining its magnitude can sometimes be rather complicated, particularly when the system contains baffle tubes. The degree of off-axis light loss can be calculated using a standard ray-tracing analysis such as the program available as an option with this book.

The method is as follows: Parameters for all the elements are fed into the program. Stops and obstructions are entered as optical surfaces having zero power. Central obstructions, which cause vignetting, are also entered. Then the ray trace is carried out for an axial beam and various oblique beams.

Our computer program has been written so that if a ray cannot pass an obstruction, or falls outside the rim of an optical element, it is not traced any further. The off-axis light loss is calculated for various off-axis distances by counting the

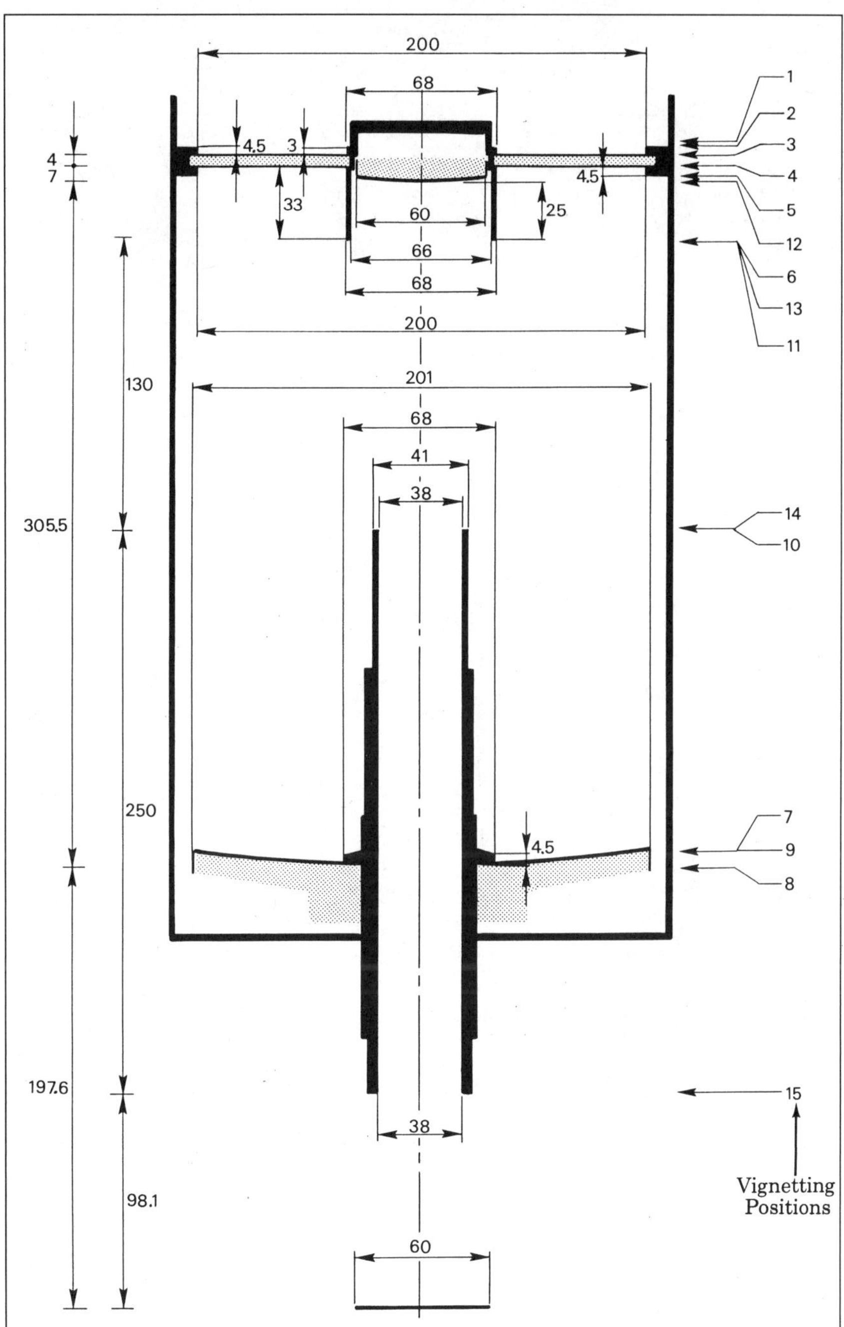

Fig. 19.11 *Layout of 200 mm f/10 Schmidt-Cassegrain Telescope.*

Table 19.2
Vignetting Parameters of a 200 mm *f*/10 SCT
(All dimensions in millimeters)

No.	Position in system	Edge diameter	Obstruction diameter
1.	Corrector, retaining ring	200 (200)	—
2.	Secondary, retaining ring	200 (200)	68 (75)
3.	Corrector, front side	200	68 (75)
4.	Corrector, back side	200	68 (75)
5.	Corrector, support	200	—
6.	Front baffle, outside rear end	—	68 (75)
7.	Retaining ring, primary	—	68
8.	Primary mirror edge	201 (211)	68
9.	Retaining ring, primary	—	68
10.	Rear baffle, outside front end	—	41 (45)
11.	Front baffle, inner rear end	66 (73)	—
12.	Secondary mirror edge	60 (67)	—
13.	Front baffle, inside back end	66 (73)	—
14.	Rear baffle, inside front end	38 (42)	—
15.	Rear baffle, inside rear end	38 (42) (60)	—

number of rays that reach the focal plane. If, for instance, 200 rays reach the focal plane on axis, but only 150 rays reach the focal plane at some off-axis distance, the light loss at that place is:

$$\frac{200 - 150}{200} \cdot 100\,\% = 25\%. \qquad \textbf{(19.4.1)}$$

The relative illumination amounts to 75% at that place.

We have performed this analysis for the 200 mm Schmidt-Cassegrain telescope described in section 9.3. Its dimensions are given in fig. 19.11. Table 19.2 shows that a vignetting analysis of this system takes into account no less than fifteen places. The number in the table corresponds with the position numbers in the figure.

In some places only edge limitations must be considered; in other places, only the central obstruction; and at other places, both. Numbers between brackets refer to alternative cases. Based on this table, five situations were analyzed with respect to the light drop-off:

1. The configuration shown in fig. 19.11 with a 68 mm (34%) central obstruction. The diameter of the primary mirror is 201 mm, which is just sufficient to catch all outer rays of an axis-parallel bundle.
2. The same configuration as (1), but with a 5% oversize mirror (211 mm).
3. A configuration with a widened baffle tube system. Instead of an inner rear baffle tube diameter of 38 mm, 42 mm is used. To stop stray light

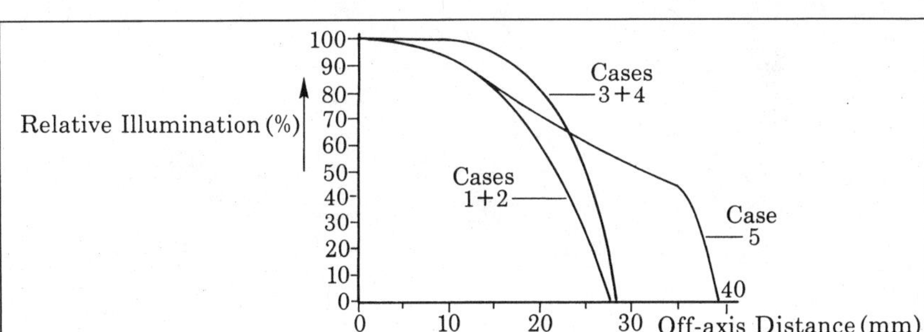

Fig. 19.12 *Relative Illumination in the Focal Plane of a 200 mm f/10 Schmidt-Cassegrain Telescope.*

from reaching the focal plane, the front baffle must then be widened to 75 mm, so that the central obstruction is increased to 37%. The diameter of the primary is held to 201 mm.

4. The same configuration as (3), but with a 5% oversize primary mirror (211 mm).

5. The same configuration as (3), but instead of a cylindrical rear baffle tube, a conical baffle tube. The inside front diameter is still 38 mm, but the back end is widened to 60 mm. This enlarges the field because the back end blocks skew bundles. The conical baffle would make focusing by moving the primary mirror either quite difficult or impossible.

In fig. 19.12, we graph the light loss curves for these five cases relative to the axial bundle. Note that in the 37% central obstruction, the illumination will be some 2% lower than in the cases of 34% obstruction.

The illumination curves of cases 1 and 2, and of 3 and 4, coincide. We conclude that the oversize primary mirror is useless in these particular cases, because the rear baffle tube determines the vignetting for off-axis light.

Curve 5 is an interesting one. For off-axis distances less than 15 mm, it coincides with cases 1 and 2, because the front end of the rear baffle tube mainly determines the vignetting. More than 15 mm off-axis, the back side of the rear baffle tube also plays its part. A considerably larger field is obtained when the rear baffle tube is conical.

The reader may have noticed that for most telescopes and astrocameras that were treated in the foregoing chapters, the diameters of the various optical components were not specified. The reason for this is that these diameters are partly dependent on the compromises that the designer wishes to make between various conflicting demands.

As soon as the designer has determined the optical data such as the axial thickness and distances, radii of curvature and kinds of glass, the diameters must be specified. As should be obvious, the minimum required diameters of the components are determined by the axial bundle.

Next, the diameters may be increased for a wider field, but this depends en-

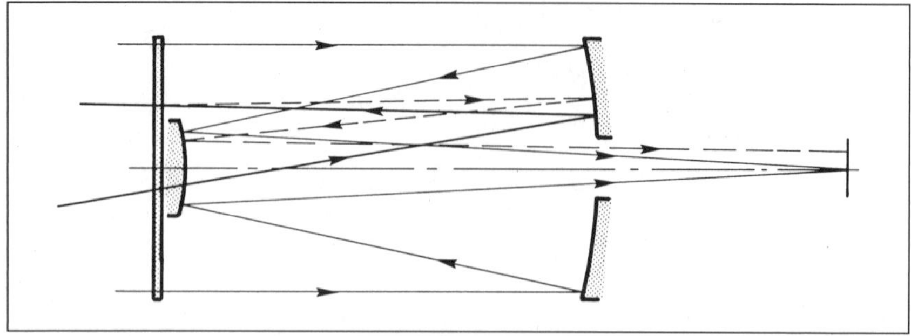

Fig. 19.13 *Stray Light in a Schmidt-Cassegrain Telescope Caused by Reflection at the Corrector.*

tirely on the degree of vignetting and off-axis light loss that the designer is willing to accept. In the presence of a secondary mirror, the maximum acceptable central obstruction plays an important role too. As we have seen, in Cassegrain-like instruments the baffle tubes may be the determining factor.

In production telescopes, economic considerations often play an important role in choosing the sizes of the optical elements. The computer program, with its capability for performing the vignetting analysis, is an excellent tool for the designer seeking the best compromise between the conflicting demands in a particular design.

19.5 Internal Reflections in Catadioptric Systems

Internal reflections in catadioptric telescopes can sometimes lead to a considerable loss of contrast. In fig. 19.13, we show a Schmidt-Cassegrain telescope. Normally, those ray bundles inclined at small angles relative to the optical axis will be imaged in the focal plane. Bundles entering at larger off-axis angles will be reflected by the primary mirror and thrown on the back side of the corrector. Most of this light passes through the corrector. Approximately 8% for a non-coated corrector and 2% for a well-coated corrector is reflected back to the primary mirror and thence to the focal plane, diminishing image contrast.

The authors have noticed that considerable loss of contrast can occur during daytime observing with a non-coated corrector, especially when the object under observation has a bright background. At night, Schmidt-Cassegrain telescopes with coated optics show a considerably darker sky than those with non-coated optics. As little off-axis light as possible should enter the telescope; this can be partially realized by providing the telescope with a long lens shade.

Yet another source of reflections that occurs in Schmidt correctors is ghosting. Ghost images result from reflections between the two surfaces of the corrector plate. Ghosts appear as halo-like effects when bright objects are observed or photographed against a dark background.

Internal reflections and ghosts are less bothersome for bent correctors, such as the Maksutov corrector. One European supplier has developed a special tech-

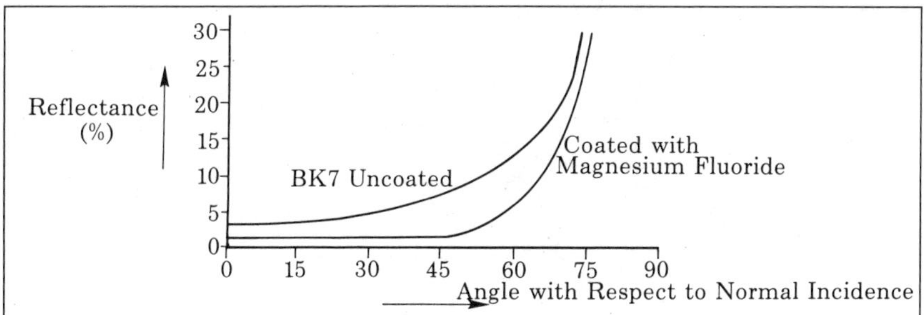

Fig. 19.14 *Reflectance of BK7 Glass.*

Table 19.3
Reflection by Optical Glasses

Refractive index:	1.4	1.5	1.6	1.7	1.8	1.9
% Reflection:	2.8	4.0	5.3	6.7	8.2	9.6

nique for generating bent Schmidt correctors. Another technique for suppressing these two sources of internal reflections is coating. Contrary to popular belief, the elimination of reflections and the resultant gain in contrast is more important for the observer than the gain in light transmission, which amounts to only 4% per surface.

19.6 Lens Coatings

The unwanted surface reflectance of lens surfaces can be reduced by an optical coating. The reflectivity of an uncoated lens surface depends on the refractive index of the glass and the angle of incidence. For normal incidence, the reflectivities are shown in table 19.3.

Note that the reflectivity becomes quite high in high-index glasses. For angles of incidence greater than roughly 30 degrees, reflectivity increases rapidly. This rise is graphed in fig. 19.14.

To reduce unwanted reflections, we can coat the surface with a layer of a substance having a refractive index equal to the square root of the index of the glass. Nowadays, magnesium fluoride, a durable substance with a refractive index of 1.38 at 550 nm, is the most commonly used coating. The optical coating thickness is 1/4 the wavelength of the light we wish to suppress.

What happens in a coating on a lens?

Instead of one reflection at the air-glass interface, there are two. One occurs at the air-coating surface, and the other at the coating-glass interface. The light from the reflections interferes destructively because it differs by half a wavelength. The shift of phase of a half wavelength occurs because the glass-reflected light has traversed the coating layer twice. The result is that the reflection is min-

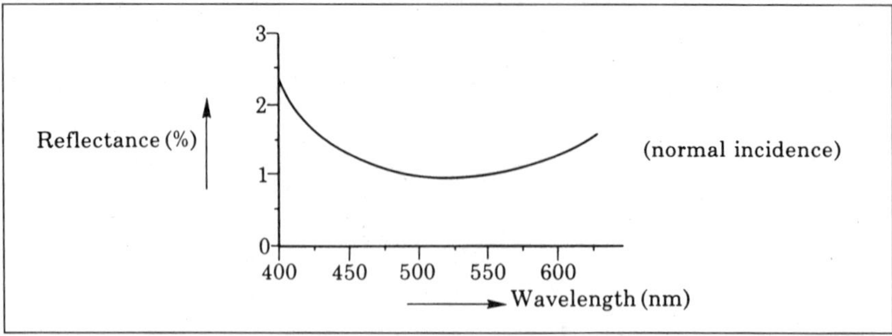

Fig. 19.15 *Reflectance of Magnesium-Fluoride-Coated BK7 Glass.*

imized.

The effectiveness of the interference requires that both reflections have the same strength. This happens when the ratios of the indexes at the two interfaces are the same. For magnesium fluoride this occurs for glass having a refractive index of 1.90 (i.e., $1.38 \cdot 1.38$). For glasses with a lower refractive index, this coating is somewhat less effective.

For BK7 glass, a magnesium fluoride coating reduces the reflectance from 4% to 1% at normal incidence. Because of the principle of energy conservation the reduction of reflectivity results in a corresponding increase of light transmission, from 92% to 98% for a lens with two surfaces.

As can be seen in fig. 19.15, the efficiency of a single-layer coating is optimum at one wavelength, and the reflectance for other wavelengths is higher. By using multiple layers and a combination of different coating materials, the reflectance can be reduced over a wider spectral range, or reduced to 0.25% or less over a narrow spectral range.

Coatings have become a necessity, especially for modern telescope eyepieces. These often contain many lenses, frequently employing high-index glass, and large angles of incidence occur in the wide-field designs.

Chapter 20

Optical Calculations

20.1 Introductory Remarks to Chapters 20 and 21

In order not to discourage our readers, we have placed chapter 20 (Optical Calculations) and 21 (Design of Optical Systems) at the end of the book. Optics, by its very nature, is a science with many and often complicated formulae, so it is unlikely that a large number of telescope makers will be inclined to become familiar with optical calculation and design. In our experience, though, telescope making provides the deepest satisfaction to the telescope maker who not only constructs, but also designs, his optics.

Chapters 20 and 21 are not intended to replace handbooks on optical design. Instead, they are written to instruct the reader in the use of the standard optical formulae. After reading these chapters carefully and patiently, the telescope maker will be able to design excellent telescopes.

The reader should not be awestruck, frightened, or daunted by the many formulae presented in these chapters. In the past, optical computing was time-consuming and burdensome, and this alone prevented amateurs from contemplating, let alone attempting, optical design. Today, however, such formulae can be programmed on a personal computer—and satisfying results produced with remarkable speed.

20.2 Methods of Optical Calculation

In chapter 3 we gave a brief account of image formation, and how the focal length for a compound system can be determined. The formulae given in that chapter are approximations: they neither take into account the thickness of the optical components nor give any indication of the quality of the images formed.

The best method for determining the performance of an optical system is to perform a skew-ray-trace calculation. This generates a set of spot diagrams for various off-axis distances and colors, as described in chapter 4. However, because running a full set of spot diagrams is still time-consuming, we need another method for evaluating designs in their early stages. Ideally, we would like to obtain a sense of the influence of small changes of the design parameters on the optical performance, the focal length, and the position of the focus. Since optical design is often not a straightforward process, but proceeds iteratively, as a series of small steps toward perfection, it is of the utmost importance that the designer maintain an overview of the changes, and possess some way telling whether each change improves or worsens the optical quality.

With a new telescope design, the designer will start with approximations that are quick and easy, but not very accurate. As he improves the system's performance, he relies more and more on the precise skew-ray trace, until the design can no longer be improved. Given a design that already performs moderately well, though, the early stages can be skipped, and the skew-ray trace used to arrive at a final design.

This is the situation readers of this book will most often encounter with catadioptric telescopes "predesigned" with the computer program which is an option with the book. It is difficult to give rules for how to arrive at a final, fully-optimized design because of the numerous types of designs and starting points. Frequently, however, the design progresses through these four stages:

The paraxial calculation. This method, called a "first-order" calculation, is carried out for the region in the pupil close to the optical axis, with small off-axis distances and angles. This gives the position of the principal planes (described in chapter 3) and the focal length of the system, but offers no information about image aberrations. Having roughed out preliminary values for the radii of curvature, thicknesses, and distances of the elements, the designer proceeds to:

The Seidel calculation. Using what is called "third-order" aberration theory, the designer explores the types and magnitudes of the image aberrations. The Seidel method offers a quick optical evaluation, and is therefore well-suited to the initial design stages. It is followed by:

The meridional ray trace. This technique gives rigorous results, but in one plane only, that of the cross-sectional, or meridional plane. In most cases this calculation is carried out for a small number of rays, typically the center ray (or "principal" ray), a rim ray, and some rays between them. This method has been used for a long time for presenting the optical quality of optical systems.

The three preceding methods can be carried out quickly with no more than a small programmable calculator. When high-speed computer equipment is available to the designer, though, the meridional ray trace can be augmented with:

The skew-ray trace. In this technique, we calculate the paths of a large number of rays covering the entire entrance pupil, to produce spot diagrams. These show what sort of image the optical system will produce, and allow the designer to study the system fully before it is built.

20.3 Optical Surfaces

Before treating methods of optical calculation, it is necessary to define the types of optical surfaces used and how these can be described mathematically. From the foregoing chapters, it is clear that plane and spherical surfaces are the most frequently used, followed by paraboloids, hyperboloids, and ellipsoids. Occasionally Schmidt surfaces are used.

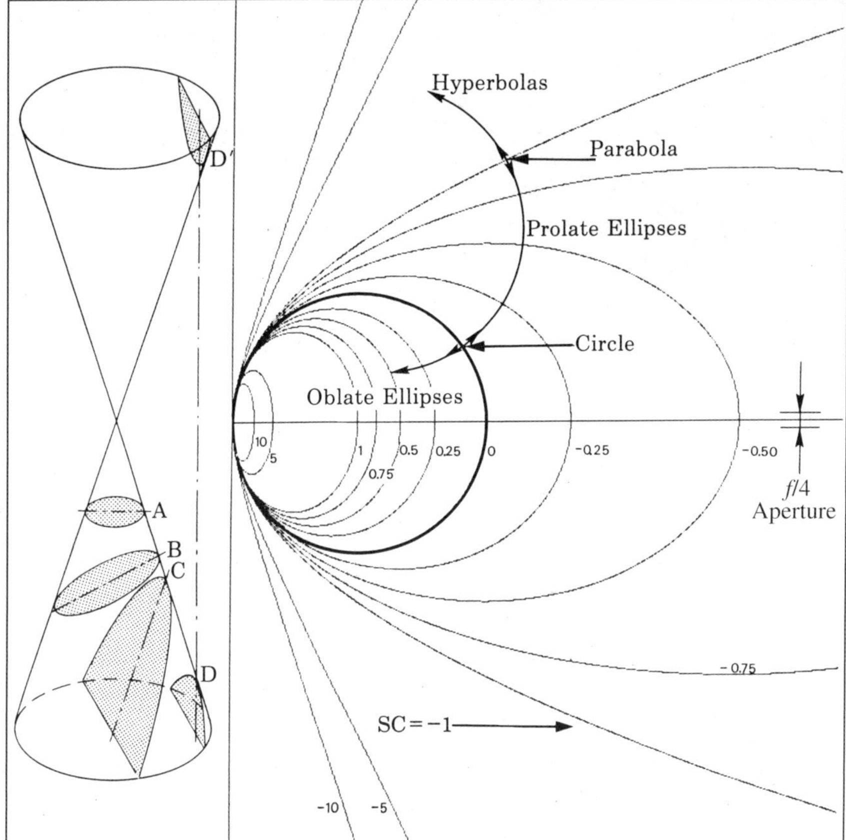

Fig. 20.1 *Conic Sections.*

As far as plane, or flat, surfaces are concerned, no further explanation is necessary.

20.3.1 Conic Sections

The circle, parabola, hyperbola, and ellipse are all conic sections. The sphere, paraboloid, hyperboloid, and ellipsoid are three-dimensional surfaces generated when these two-dimensional curves are rotated about their axes.

The origin of a conic section is shown in fig. 20.1. The circle is a section of a cone that lies on a plane perpendicular to the axis of the cone. For an ellipse, the intersecting plane lies between perpendicular and parallel to a side of the cone. When the axis of the plane containing the section is parallel to one side of the cone, we have a parabola. When the angle is still larger, we have the hyperbola. Note that in the opposite cone another hyperbola exists.

There are two types of ellipses: the prolate ellipse, with both focal points on the optical axis, and the oblate ellipse, in which the focal points lie on opposite

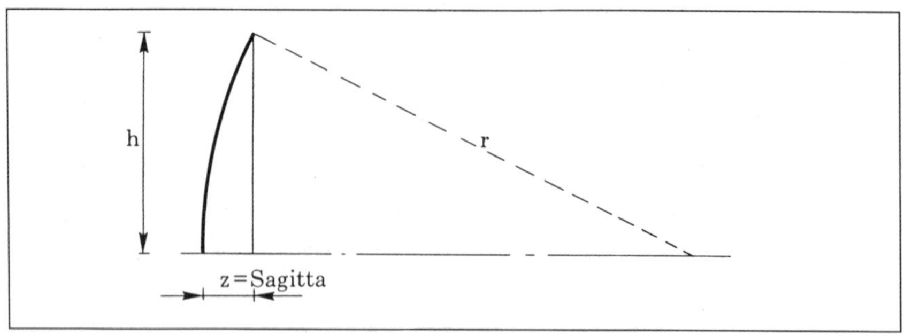

244. 20.2 *Some Parameters of the Circle.*

Table 20.1
Eccentricity and Schwarzschild Constant

Surface	Eccentricity (e)	Schwarzschild Constant (SC)
Circle	0	0
Parabola	1	–1
Hyperbola	> 1	< –1
Prolate ellipse	< 1 and > 0	< 0 and > –1
Oblate ellipse	—	> 0

sides of the optical axis. It should be kept in mind that only the circle and the parabola are strictly defined, because only one circle and one parabola exist. However, an indefinite number of ellipses and hyperbolas can be drawn.

There are a variety of mathematical means of describing conic sections. We have chosen to use the deformation parameter, also called the Schwarzschild constant. The eccentricity, e, is related to the Schwarzschild constant (SC) as $-e^2$. The values of e and SC (Note: SC is frequently called K in the optical industry) are shown in table 20.1.

The general shape is then defined by:

$$z = \frac{h^2}{r(1 + \sqrt{1 - (h^2/r^2)(SC + 1)})} \quad \textbf{(20.3.1)}$$

in which z is the "sagitta" or height with respect to the (X–Y) plane, h is the incidence height, SC is the Schwarzschild constant, and r is the paraxial radius of curvature. You may see these equations with the variable C in them, where C is the reciprocal of the radius and is called the curvature.

For every conic section, there exists a reference circle (and in three dimensions, a reference sphere) which has the same radius as the paraxial radius of curvature of the conic section. fig. 20.1 shows a set of conic sections with different deformations sharing a common reference circle. Note that even an aperture as large as *f*/4 occupies only a small part of the curve.

For a circle (see fig. 20.2), the equation above reduces to:

$$z = r - \sqrt{r^2 - h^2}. \tag{20.3.2}$$

For a parabola, the sagitta is:

$$z = \frac{h^2}{2r} \tag{20.3.3}$$

where r is the paraxial radius of curvature.

These formulae are important for the grinding of mirrors. As we have seen in chapter 5, the difference between the shape of a parabola and that of a circle is quite small. In a 200 mm *f*/4 mirror, for instance, the edge difference, Δz, is only 0.003 mm.

20.3.2 Higher-Order Surfaces

These are surfaces that are neither flat nor conic sections but nevertheless rotationally symmetric. The most important optical higher order surface is the Schmidt shape, sometimes called the "doughnut" profile. The general equation for a higher order surface is:

$$z = A_1h^2 + A_2h^4 + A_3h^6 + A_4h^8 + A_5h^{10} \tag{20.3.4}$$

In this equation, we start with the X–Y plane where $h = \sqrt{X^2 + Y^2}$. While this method is perfectly general, it is not necessarily the most useful way to represent conic sections because faster methods of calculation exist. The exponents are even because the surface is rotationally symmetric.

For generalized aspherics, the first term in eq. 20.3.4 may be replaced by eq. 20.3.1, $1/r$ by C, and SC by K, thusly:

$$z = \frac{Ch^2}{1 + \sqrt{1 - (K+1)C^2h^2}} + A_2h^4 + A_3h^6 + \ldots$$

See ref. 20.3. Note that A_2, A_3, etc. in this equation *do not* have the same value as those in eq. 20.3.4.

Because this general equation is important, we will look at several examples. Consider first, that it contains the equation of the parabola if the constant A_1 is set equal to $1/2r$ and the other constants, A_2, A_3, A_4, and so on, are equal to zero. For the circle, we use the series expansion yielding $A_1 = 1/2r$, $A_2 = 1/8r^3$, $A_3 = 1/16r^5$, and so on.

Consider now a Schmidt corrector. As we saw in chapter 8, the general equation of a Schmidt surface is:

$$z = Ah^2 + Bh^4 \tag{20.3.5}$$

and even more accurately:

$$z = Ah^2 + Bh^4 + Ch^6 + \ldots \tag{20.3.6}$$

As the number of terms in this equation increases, the shape will be described more accurately. What do these terms mean? Suppose, for the sake of argument, that we restrict ourselves to the paraxial region, and consider the first term as representing a sphere of radius r. In this region, the higher terms represent the surface deviation relative to the radius of that sphere. Outside the paraxial region, the first term represents a parabola, and the higher-order terms represent deviations from that parabola.

Note that polynomial aspherics do not scale linearly. For a scale factor, SF, then a new value of z, z', must be used:

$$z' = \frac{A_1}{SF}h^2 + \frac{A_2}{SF^3}h^4 + \frac{A_3}{SF^5}h^6 + \frac{A_4}{SF^7}h^8 + \frac{A_5}{SF^9}h^{10} + \ldots$$

Note that h is not scaled when the scale factor is applied.

In chapter 8 we saw that for a Schmidt camera, the constants in the two-term equation are:

$$A = \frac{3D^2}{32(n-1)r^3} \tag{20.3.7}$$

and

$$B = \frac{-1}{4(n-1)r^3} \tag{20.3.8}$$

for a neutral zone at 86.6% of the radius. These formulae play an important role in the paraxial and third-order calculations to be discussed later. The factors in the three-term Schmidt equation are discussed in section 21.11.

In chapters 9 (on the Schmidt-Cassegrain) and 13 (on the Wright camera), we introduced the term "relative power" for a Schmidt corrector. The relative power factor, g, is defined as the optical power of the Schmidt corrector in a system divided by the power required to make a true Schmidt camera with a mirror having the same paraxial radius as the mirror used in the Schmidt-Cassegrain or Wright camera. The relative power of the corrector in a Schmidt-Cassegrain is often less than 1, but it is considerably greater than 1 in the Wright camera. This relative power factor must be taken into account in the equations given above. For the simple paraxial calculation the aspheric shape of the Schmidt corrector need not be taken into account. For such a calculation only the paraxial radius of the Schmidt shape is used, and this equals:

$$r_{\text{par}} = \frac{4(n-1)r^3}{(NZ)^2D^2g} \tag{20.3.9}$$

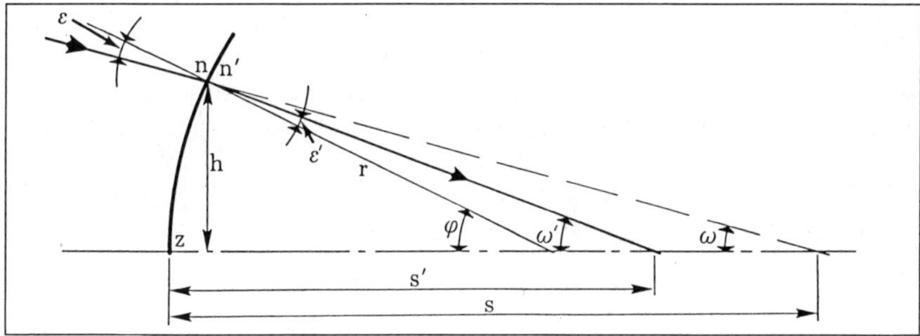

Fig. 20.3 *Terms Used in Paraxial Calculations.*

where g is the relative power factor and NZ is the relative position of the neutral zone. NZ is 0.866 for a corrector having the minimum of chromatic aberration.

20.4 Sign Conventions

Before starting optical calculations and derivations, we must be sure to understand the sign conventions we will use, namely:

- Light entering the optical system travels from left to right.
- Distances from left to right are signed positive; those from right to left, negative.
- Curvatures with the convex side to the left are signed positive; otherwise they are negative.
- Intersection points above the optical axis are positive; those below the axis are negative.
- An angle between a ray and the optical axis is measured in the direction of the ray. When this angle is up from the axis, it is positive.
- The sign of the refractive index is the same as the sign of the direction in which light travels in the medium.
- In the case of reflection, therefore, the signs of the refractive indices are reversed, so that $n' = -n$.
- Surfaces are numbered in the sequence that they are hit by the rays.

20.5 The Paraxial Calculation

As mentioned before, the paraxial calculation is used to obtain an overview of the geometry of a system—to find the distances between optical elements, the lens thicknesses, and the radii of curvature—and, derived from that, the focal length and the position of the focal plane.

The paraxial calculation is carried out in a region very close to the optical axis. Fig. 20.3 shows the geometry of the paraxial calculation. In the paraxial region, angles between the rays and the optical axis are so small that the sines and tangents of these angles are very close to the angles themselves, and the cosines of the angles are very close to unity. Moreover, the off-axis distance h is so small that the value of the sagitta, z, is approximately zero.

Angles and off-axis distances in fig. 20.3 are shown considerably enlarged for the sake of clarity. The following terms are defined:

- h = height of the incident ray on the optical surface.
- z = sagitta of the sphere at incidence height h.
- r = paraxial radius of curvature of the optical surface.
- d = axial distance to the next plane.
- S = axial distance between the optical surface and the intersection point along the light ray without refraction or reflection.
- S' = axial distance between the optical surface and the intersection point along the light ray after refraction or reflection.
- ε = angle of incidence at the optical surface before refraction or reflection.
- ε' = angle of refraction or reflection at the optical surface after refraction or reflection.
- ω = angle between the ray and the optical axis before refraction or reflection.
- ω' = angle between the ray and the optical axis after refraction or reflection.
- φ = angle between the normal and the optical axis at the point of intersection.
- n = refractive index of the medium before refraction or reflection.
- n'= refractive index of the medium after refraction or reflection.

The formulae for the paraxial calculations are as follows:

$$\left.\begin{aligned}\omega + \varepsilon &= \varphi \\ \omega' + \varepsilon' &= \varphi\end{aligned}\right\}\omega + \varepsilon = \omega' + \varepsilon' \qquad \textbf{(20.5.1)}$$

$$\omega = \frac{h}{s}$$

$$\omega' = \frac{h}{s'}$$

$$\varphi = \frac{h}{r}.$$

For small angles:

$$\frac{\varepsilon}{\varepsilon'} = \frac{n'}{n}$$

$$\left.\begin{aligned} \varepsilon &= \varphi - \omega = \frac{h}{r} - \frac{h}{S} = \left(\frac{1}{r} - \frac{1}{S}\right) h \\ \varepsilon' &= \varphi - \omega' = \frac{h}{r} - \frac{h}{S'} = \left(\frac{1}{r} - \frac{1}{S'}\right) h \end{aligned}\right\} \begin{aligned} nh\left(\frac{1}{r} - \frac{1}{S}\right) &= n'h\left(\frac{1}{r} - \frac{1}{S'}\right) \\ n\left(\frac{1}{r} - \frac{1}{S}\right) &= n'\left(\frac{1}{r} - \frac{1}{S'}\right). \end{aligned}$$

From these equations, we then derive:

$$S' = \frac{n'}{\frac{n}{S} + \frac{n' - n}{r}}$$

or

$$S' = \frac{1}{\frac{1}{r} - \frac{n}{n'}\left(\frac{1}{r} - \frac{1}{S}\right)}. \qquad \textbf{(20.5.2)}$$

From S' and d, we find a new value of S for the next surface:

$$S_{s+1} = S'_s - d_s$$

where s is the number of the surface.

In this way, the location of the paraxial focus can be determined. As we saw from fig. 3.5, the focal length of a compound optical system is the length of the light cone which has a diameter equal to that of the entering bundle. In a paraxial calculation, however, the height of the incident bundle is not considered, so we must derive the focal length from the quantities S and S'. Since the definition of the focal length is always based on an infinite object distance, the distance S of the first optical surface will be infinite.

Fig. 20.4 shows the geometry of a ray passing a single lens. If only the first surface is taken into account, the focal length equals S_1', or:

$$\frac{S'_1}{h} = \frac{f}{h}.$$

When more surfaces are present, we repeat this procedure:

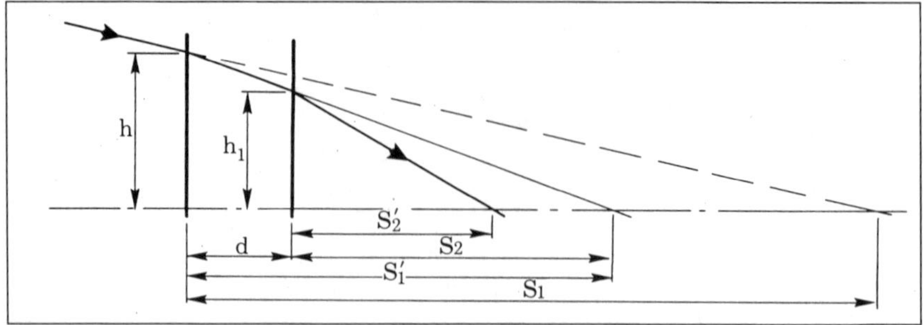

Fig. 20.4 *First-Order Geometry of a Simple Lens.*

$$\frac{f}{h} = \frac{S'_2}{h_1}.$$

We calculate the next ray height, h_1, from the cone defined by h and S_1', and d, the thickness of the lens:

$$h_1 = \frac{S'_1 - d}{S'_1} \cdot h \text{ or: } h_1 = \frac{S_2}{S'_1} \cdot h.$$

Substituting these results:

$$\frac{f}{h} = \frac{S'_2}{h_1} = \frac{S'_2}{\left(\frac{S_2}{S'_1}\right)h} = \frac{S'_1 S'_2}{S_2 h}.$$

Hence we find the focal length:

$$f = \frac{S'_1 S'_2 \ldots}{S_2 \ldots}. \qquad \textbf{(20.5.3)}$$

From this value of f and the last value of S', we can calculate the positions of the principal points of a lens, as illustrated in fig. 20.5.

Consider a biconvex lens with the following characteristics:

$$R_1 = +100 \text{ mm}$$

$$R_2 = -600 \text{ mm}$$

$$n_{\text{glass}} = 1.5$$

$$\text{thickness} = 10\text{mm}.$$

The position of the foci, the principal points, and the focal length are determined as follows. For the first surface:

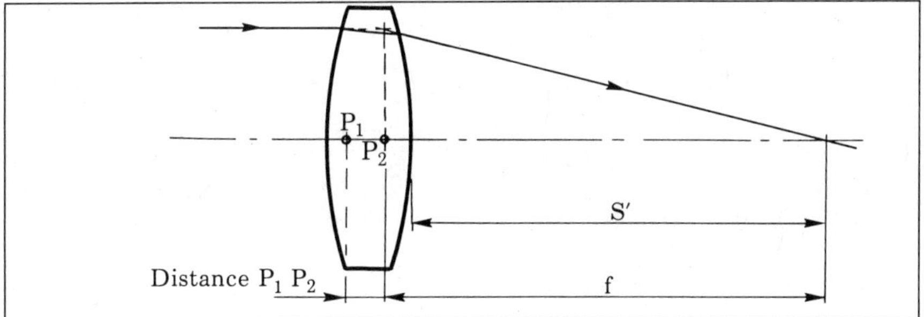

Fig. 20.5 *The Paraxial Calculation of a Principal Point.*

$$\left.\begin{aligned} S_1 &= \infty \\ r_1 &= 100\text{mm} \\ n_1 &= 1 \\ n'_1 &= 1.5 \end{aligned}\right\} S'_1 = \frac{n'_1}{\dfrac{n_1}{S_1} + \dfrac{n'_1 - n_1}{r_1}} = \frac{1.5}{\dfrac{1}{\infty} + \dfrac{1.5 - 1}{100}} = 300\text{mm}.$$

For the second surface,

$$\left.\begin{aligned} S_2 &= (S'_1 - d) \\ = 300 - 10 &= 290\text{mm} \\ r_2 &= -600\text{mm} \\ n_2 &= 1.5 \\ n'_2 &= 1 \end{aligned}\right\} S'_2 = \frac{n'_2}{\dfrac{n_2}{S_2} + \dfrac{n'_2 - n_2}{r_2}} = \frac{1}{\dfrac{1.5}{290} + \dfrac{1 - 1.5}{-600}} = 166.507\text{mm}.$$

This places f_1 166.507 mm behind the lens.

Calculating backwards, we then determine the position of f_2. For the first surface:

$$\left.\begin{aligned} S_1 &= -\infty \\ r_1 &= -600\text{mm} \\ n_1 &= -1 \\ n'_1 &= -1.5 \end{aligned}\right\} S'_1 = \frac{n'_1}{\dfrac{n_1}{S_1} + \dfrac{n'_1 - n_1}{r_1}} = \frac{-1.5}{\dfrac{-1}{-\infty} + \dfrac{-1.5 + 1}{-600}} = -1800\text{mm}.$$

For the second surface, we find

$$\left.\begin{aligned} S_2 &= (S'_1 - d) \\ &= -1800 + 10 = 1790\text{mm} \\ r_2 &= 100\text{mm} \\ n_2 &= -1.5 \\ n'_2 &= -1 \end{aligned}\right\} S'_2 = \frac{-1}{\frac{-1.5}{-1790} + \frac{-1+1.5}{100}} = -171.291\text{mm}.$$

Therefore, f_2 is located 171.291 mm in front of the lens.

To determine the focal length, we can use either the left-to-right or the right-to-left calculation. From left to right, it is:

$$f = \frac{S'_1 \cdot S'_2}{S_2} = \frac{300 \cdot 166.507}{290} = 172.248\text{mm}$$

and from right to left, it is:

$$f = \frac{S'_1 \cdot S'_2}{S_2} = \frac{-1800 \cdot (-171.291)}{-1790} = -172.248\text{mm}$$

The negative sign in the last value does not mean that the lens has a negative power, but only that the focal length is described as a distance measured from left to right.

The distance from a principal point P to the lens surface S_p is defined by:

$$S_p = S' - f$$

as shown in fig. 20.5. The position of the principal points is calculated as follows:

$$S_{p1} = S' - f = 166.507 - 172.248 = -5.741\text{mm}$$

$$S_{p2} = S' - f = -171.291 + 172.248 = 0.957\text{mm}$$

The distance $P_1 - P_2$ is equal to 10 – 5.741– 0.957, or 3.301 mm.

20.6 The Seidel Calculation

The Seidel, or third-order, calculation was derived in 1856 from the paraxial calculation by Seidel to gain insight into the magnitude of the image aberrations of an optical system without making a complete ray-tracing analysis. In the middle of the 19th century, optical calculations were carried out entirely by hand, and exact ray tracing was very time-consuming. Seidel's method can still save a lot of time in the early stages of roughing in a new design.

The term "third-order," sometimes used for the Seidel calculation, and the term "first-order," used for the paraxial calculation, refer to certain properties of the trigonometric functions. Image aberrations are highly dependent on the image angle, particularly the sine and tangent functions. The sine may be expanded in a series,

$$\sin x = x - \frac{x^3}{3!} + \frac{x^5}{5!} - \frac{x^7}{7!} + \ldots \qquad \textbf{(20.6.1)}$$

where x is in radians.

In paraxial calculations, the angles involved are so small that the sine of the angle equals the angle itself, or $\sin x = x$, which means that the higher-order terms of the expansion are neglected. Only the first-order term in x is taken into account, hence the name of the paraxial calculation.

For systems with larger angles, however, we must use more terms in the expansion in order to describe the imaging properties of the system accurately. Seidel took the third-order term in x into account, neglecting the fifth, seventh, and higher orders. The Seidel calculation takes its name from this fact. This method is still frequently used in the initial design stage, especially when many optical components are involved. More sophisticated schemes utilizing the fifth- and higher-order terms are used by professional optical designers, but they involve very complicated formulae.

For the practical optical designer, it is not necessary to give the derivation of the Seidel formulae; anyway, modifications of the original theory have been adopted over the years. For instance, H. Köhler derived simpler formulae than the original Seidel formulae. We use the Köhler formulae here, though we still call the calculation the Seidel method.

Unlike exact ray-trace methods, in which we find actual blur patterns, the Seidel method treats every individual monochromatic aberration (spherical aberration, coma, astigmatism, curvature of field, and distortion) as a separate calculation. The power of the method is that each optical surface makes its own contribution to each of these aberrations, so that it is immediately evident which of the surfaces has made the largest contribution to each aberration. In using the Seidel method, the designer should recognize that the formulae are accurate only for rather slow systems, with rather weak curves, and small ray angles.

The following method is used in making Seidel calculations. We first determine the focal length and the values of S, S', and so on for each of the optical surfaces. Then all values of the radii of curvature r, axial thicknesses, and axial distances, and the values of S and S' are divided by the system focal length, so the resulting numbers are correct for a unit focal length. The incident ray heights for the system are computed next, likewise scaled so that the incidence height of the entering bundle is equal to 1. As in the paraxial calculation, the incidence height at the second surface is then:

$$h_2 = \frac{S_2}{S'_1}.$$

And for the third surface:

$$h_3 = \frac{h_2 \cdot S_3}{S'_2}.$$

For each surface, in fact, we find that:

$$h_s = \frac{h_{s-1} \cdot S_s}{S'_{s-1}} \tag{20.6.2}$$

where s is the number of the surface.

We next compute the following auxiliary quantities:

$$Q_s = n_s\left(\frac{1}{r_s} - \frac{1}{S_s}\right) = n'_s\left(\frac{1}{r_s} - \frac{1}{S'_s}\right) \tag{20.6.3}$$

$$\Delta\left(\frac{1}{n \cdot S}\right)_s = \frac{1}{n'_s \cdot S'_s} - \frac{1}{n_s \cdot S_s} \tag{20.6.4}$$

$$TH_s = \sum_{1 \to S} \frac{d_{s-1}}{n_s \cdot h_{s-1} \cdot h_s} - d_p. \tag{20.6.5}$$

The summation sign Σ means the sum of all previous terms of the fraction after this sign, and d_p is the distance from the entrance pupil to the first optical surface. This is taken into account only for the first surface.

$$p_s = \frac{1}{h_s^2 \cdot Q_s} + TH_s \tag{20.6.6}$$

$$P_s = \frac{1}{r_s}\left(\frac{1}{n'_s} - \frac{1}{n_s}\right). \tag{20.6.7}$$

Using these auxiliary quantities, we then calculate the aberration contribution for each spherical surface in this optical system:
Spherical aberration:

$$A_s = h_s^4 \cdot Q_s^2 \cdot \Delta\left(\frac{1}{nS}\right)_s. \tag{20.6.8}$$

Coma:

$$B_s = p_s \cdot A_s. \tag{20.6.9}$$

Astigmatism:

$$C_s = p_s^2 \cdot A_s. \tag{20.6.10}$$

Curvature of field:

$$P_s. \tag{20.6.11}$$

Distortion:

$$V_s = p_s^3 \cdot A_s + p_s \cdot P_s . \tag{20.6.12}$$

A similar method can be used for aspheric surfaces. Conic sections can all be derived from a sphere; the deviation of the calculation method is small. To the coefficient for the reference sphere (defined in section 20.3.1), we add the extra term K:

$$K_s = \frac{SC_s}{8r_s^3} \tag{20.6.13}$$

in which r_s is the curvature ratio of the reference sphere, that is, the actual radius divided by the system focal length.

Other aspheric surfaces cause considerably more trouble. While it is beyond the scope of this book to treat all possible aspheric surfaces, we have included one extremely important aspheric surface for astronomical optics: the Schmidt surface. The profile is described using a paraxial radius of curvature with aspheric factors, just as we did in the paraxial calculation. The aspheric factor is:

$$K_s = \frac{g_s}{4(n_g - 1)r_c^3} \tag{20.6.14}$$

in which n_g is index of refraction of the glass for the design color, r_c is the curvature ratio of primary (i.e., the radius of curvature divided by the focal length), and g_s is relative power of the Schmidt corrector. The paraxial radius of curvature of the Schmidt corrector r_{par} (derived in section 20.3.2) is substituted for r_s in the formulae above. The surface coefficients are then calculated with the aid of following auxiliary quantity:

$$\Delta n_s = n'_s - n_s .$$

The surface coefficients for the aspheric surfaces (designated with *) are as follows:

$$A_s^* = 8h_s^4 \cdot K_s \cdot \Delta n_s \tag{20.6.15}$$

$$B_s^* = TH_s \cdot A_s^* \tag{20.6.16}$$

$$C_s^* = TH_s^2 \cdot A_s^* \tag{20.6.17}$$

$$P_s^* = 0 \tag{20.6.18}$$

$$V_s^* = TH_s^3 \cdot A_s^* . \tag{20.6.19}$$

The Seidel parameters for the whole system are the sums of the individual surface contributions. So the Seidel sums are:

$$\sum A = \sum_{1 \to s} A + \sum_{1 \to s} A^* \qquad (20.6.20)$$

$$\sum B = \sum_{1 \to s} B + \sum_{1 \to s} B^* \qquad (20.6.21)$$

$$\sum C = \sum_{1 \to s} C + \sum_{1 \to s} C^* \qquad (20.6.22)$$

$$\sum P = \sum_{1 \to s} P \qquad (20.6.23)$$

$$\sum V = \sum_{1 \to s} V + \sum_{1 \to s} {}^* . \qquad (20.6.24)$$

With the formulae given above, nearly all amateur optical systems can be analyzed.

Now let us see how to use the Seidel sums to calculate the size of the image aberrations. In the calculations above, we did not consider the focal ratio, the focal length, or the image angle. The actual image aberrations depend on these quantities, and are calculated as shown below. In these equations, F is the effective focal length, N is the focal ratio, and ω is off-axis image angle for which the aberrations are calculated.

SPHERICAL ABERRATION

Longitudinal Aberration:

$$-\frac{F}{8N^2} \cdot \sum A \qquad (20.6.25)$$

Transverse Aberration:

$$\frac{F}{16N^3} \cdot \sum A \qquad (20.6.26)$$

COMA

Transverse Aberration:

$$\frac{3F}{8N^2} \tan\omega \cdot \sum B \qquad (20.6.27)$$

ASTIGMATISM

Transverse Aberration (Tangential):

$$\frac{-F}{4N}\tan^2\omega\left(\sum 3\cdot C+\sum P\right) \quad \textbf{(20.6.28)}$$

Transverse Aberration (Sagittal):

$$\frac{-F}{4N}\tan^2\omega\left(\sum C+\sum P\right) \quad \textbf{(20.6.29)}$$

Transverse Aberration (Average Curved Surface):

$$\frac{-F}{4N}\tan^2\omega\sum C \quad \textbf{(20.6.30)}$$

CURVATURE OF FIELD

Tangential Radius of Curvature:

$$\frac{-F}{3\cdot\Sigma C+\Sigma P} \quad \textbf{(20.6.31)}$$

Sagittal Radius of Curvature:

$$\frac{-F}{\Sigma C+\Sigma P} \quad \textbf{(20.6.32)}$$

Average Radius of Curvature:

$$\frac{-F}{2\cdot\Sigma C+\Sigma P} \quad \textbf{(20.6.33)}$$

DISTORTION

$$\frac{\tan^3\omega}{2}\cdot\Sigma V \quad \textbf{(20.6.34)}$$

When the spherical aberration is not equal to zero and coma is present, the coma can be eliminated by changing the location of the entrance pupil. The position for zero coma is:

$$x = -F\cdot\frac{\Sigma B}{\Sigma A}. \quad \textbf{(20.6.35)}$$

It must be emphasized again that all of these formulae are approximations. Apart from the third-order aberrations we have calculated, there are fifth-, seventh-, and higher-order aberrations which can occur. The higher-order aberrations are completely neglected in the Seidel method, but can be quite important in systems with large image angles and fast focal ratios. Exact ray-trace analysis takes all orders into account.

Third-order Seidel analysis is very useful in the initial stages of a design, when the designer is still trying to establish parameters such as the axial lens thicknesses and distances, the types of glass, the radii of curvature, and the lens bendings. Despite the complicated appearance of the formulae, the sums can be calculated very quickly with a programmable calculator or a computer, and the influence of small changes in the system parameters analyzed in a short period of time.

When a calculation of the aberrations is made with the help of the Seidel method, it is clear that criteria for the maximum acceptable aberrations must be defined. The criterion is expressed as the diameter of the admissible spread figure, in seconds of arc. The Seidel sums can be converted by multiplying by $f/206{,}265$, where f is the effective focal length of the system.

Since coma and astigmatism only occur away from the optical axis, an off-axis angle must be specified for them. When θ is the maximum admissible unsharpness criterion, and ω is the off-axis angle, then the maximum allowable Seidel sums are:

$$A_{max} = \theta \cdot 16 \cdot \frac{N^3}{F} \tag{20.6.36}$$

$$B_{max} = \theta \cdot 8 \cdot \frac{N^2}{3 \cdot F \cdot \tan\omega} \tag{20.6.37}$$

$$C_{max} = \frac{\theta \cdot 4 \cdot N}{F \tan^2 \omega} . \tag{20.6.38}$$

For example, consider a 200 mm $f/10$ telescope for which we have established an imaging criterion of 1 second of arc 15 minutes of arc off axis. The criteria are 1 arcsecond = 0.0097 mm at $\tan 15' = 4.36 \cdot 10^{-3}$. The maximum Seidel sums are then:

$$\Sigma A_{max} = 0.078$$

$$\Sigma B_{max} = 0.296$$

$$\Sigma C_{max} = 10.185 .$$

Examine the tables on the next page as examples of the calculation of the Seidel sums. These systems are also analyzed by skew-ray tracing in this book, so that a direct comparison between the Seidel method and the exact ray trace is possible.

Note in the second table that the values of ΣA and ΣB are not exactly zero. This results from the fact that the construction parameters in table 7.2 are rounded-off values.

Because the Schmidt-Cassegrain has strong curves, and also because the Schmidt corrector is treated as a spherical surface with a strong aspheric deformation, the third order results will deviate somewhat from the exact ray trace. For ex-

Parabolic Mirror (R = 2000, F = 1000, SC = −1)							
Surface	R	K	A	B	C	P	V
1	−2.000		−0.2500	−0.5000	−1.000	−1.0000	0
1*		0.015625	0.2500	0	0		0
Σ			0	−0.5000	−1.000	−1.0000	

Ritchey-Chrétien Cassegrain (see Table 7.2)							
Surface	R	K	A	B	C	P	V
1	−0.75		4.7404	−3.5554	2.6667	−2.6667	0
1*		0.336814	−5.3890	0	0		0
2	−0.3927		−1.8332	1.6423	−1.4712	5.0925	−3.2441
2*		13.583227	2.4930	1.9218	1.4815		1.1420
Σ			0.0112	0.0087	2.6770	2.4258	−2.1021

Schmidt-Cassegrain (see Table 9.1)							
Surface	R	K	A	B	C	P	V
1			0	0	0	0	0.5664
2	−22.2230		0.001	−0.0016	0.0233	0.0154	−0.5663
2*		6.122664	−25.4076	−0.0330	0.0000		0
3	−0.4034		30.5914	−7.5399	1.8584	−4.9574	0.7638
4	−0.1252		−8.7408	5.1205	−2.9996	15.9703	−7.5984
4*		56.370156	3.4194	2.6189	2.0058		1.5362
Σ			−0.1366	0.1649	0.8879	11.0283	−5.2983

ample, if the design were optimized using only the Seidel method, so that ΣA and ΣB were both zero, the design would still show aberrations when it was ray traced. The reason is that the Seidel calculation is an approximation.

20.7 The Meridional Calculation

Unlike the two previous approximation methods, the meridional calculation is an exact method. It is an exact ray trace in which the path of each ray is computed through the optical elements using the laws of refraction and reflection. The meridional ray trace is restricted to a plane passing through the center of the elements often called the meridional or tangential plane. It is convenient to think of this as "the plane of the drawing."

The equations describing the passage of a ray through a spherical surface are given below, and shown in fig. 20.6:

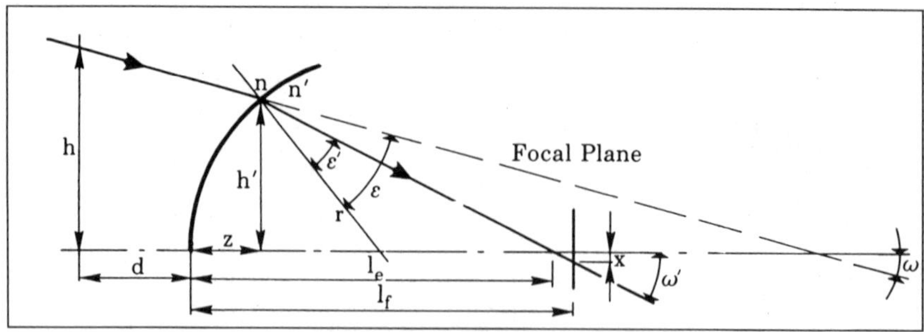

Fig. 20.6 *Terms Used in the Meridional Calculation.*

$$\sin\varepsilon = \frac{h}{r}\cos\omega + \left(\frac{d}{r} + 1\right)\sin\omega \tag{20.7.1}$$

$$h' = r\sin(\omega + \varepsilon) \tag{20.7.2}$$

$$z = r - r\cos(\omega + \varepsilon) \tag{20.7.3}$$

$$\sin\varepsilon' = \frac{n}{n'}\sin\varepsilon \tag{20.7.4}$$

$$\omega' = \omega + \varepsilon - \varepsilon' . \tag{20.7.5}$$

The ray intersects the optical axis at distance l_e after the last surface:

$$l_e = \frac{h'}{\tan\omega'} + z . \tag{20.7.6}$$

The incident height in the focal plane at a distance l_f after the last surface is:

$$x = h' - (l_f - z)\tan\omega' . \tag{20.7.7}$$

These formulae are valid only for spherical surfaces, and can be used quite easily with a small programmable calculator. Employing a meridional trace for aspheric surfaces is rather cumbersome because an iterative method must be used to determine the point of intersection. A more effective method for calculating the intersection points for aspheric surfaces is discussed in the following section.

20.8 The Skew-Ray Trace

20.8.1 Introduction

The best method for evaluating the performance of an optical system is to calculate a series of spot diagrams; the skew-ray trace is used to do this. Skew rays are rays that lie outside the meridional plane. Skew rays generally do not remain in the same plane during their passage through an optical system, complicating the calculations and requiring considerably more computing time. This explains why, before the advent of programmable computing equipment, the method of skew-

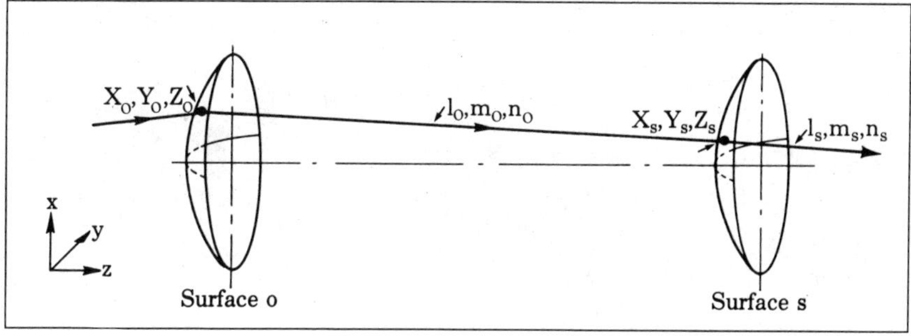

Fig. 20.7 *Skew-Ray Tracing.*

ray tracing was hardly used.

The trigonometric methods described in section 20.7 will not be used for the skew-ray trace because the calculations are slow and not very accurate on small computers. Instead, we have employed algebraic methods.

The skew-ray trace uses a different method to calculate each of the following types of surface:

- plane surfaces
- spherical surfaces
- non-spherical conic sections
- arbitrarily defined rotationally symmetric surfaces.

In the skew-ray trace, the position of a light ray is defined by the *X*, *Y*, and *Z* coordinates of the intersection point in a certain reference or optical surface, and by the direction cosines *l*, *m*, and *n*. In this scheme, *l* is the cosine of the angle between the ray and the *X*-axis; *m*, of the angle with the *Y*-axis; and *n*, of the angle with the *Z*-axis (see fig. 20.7). Note that the *X–Z* plane is considered the meridional or tangential plane ("the plane of the drawing"), while the *Y–Z* plane is the sagittal plane ("out of the page").

When the calculation is started, the ray has the coordinates (X_o, Y_o, Z_o), and travels in the direction (l_o, m_o, n_o).

The first step is the calculation of the intersection point at the next surface, with the coordinates (X_s, Y_s, Z_s). After we calculate the refraction or reflection at the surface, the ray has the new direction cosines (l_s, m_s, n_s).

Continuing the calculation, we treat the new surface as a new starting point, so that:

$$X_o = X_s \quad l_o = l_s$$
$$Y_o = Y_s \quad m_o = m_s$$
$$Z_o = Z_s \quad n_o = n_s$$

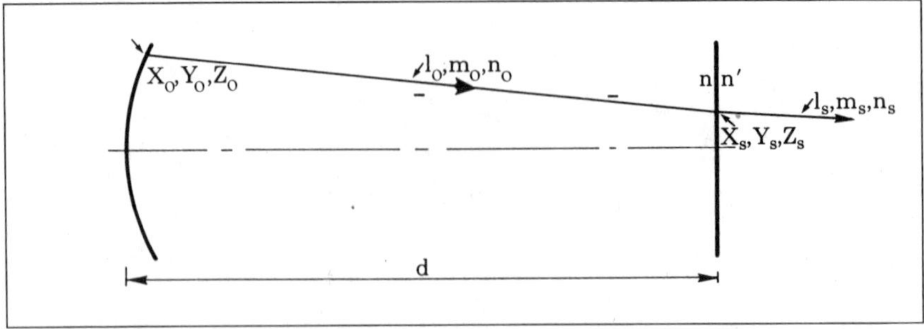

Fig. 20.8 *Skew-Ray Tracing Through a Plane Surface.*

In the following sections, we have shown rays in the meridional plane for the sake of clarity, but it should be understood that the equations apply to skew rays.

20.8.2 Flat Surfaces

We first compute these auxiliary quantities:

$$\mu_s = \frac{n}{n'}$$

$$t_o = Z_o - d$$

$$\mu_v = \frac{t_o}{n_o}.$$

The coordinates of the new intersection point are:

$$X_s = X_0 - \mu_v \cdot l_0$$

$$Y_s = Y_o - \mu_v \cdot m_o.$$

(Note that $Z_s = 0$ because a plane surface is two-dimensional.)

Finally, the new direction cosines of the ray are:

$$l_s = \mu_s \cdot l_o$$

$$m_s = \mu_s \cdot m_o$$

$$n_s = \sqrt{\mu_s^2 \cdot (n_o^2 - 1) + 1}.$$

20.8.3 Spherical Surfaces

We first compute the auxiliary quantities:

$$|r| = \text{abs}(r)$$

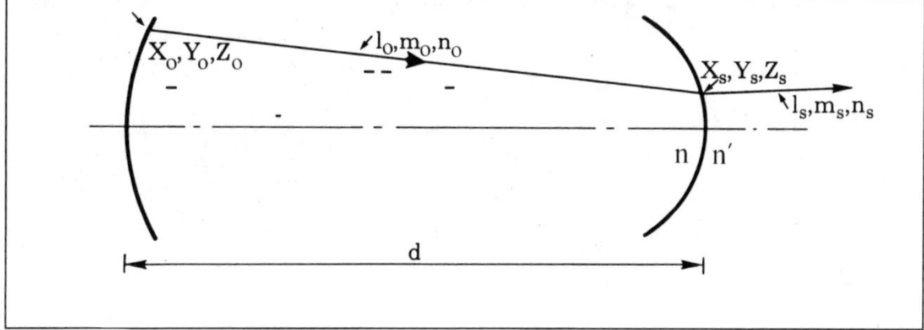

Fig. 20.9 *Skew-Ray Tracing through a Spherical Surface.*

$$\frac{r}{|r|} = \operatorname{sgn}(r)(r \neq 0)$$

$$\mu_s = \frac{n}{n'}$$

$$t_o = Z_o - d$$

$$Z_r = t_o - r$$

$$Zr = t_o - r$$

$$A = r^2 - X_o^2 - Y_o^2 - Z_r^2$$

$$B = l_o \cdot X_o + m_o \cdot Y_o + n_o \cdot Z_r$$

$$C = B + \frac{r}{|r|}\sqrt{A + B^2}. \tag{20.8.1}$$

The coordinates of the new point of intersection are:

$$X_s = X_o - C\ \ l_o$$

$$Y_s = Y_o - C \cdot m_o$$

$$Z_s = t_o - C \cdot n_o.$$

Next:

$$C_i = \frac{\sqrt{A + B^2}}{|r|}$$

$$C_r = \sqrt{\mu_s^2 \cdot (C_i^2 - 1) + 1}$$

$$P = C_r - \mu_s \cdot C_i.$$

Finally, the new direction cosines of the ray are:

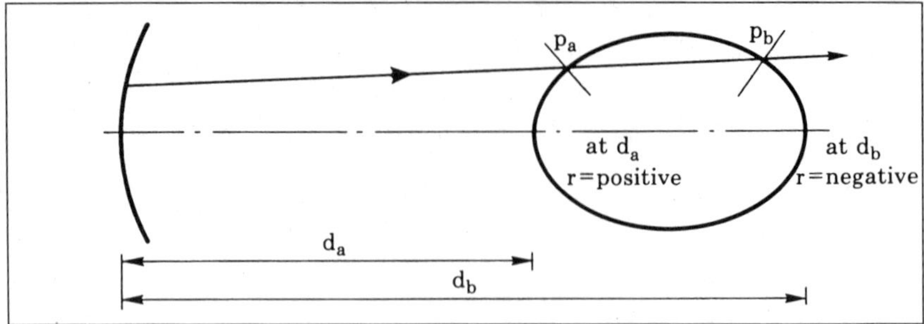

Fig. 20.10 *A Ray Intersects a Conic Section Twice.*

$$l_s = \mu_s \cdot l_o - \frac{P}{r} \cdot X_s$$

$$m_s = \mu_s \cdot m_o - \frac{P}{r} \cdot Y_s$$

$$n_s = \mu_s \cdot n_o - \frac{P}{r} \cdot (Z_s - r) .$$

20.8.4 Conic Sections

The calculation of conic sections is based on the paraxial radius of curvature, or reference sphere. The method of calculating the point of intersection of the ray and the conic section is based on methods used in matrix algebra. There are two points of intersection between a conic surface and a straight line (with the exception of a parabola and a line parallel to its axis); only the intersecting point nearest the vertex of the surface is relevant (figs. 20.10 and also 20.11). It is, therefore, necessary to check only this point. Determining the new direction cosines is treated differently for reflecting and refracting surfaces, but in determining the point of intersection, it does not matter whether the surface is reflecting or refracting.

We begin by calculating the intersection of the ray with an auxiliary plane. We calculate the following quantities, just as we did in section 20.8.2, above:

$$\mu_s = \frac{n}{n'}$$

$$t_o = Z_o - d$$

$$\mu_v = \frac{t_o}{n_o} .$$

We then find the point of intersection (X_v, Y_v, Z_v) between the ray and the plane:

$$X_v = X_o - \mu_v \cdot l_o$$

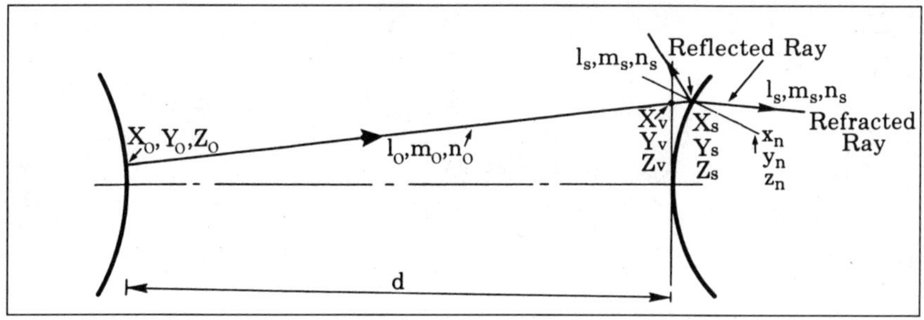

Fig. 20.11 *Skew Rays Reflected and Refracted at a Conic Surface.*

$$Y_v = Y_o - \mu_v \cdot m_o$$

$$Z_v = 0.$$

Starting from the plane, we determine the intersection of the ray and the conic surface:

$$A = l_o^2 + m_o^2 + n_o^2(SC + 1)$$

$$B = 2(X_v \cdot l_o + Y_v \cdot m_o - r \cdot n_o)$$

$$C = X_v^2 + Y_v^2.$$

A, *B*, and *C* are the three constants of the general quadratic equation. When $A = 0$ there is only one solution. This is:

$$h_v = -\frac{C}{B}.$$

When $A \neq 0$, then the nearest point of intersection is determined from:

$$h_v = \frac{-2C}{B - (\sqrt{B^2 - 4AC})\frac{-B}{|B|}}.$$

The intersection point between the ray with the conic section is then:

$$X_s = X_v + h_v \cdot l_o$$

$$Y_s = Y_v + h_v \cdot m_o$$

$$Z_s = h_v \cdot n_o.$$

In order to be able to calculate the refraction or reflection, we must first know the direction of the normal to the surface at the intersection point. This is calculated with the aid of the following auxiliary quantities:

$$K_c = (SC + 1) \cdot Z_s - r$$

$$C = X_s^2 + Y_s^2$$

$$n_v = \sqrt{C + K_c^2}\,.$$

The direction cosines of the normal are:

$$X_n = \frac{X_s}{n_v}$$

$$Y_n = \frac{Y_s}{n_v}$$

$$Z_n = \frac{K_c}{n_v}$$

$$V_h = X_n \cdot l_o + Y_n \cdot m_o + Z_n \cdot n_o\,.$$

If $Z_n \cdot n_o < 0$, the normal points in the wrong direction. This is corrected by reversing the signs of the direction cosines:

$$X_n = -X_n$$

$$Y_n = -Y_n$$

$$Z_n = -Z_n\,.$$

All the necessary quantities are now known. For reflecting conic surfaces the new direction cosines are:

$$l_s = 2 \cdot V_h \cdot X_n - l_o$$

$$m_s = 2 \cdot V_h \cdot Y_n - m_o$$

$$n_s = 2 \cdot V_h \cdot Z_n - n_o\,.$$

Although refracting conic surfaces are seldom used, we have included the equations for refraction. We begin by calculating the intersection of the ray with the auxiliary plane, as we did above, then find the intersection of the ray with the conic surface:

$$C_o = \mu_s^2 V_h^2 - 1) + 1$$

$$A_a = (V_h - 1) \cdot (V_h - 1 + 2 \cdot C_o)$$

$$B_b = 2 \cdot (1 - C_o) \cdot (V_h - 1)$$

$$C_c = 1 - C_o\,.$$

If $A_a = 0$, only one solution is possible:

$$l_v = -\frac{C_c}{B_b}$$

For all other cases:

$$lv_1 = \frac{-2C_c}{B_b - (\sqrt{B_b^2 - 4 \cdot A_a \cdot C_c}) \cdot \frac{-B_b}{|B_b|}}.$$

This quantity should be positive. If it is not, then we have found the negative root of the equation. The positive root is:

$$l_{v2} = \frac{-B_b}{A_a} - l_{v1}.$$

We next find the direction cosines of the normal to the surface:

$$l_n = X_n + l_v(l_o - X_n)$$

$$m_n = Y_n + l_v(m_o - Y_n)$$

$$n_n = Z_n + l_v(n_o - Z_n)$$

$$V_n = \sqrt{l_n^2 + m_n^2 + n_n^2}.$$

Finally, the new direction cosines of the ray are:

$$l_s = l_n / V_n$$

$$m_s = m_n / V_n$$

$$n_s = n_n / V_n.$$

20.8.5 Higher-Order Surfaces

It is not possible to determine the intersection point of a ray with a deformed surface directly; instead we use an iterative procedure. This means that we use a trial value, then check whether the height of the calculated intersecting point coincides with the height of the real profile (see fig. 20.12). As long as this aim has not yet been achieved, we repeat the procedure using the previous value as the starting point. Because of round-off errors, an iterative process must be stopped at some point, otherwise the results begin to oscillate. A value for the error that is suitable for microcomputers using double-precision variables is 10^{-12} millimeters.

We first calculate the curvature of the surface:

$$C_a = \frac{1}{r}.$$

When r is ∞, $C_a = 0$.

The process works as follows: we determine the point of intersection with a reference surface, which can be either a plane or a spherical surface. This point has the coordinates (X_v, Y_v, Z_v). For a plane reference surface, $Z_v = 0$ and $R_h = X_v^2 + Y_v^2$. Using the deformation constants A_1, A_2, ..., we find the height of the profile:

$$Z_a = \frac{C_a \cdot R_h}{1 + \sqrt{1 - C_a^2 \cdot R_h}} + A_1 \cdot R_h + A_2 \cdot R_h^2 + A_3 \cdot R_h^3 + A_4 \cdot R_h^4 + A_5 \cdot R_h^5 .$$

We then compute the auxiliary quantities:

$$A_{2z} = 2 \cdot A_1 + 4A_2 \cdot R_h + 6A_3 \cdot R_h^2 + 8A_4 \cdot R_h^3 + 10A_5 \cdot R_h^4$$

$$n_a = \sqrt{1 - R_h \cdot C_a^2}$$

$$l_a = X_v \cdot -(C_a + n_a \cdot A_z)$$

$$m_a = Y_v \cdot -(C_a + n_a \cdot A_z)$$

$$F_a = l_o \cdot l_a + m_o \cdot m_a + n_o \cdot n_a$$

$$G_s = n_a \cdot \frac{Z_a - Z_v}{F_a}$$

from which we obtain a new intersection point:

$$X_s = G_s \cdot l_o + X_v$$

$$Y_s = G_s \cdot m_o + Y_v$$

$$Z_s = G_s \cdot n_o + Z_v$$

The new value for R_h is then:

$$R_h = X_s^2 + Y_s^2 . \tag{20.8.2}$$

The height of the surface at that point is:

$$Z_p = \frac{C_a \cdot R_h}{1 + \sqrt{1 - C_a^2 \cdot R_h}} + A_1 \cdot R_h + A_2 \cdot R_h^2 + A_3 \cdot R_h^3 + A_4 \cdot R_h^4 + A_5 \cdot R_h^5 .$$

This process continues until the difference in the profile heights ($Z_p - Z_a$) is less than the difference criterion. If this value is larger than the criterion, then we treat the new point as a new starting point:

$$X_v = X_s$$

$$Y_v = Y_s$$

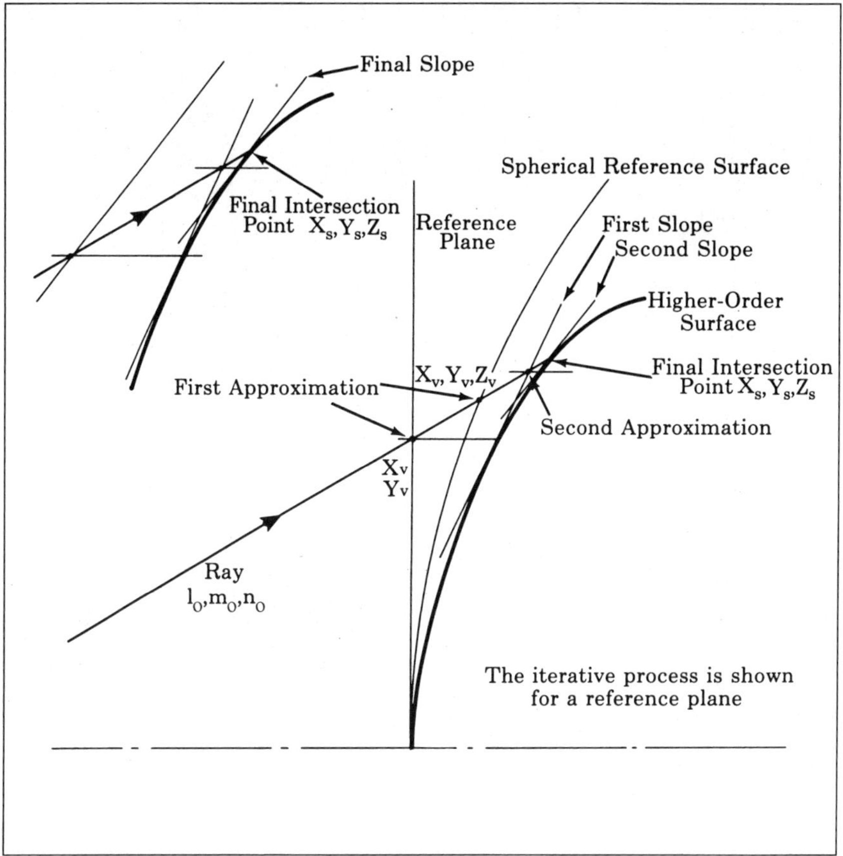

Fig. 20.12 *The Calculation of a Higher-Order Surface.*

$$Z_v = Z_s$$

and the procedure is repeated.

If the criterion is satisfied, then we continue the calculation as follows:

$$P_a = l_a^2 + m_a^2 + n_a^2$$

$$F_s = \frac{\sqrt{P_a \cdot (n^2 - n'^2) + n'^2 \cdot F_a^2}}{|n|}$$

$$G_a = \frac{F_s - \mu_s \cdot F_a}{P_a}.$$

Finally, the new direction cosines of the ray are:

$$l_s = \mu_s \cdot l_o + G_a \cdot l_a$$

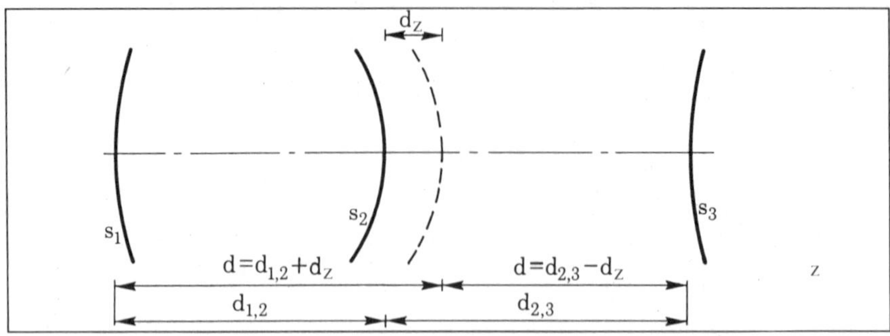

Fig. 20.13 *Terms Used for an Axial Deviation.*

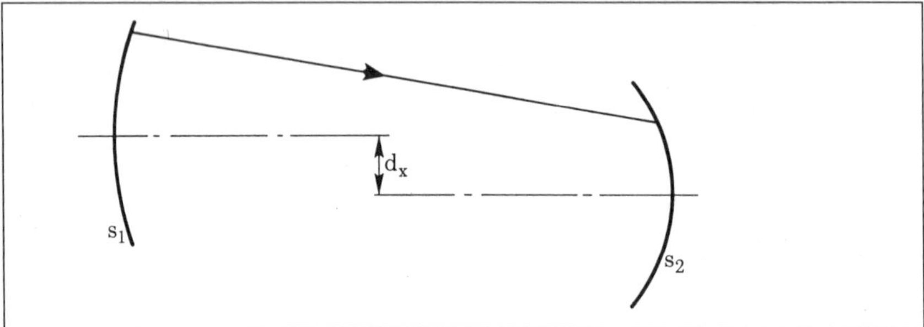

Fig. 20.14 *Terms Used for a Transverse Deviation.*

$$m_s = \mu_s \cdot m_o + G_a \cdot m_a$$

$$n_s = \mu_s \cdot n_o + G_a \cdot n_a .$$

In our computer program, the variables X_v, Y_v, and Z_v are called X_s, Y_s, and Z_s, respectively. This is possible because the two sets of variables do not appear simultaneously.

20.9 Calculation of Non-Centered Systems

As is true in the calculation of centered systems, there are a number of ways to compute non-centered systems. The method presented here is quite simple, and has been chosen because it leads directly to the intersection points for each optical element. This is extremely useful in making size or vignetting calculations.

An optical element can have three deviations with respect to another element; this is demonstrated in chapter 17. The simplest deviation is an axial shift of one element. In principle, nothing in the calculation is changed except that the distance from the preceding surface is increased or decreased (fig. 20.13), and the distance to the following surface is decreased or increased by the same amount.

For a transverse shift (see fig. 20.14), the calculation remains relatively simple:

In the *X*-direction:

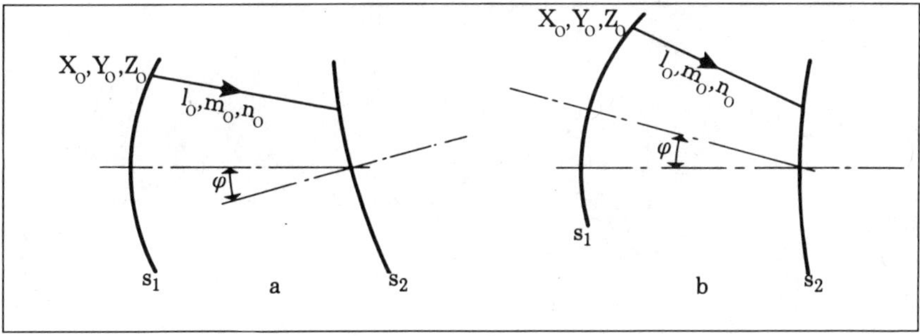

Fig. 20.15 *Terms Used with Tilted Elements.*

$$X_o = X_o + d_x .$$

In the Y-direction:

$$Y_o = Y_o + d_y .$$

And after passing the shifted surface:
In the X-direction:

$$X_s = Y_s - d_x .$$

In the Y-direction:

$$Y_s = Y_s - d_y .$$

A more complex situation occurs when elements or surfaces are tilted with respect to the optical axis (see fig. 20.15a). The simplest method of calculating is achieved by rotating the preceding surface around the vertex of the tilted element (fig. 20.15b). The values of X_o, Z_o, l_o, and n_o change with respect to the tilted surface, but the Y and m values remain unchanged. We first assign values to the temporary variables:

$$X_i = X_o$$

$$t_i = t_o$$

$$l_i = l_o$$

$$n_i = n_o .$$

We then find:

$$P_h = \sin\varphi$$

$$h_p = \cos\varphi .$$

New values are then:

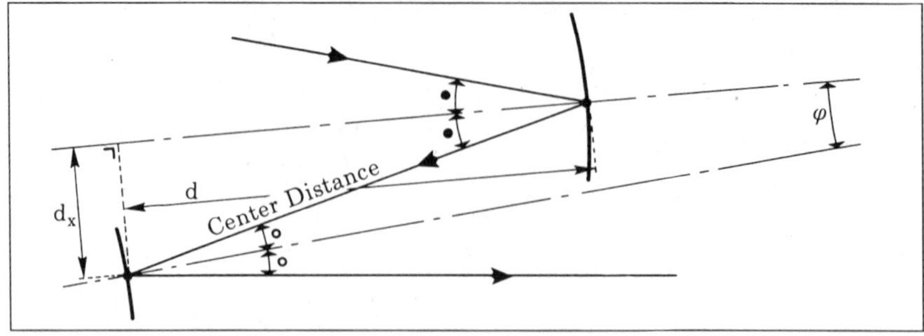

Fig. 20.16 *Terms Used in a Schiefspiegler.*

$$X_o = X_i \cdot h_p + t_i \cdot P_h$$

$$t_o = t_i \cdot h_p - X_i \cdot P_h$$

$$l_o = l_i \cdot h_p + n_i \cdot P_h$$

$$n_o = n_i \cdot h_p - l_i \cdot P_h .$$

If the rest of the system is tilted, nothing changes, but if this is the only tilted surface, then it must be tilted back after the new values of X_s, Y_s, Z_s and l_s, m_s, n_s are calculated. When the rest of the system is tilted, this is not necessary.

For a lens with an axial and transverse shift, there is no difficulty because this calculation can be done simultaneously. When a surface is both tilted and shifted, the axial shift must be calculated first, next the transverse, and lastly the tilt calculated. The sign conventions are of the utmost importance in such work.

When an optical system has tilted surfaces, as, for example, in the Schiefspiegler, use the method illustrated in fig. 20.16. Treat the main beam as an oblique entering bundle for the first surface encountered. For each of the next surfaces, project the distance between centers onto the axis of the previous surface. In this way, you can find the transverse distance between the centers of both surfaces. The tilt angle is the angle between the axes of the surfaces. The calculation is then carried out as described above.

20.10 Using Ray-Trace Results

In sections 20.7 through 20.9, we treated methods of meridional and skew-ray tracing of various optical surfaces. The results of these calculations have two primary applications: first, determining the magnitude of image aberrations, and second, determining the intercept height of the rim rays on the various optical surfaces to establish the necessary diameter and, if required, making vignetting calculations.

20.10.1 Magnitude of the Image Aberrations

We must clearly distinguish between the longitudinal and transverse presentation of image aberrations. The best-known longitudinal presentation is the longitudinal spherical aberration (LA). To calculate this, we trace parallel rays entering the system at various entrance heights (also called zones) until they intersect the optical axis. The spread of the intersecting points along the axis with respect to the paraxial focus is the LA. The longitudinal presentation is interesting in the beginning stage of a new design.

Transverse aberration (TA), in contrast, shows the spread of the intersection points of the rays on a plane perpendicular to the optical axis. It is particularly informative for the designer because it gives the size of the image blur in the focal plane. Because the size of this image blur can be directly related to the sharpness criteria (section 4.4), the designer can easily see whether or not the optical performance is adequate. Such a conclusion cannot be easily drawn from the longitudinal presentation.

The major problem with the transverse display is that the optimum position of the plane of the presentation, or focal plane, may not be well-defined. In the case of spherical aberration, the best focus (i.e., the sharpest image) lies farther along the axis than the point where the image blur is smallest. This is illustrated in section 5.2., for the spherical mirror. Particularly when the image aberrations are large, as is often the case when a mixture of coma and astigmatism is present, it is difficult to find the position of the best focal plane.

Our ray-trace program permits the user to find a best-fitting focal plane using a process called refocusing. The computer holds the positions of all exit rays computed in its memory. The user chooses a plane he expects to be close to the best focal plane. From the size, shape, and distribution of light in the spot diagram, he decides whether to shift the focal plane, and repeats this until the best focal plane has been found. This procedure can be carried out for axial and off-axis images, which also allows the designer to determine system's field curvature.

20.10.2 Determining the Diameters of Optical Elements

To determine the diameters required for the optical components in a system, the designer must know the intercept heights of the edge rays for a beam parallel to the axis and for the most oblique beam, for every optical surface. Often we see that vignetting occurs somewhere in the system, so that the full oblique beam cannot pass. This was, for instance, the case for the Schmidt-Cassegrain discussed in section 19.4, because of the presence of a baffle tube system. In such a case, it is not necessary for all the other components to have the full diameter required to pass the edge rays of the most oblique beam. The minimum size required for optical components is determined by the height of the edge rays for a beam entering the system parallel to the axis, having a diameter which is equal to the entrance pupil. Often, the designer must compromise, and select optical component diameters somewhere between the maximum and minimum required values.

20.11 Other Optical Calculations

Many methods for the calculation and presentation of image aberrations exist. One method of presentation which bears mention is the H' tanU' plot (ref. 20.4). H' means the intercept height of the exit ray with respect to the optical axis at some reference plane. Its calculation is described in this chapter. The reference plane normally chosen is the paraxial focal plane. U' means the exit angle with respect to the optical axis. From the slope and the shape of the resulting plots, the designer can draw conclusions about the aberrations present in the image.

Another common analytical tool is the optical path difference (OPD). In this method, rays at various heights are followed through the system until they arrive at the focal plane. The total length of the optical path of the ray, expressed in wavelengths, is then computed. Where light rays pass through refractive elements, the transfer lengths are multiplied by the refractive index of the glass. In a perfect optical system, the path lengths of all the rays will be exactly the same; the light will arrive in phase. If the rays arrive at the focal plane out of phase, aberrations will be present. When the optical path length variation differs less than 1/4 of the wavelength of the light, the system is normally considered diffraction-limited. Readers interested in these subjects are referred to refs. 20.1 and 20.2.

Chapter 21

Designing Telescope Optical Systems

21.1 Introduction

In this chapter we describe design procedures for ten telescope types. These are Cassegrains, Schmidt-Cassegrains, Houghton-Cassegrains, Maksutov-Cassegrains, Schmidt cameras, Wright cameras, Houghton cameras, Maksutov cameras, doublet refractors, and triplet refractors.

Most of these systems are not difficult to design. Only the refractors and the variants on the Maksutov pose any real difficulty. The reason for this is that a system like a two-mirror Cassegrain is, in fact, a very simple optical system, and offers just a few degrees of freedom. As soon as the configuration has been chosen, the system is optically almost fixed. The only changes that can still be made are the asphericities of the optical surfaces, and these are readily derived from third-order Seidel theory. Thus, designing is a straightforward process of applying established formulae.

The same is true for the Schmidt and Houghton derivatives, because the optical power and curvatures of the correctors are relatively weak. These catadioptric Cassegrains are simply modifications of a two-mirror Cassegrain, and can be described with good accuracy with a number of formulae. This greatly facilitates the design procedure.

On the other hand, refractor objectives and Maksutov systems offer the designer more degrees of freedom At the outset, it will not be immediately clear to the designer what starting values he should choose for the optical parameters, and what optimization procedure he should follow. In most cases, then, the design procedure for refractors must be iterative. The same holds true for Maksutov systems because the corrector has highly curved surfaces in relation to its power. A straightforward design process is possible, but would require very complex formulae.

In the 1970s, Robert D. Sigler studied extensively various catadioptric Cassegrain systems (refs. 21.1, 21.2, and 21.3). His work opened up a large area of telescope systems which are particularly interesting to the amateur. It is due to the work of this researcher that we can present the design procedures for these systems here.

The telescope design program available as an option with this book performs the predesign for all the mirror systems listed above; refractors have a separate de-

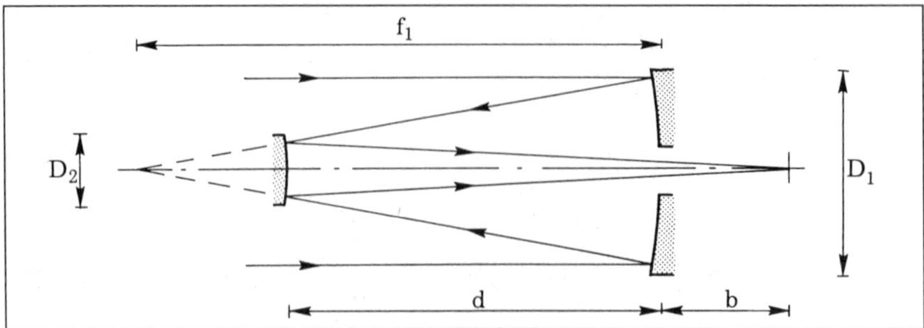

Fig. 21.1 *Characteristics of the Cassegrain Telescope.*

sign program of their own. For doing a design of Cassegrain and catadioptric systems, optimization by means of a skew-ray trace is required. This final "touch-up" process takes only a little time because the predesign methods produce systems that are already quite good.

21.2 Designing a Cassegrain

Cassegrain telescopes are preferred by many amateurs when a combination of a short tube and a long effective focal length is desired. We described the general layout of a Cassegrain in chapter 7, and discussed the relations among various parameters. Insight into the optical performance of four types of Cassegrains was given there.

We present the following dimensionless quantities which were algebraically defined by Schwarzwald (see fig. 21.1):

$$S = \frac{f_2}{f_1} = \frac{r_2}{r_1}$$

$$T = \frac{D_2}{D_1}$$

$$R = \frac{d}{f_1}$$

$$E = \frac{b}{f_1}$$

$$M = \frac{f}{f_1} \text{ (secondary magnification)}$$

$$B = \frac{b}{f}.$$

Note that f_1, f_2, r_1, and r_2 are two focal lengths and radii of the primary and secondary respectively; f is the focal length of the system.

Designing begins by determining the demands on the system. The basis of the design is the geometric layout. The value of b, for example, is not fixed, but depends on the thickness of the primary mirror, the type of the mirror support, and the length of the focusing mechanism; it is the first design value to be defined. The value of E will almost always be larger for small telescopes than for large telescopes.

The secondary magnification, M, may be a design condition. This often occurs when the designer wants to build a system having focal length f using an existing primary mirror with focal length f_1. M is then fixed by f and f_1.

The maximum admissible obstruction, T_{max} is often an independent design condition. If this is the case, the secondary magnification will depend on the value of T_{max} chosen. Allowance must be made, in setting the size of the secondary mirror, for oblique beams. The obstruction for only the axial parallel beam is usually some 20% smaller. For example, if $T_{max} = 0.3$, the design should specify T as (0.3–(0.2 × 0.3)), or 0.24.

From this value, we find the secondary magnification:

$$M = \frac{1+E}{T} - 1 . \quad \textbf{(21.2.1)}$$

The maximum length of the instrument is often another independent design condition. In that case, the distance between the mirrors, d, is an important parameter. The secondary magnification can be determined as follows:

$$M = \frac{E+R}{1-R} . \quad \textbf{(21.2.2)}$$

The parameters of the system are then calculated using the following formulae, at least to the extent that they are not yet defined as design conditions.

$$R = \frac{M-E}{M+1} \quad \textbf{(21.2.3)}$$

$$E = M - (M+1) \cdot R \quad \textbf{(21.2.4)}$$

$$T = \frac{1+E}{M+1} \quad \textbf{(21.2.5)}$$

$$S = \frac{M(1-R)}{M-1} . \quad \textbf{(21.2.6)}$$

At this stage, the type of Cassegrain has not yet been chosen. The formulae above apply equally well for the classical Cassegrain, the Dall-Kirkham, the Ritchey-Chrétien, and the Pressmann-Camichel. The type of Cassegrain is determined by the aspheric deformation of the mirrors.

The third-order Seidel coefficients for a two-mirror Cassegrain are:

Spherical aberration:

$$A_{\text{cass}} = 1 + SC_1 - \left[SC_2 + \left(\frac{M+1}{M-1}\right)^2\right]\frac{(M-1)^3 \cdot (1-R)}{M^3} \qquad \textbf{(21.2.7)}$$

Coma:

$$B_{\text{cass}} = \frac{2}{M^2} + \left[SC_2 + \left(\frac{M+1}{M-1}\right)^2\right]\frac{(M-1)^3 \cdot R}{M^3} \qquad \textbf{(21.2.8)}$$

Astigmatism:

$$C_{\text{cass}} = \frac{4(M-R)}{M^2(1-R)} - \left[SC_2 + \left(\frac{M+1}{M-1}\right)^2\right]\frac{(M-1)^3 \cdot R^2}{M^3(1-R)} . \qquad \textbf{(21.2.9)}$$

In order to simplify the formulae above, we introduce several auxiliary quantities. These will also be used in the derivations of the Schmidt-Cassegrain systems, in section 21.4.

$$\alpha = \left(\frac{M+1}{M-1}\right)^2 \qquad \textbf{(21.2.10)}$$

$$\beta = \frac{(M-1)^3(1-R)}{M^3} \qquad \textbf{(21.2.11)}$$

$$\gamma = \frac{(M-1)^3 R}{M^3} \qquad \textbf{(21.2.12)}$$

$$\delta = \frac{2}{M^2} \qquad \textbf{(21.2.13)}$$

$$\varepsilon = \frac{4(M-R)}{M^2(1-R)} \qquad \textbf{(21.2.14)}$$

$$\vartheta = \frac{(M-1)^3 R^2}{M^3(1-R)} . \qquad \textbf{(21.2.15)}$$

Spherical aberration must be zero; that is, $A_{\text{cass}} = 0$. From this condition we can determine the deformation constants (Schwarzwald constants) SC_1 and SC_2 of the primary and secondary mirrors for the various types of Cassegrains.

For the classical Cassegrain, the primary mirror is parabolic (i.e., $SC_1 = -1$). Eq. 21.2.7 may be reduced to:

$$SC_2 = -\alpha . \qquad \textbf{(21.2.16)}$$

In the Dall-Kirkham, the secondary mirror is spherical (i.e., $SC_2 = 0$), so eq. 21.2.7 reduces to:

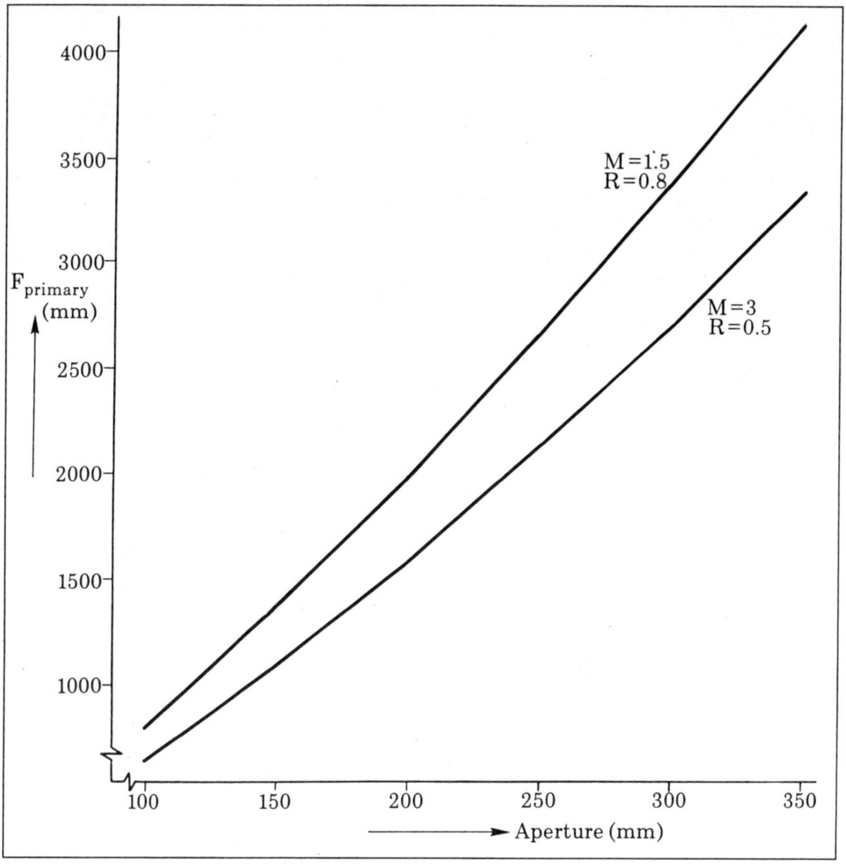

Fig. 21.2 *Parameters for a Cassegrain Telescope with Spherical Mirrors.*

$$SC_1 = \alpha \cdot \beta - 1 . \qquad \textbf{(21.2.17)}$$

For the Ritchey-Chrétien, both spherical aberration and coma are eliminated, so $A = 0$ and $B = 0$. The resulting two equations with two unknowns are solved thus:

$$SC_1 = -\left(1 + \frac{\beta \cdot \delta}{\gamma}\right) \qquad \textbf{(21.2.18)}$$

$$SC_2 = -\left(\alpha + \frac{\delta}{\gamma}\right) . \qquad \textbf{(21.2.19)}$$

Finally, in the Pressmann-Camichel, with its spherical primary, $SC_1 = 0$, so eq. 21.2.7 reduces to:

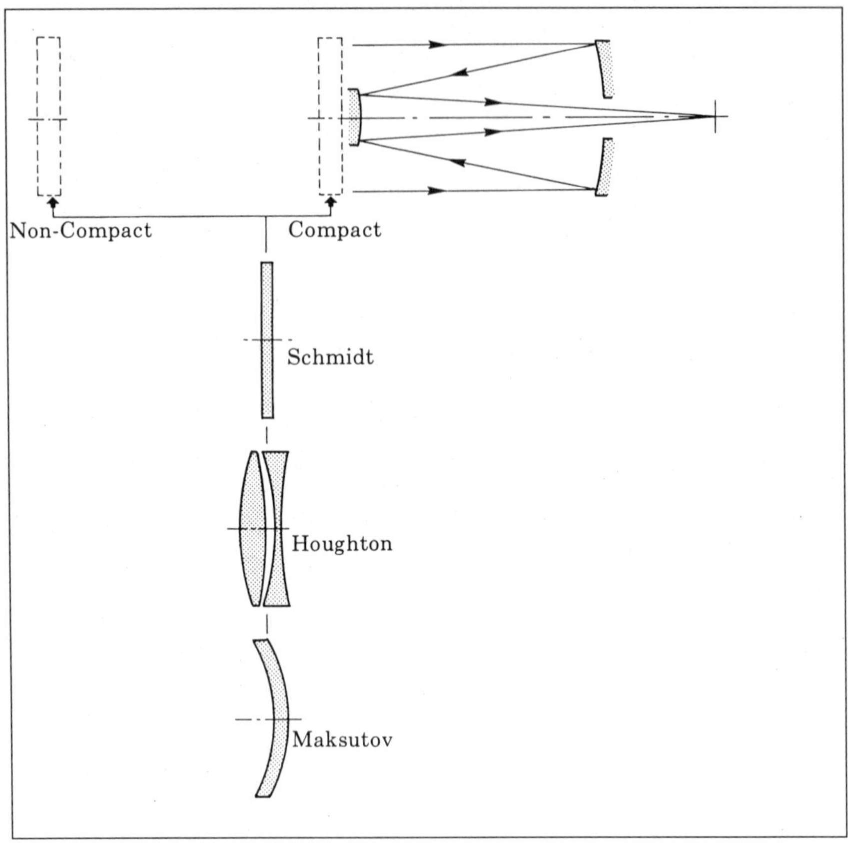

Fig. 21.3 *Compact and Non-Compact Catadioptric Cassegrains.*

$$SC_2 = \frac{1 - \alpha \cdot \beta}{\beta} = \frac{1}{\beta} - \alpha . \tag{21.2.20}$$

When an instrument is intended only for observation of objects of small angular size (planets), both mirrors can sometimes remain spherical. Considering that the spherical aberration of the primary mirror is partially corrected by the secondary mirror, above a certain focal ratio combinations in which the diameter of the spread figure is smaller than the Airy disk must exist. Analysis shows that for a given aperture the primary must exceed a minimum focal length. This depends only slightly on the secondary magnification, *M*, and the value of *R* (fig. 21.2). For apertures from 100 to 350 mm, the minimum allowable focal ratio for the primary mirror rises from 7 to 10.5. For any given aperture, the minimum system focal ratio occurs when *M* is chosen to be 3.0. Fig. 21.2 shows clearly that it is not possible to build an all-spherical Cassegrain with a short tube unless at least one mirror is aspheric.

21.3 Designing a Catadioptric Cassegrain

In the following sections we will discuss the design procedures for various catadioptric Cassegrain systems. These systems consist of a Cassegrain system plus a weak refractive corrector. The corrector may be a Schmidt plate, a two-element Houghton corrector, or a Maksutov corrector. These instruments were described in chapter 9 for the Schmidt, chapter 11 for the Maksutov, and chapter 13 for the Houghton. We saw that all of them are capable of good to excellent performance.

Unlike two-mirror Cassegrain systems, catadioptric systems can produce flat fields with good off-axis performance. Consider the catadioptric Schmidt-Cassegrain systems discussed in chapter 9. In the compact system, the corrector is placed close to the secondary mirror so that it can support the secondary. In the non-compact design, the corrector is placed farther from the primary, and the secondary mirror must be supported with a spider (fig. 21.3). However, non-compact designs generally offer better possibilities for correcting off-axis aberrations.

Since the starting point in designing a catadioptric Cassegrain is a two-mirror Cassegrain, the equations for the system geometry (eqs. 21.2.1 through 21.2.6) remain valid. These formulae are valid for both curved- and flat-field designs. However, flat-field designs result only when the two mirrors have approximately the same radius of curvature, provided the system is also free of astigmatism. When a system suffers from astigmatism, the secondary mirror can be more strongly curved to flatten the field. This may be carried out afterwards, during optimization of the designs, or when some design experience has been gained, taken into account in the design as the ratio of the curvatures of the primary and secondary mirrors.

The starting point for a flat-field design is the system focal length, the distance from the front surface of the primary mirror to the focal plane, and the knowledge that the two mirror curvatures must be approximately the same. Combining $B = b/f$ and eq. 21.2.4, we find:

$$B = 1 - R\left(1 + \frac{1}{M}\right). \tag{21.3.1}$$

Combining this with eq. 21.2.6, we find the secondary magnification:

$$M = \frac{1 + \sqrt{1 + 4S(S - B)}}{2(S - B)}. \tag{21.3.2}$$

For a flat-field system, the ratio of the focal lengths of the primary and secondary mirrors, S, is approximately 1. The value of M is used to calculate the focal length of the primary mirror, and also to evaluate eqs. 21.2.3 and 21.2.5.

As noted above, the starting point for a catadioptric Cassegrain is the two-mirror Cassegrain in which the refractive elements are considered to have zero power. In reality, the zero power assumption is not always true. Therefore, after a predesign has been completed, the system must be skew-ray traced to analyze and minimize residual aberrations. Optimization procedures are described in section 21.12, and some results are shown in chapter 22.

When you introduce the real refractive element, watch for a shift in the focal plane caused by the element's optical power. It is often difficult to guess the magnitude of this shift beforehand, because it depends on the shape and kind of the corrector used.

The shift of the focal plane is most easily understood for Schmidt correctors. In this type of corrector, the power depends on the paraxial curvature of its optical surface. When the neutral zone of the Schmidt corrector lies on the optical axis, the surface is paraxially flat and has no optical power. As the neutral zone is moved towards the edge of the corrector, the paraxial curvature increases, and so does the power. Thus the change in focal length increases as the neutral zone is moved towards the edge.

The situation is different for the Houghton and Maksutov correctors. In the Houghton system, the thickness of the lenses can initially be ignored.

When the design is optimized by ray tracing, the slight shift in the focus can be corrected in one of two ways: by moving the secondary mirror and corrector simultaneously, or by changing the curvature of the secondary mirror. Both methods proceed by trial and error until the original focus position is achieved. It should be obvious, though, that in flat-field designs, changing the radius of the secondary mirror involves some risk that the field will no longer be flat. This means that considerable trial and error work may be necessary to determine the best position and curvature for the secondary mirror, once the power of the corrector or the field curvature is taken into account.

In the following sections, we will treat only the correction of aberrations, and will not further discuss the two-mirror part of the design because the process is the same for all of the telescope types. We wish to emphasize that the techniques are based on Sigler's formulae (refs. 21.1, 21.2, and 21.3), and that his design formulae are derived from Seidel theory.

In discussing catadioptric designs, we will be using an additional dimensionless number, namely, the ratio between the distance to the corrector and the focal length of the primary mirror:

$$D = \frac{d_c}{f_1}. \quad \textbf{(21.3.3)}$$

21.4 Designing a Schmidt-Cassegrain

As we saw in chapter 9, the Schmidt-Cassegrain telescope can provide excellent optical performance for a system with only three optical elements. Although the optical literature has paid a great deal of attention to the Schmidt-Cassegrain, this type of instrument is scarcely ever found in professional observatories. The reason for this is that large-diameter Schmidt correctors are expensive and difficult to make. Observatories have instead concentrated their efforts on field correctors to enlarge the field of their telescopes. For amateur size instruments, however, the situation is entirely different because the cost of manufacturing a Schmidt corrector up to 500 mm in diameter is moderate. Above this size, the cost of manufacture rises sharply.

Important constants for the Schmidt-Cassegrain are g, the relative power of the Schmidt corrector (that is, the actual power divided by the power that would be necessary to fully suppress the spherical aberration of the spherical primary mirror), and SC_1 and SC_2, the deformations of the primary and secondary mirrors. A sphere has a deformation of zero, and a parabola's is –1.

Sigler gives the following third-order formulae for the Seidel coefficients for the Schmidt-Cassegrain (refs. 21.1 and 21.2).

Spherical aberration:

$$A = 1 + SC_1 - \left[SC_2 + \left(\frac{M+1}{M-1}\right)^2\right]\frac{(M-1)^3(1-R)}{M^3} - g \tag{21.4.1}$$

Coma:

$$B = \frac{2}{M^2} + \left[SC_2 + \left(\frac{M+1}{M-1}\right)^2\right]\frac{(M-1)^3 \cdot R^2}{M^3} - g \cdot D \tag{21.4.2}$$

Astigmatism:

$$C = \frac{4(M-R)}{M^2(1-R)} - \left[SC_2 + \left(\frac{M+1}{M-1}\right)^2\right]\frac{(M-1)^3 \cdot R^2}{M^3(1-R)} - g \cdot D^2 \tag{21.4.3}$$

Before starting the design procedure, the geometrical layout must be determined using the same formulae as for a two-mirror Cassegrain (eqs. 21.2.1 through 21.2.6). Once all main parameters have been calculated, the designer determines the deformations of the mirrors and the power of the Schmidt corrector. In the Schmidt-Cassegrain, the position of the Schmidt corrector is a free parameter. In most compact designs, the corrector distance, D, lies between R and 1.1 R.

In order to be able to make derivations from the formulae, the auxiliary quantities (given in eqs. 21.2.10 through 21.2.15) may be substituted into eqs. 21.4.1 through 21.4.3 above. The calculations can then be simplified using following equations:

Spherical aberration:

$$A = 1 + SC_1 - (SC_2 + \alpha) \cdot \beta - g \tag{21.4.4}$$

Coma:

$$B = \delta + (SC_2 + \alpha)\gamma - g \cdot D \tag{21.4.5}$$

Astigmatism:

$$C = \varepsilon - (SC_2 + \alpha)\vartheta - g \cdot D^2 . \tag{21.4.6}$$

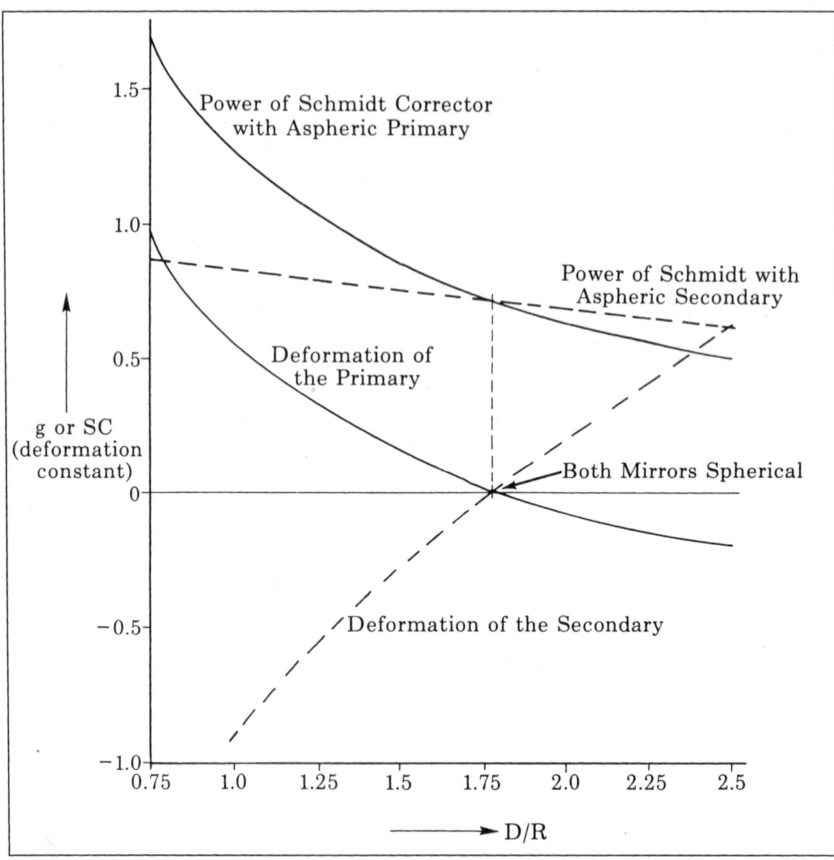

Fig. 21.4 *Influence of Position of the Corrector for an f/10 Aplanatic Curved-Field Schmidt-Cassegrain Telescope with one Aspheric Mirror.*

These formulae give us the opportunity to find the conditions for which spherical aberration, coma, and astigmatism are eliminated, because the values for A, B, and C will then equal zero.

Before the design procedure is started, it is important to define what demands are to be made with respect to off-axis aberrations. When a sharp axial image is the only requirement (i.e., no correction of coma and astigmatism is necessary), then both mirrors can remain spherical because the corrector lens can completely eliminate spherical aberration. By substituting $A = 0$ and $SC_1 = SC_2 = 0$ into eq. 21.4.4, we obtain the following simple condition for such a system:

$$g = 1 - \alpha \cdot \beta . \qquad \textbf{(21.4.7)}$$

In the case of two spherical mirrors, coma can be eliminated by moving the corrector farther from the primary mirror. The distance, D, is found by setting $B = 0$ and $SC_2 = 0$, and substituting eq. 21.4.7 in 21.4.5:

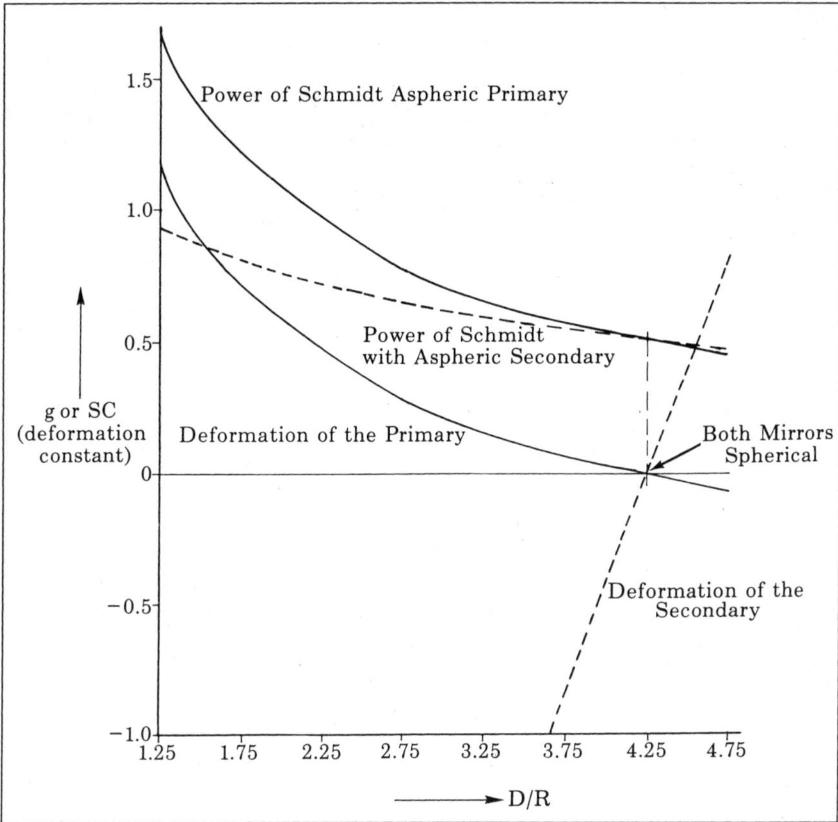

Fig. 21.5 *Influence of the Position of the Corrector for an f/4 Aplanatic Flat-Field Schmidt-Cassegrain Camera with one Aspheric Mirror.*

$$D = \frac{\delta + \alpha\gamma}{1 - \alpha \cdot \beta}. \qquad \textbf{(21.4.8)}$$

Because the corrector has been moved, the system is no longer a compact design

Equations 21.4.4, 21.4.5, and 21.4.6 can also be used for finding design conditions for other modifications. To maintain a compact design, for example, we might choose to correct coma by aspherizing one of the two mirrors. If we choose to aspherize the secondary while keeping the primary spherical, we find the following conditions from eqs. 21.4.4 and 21.4.5:

$$g = \frac{\gamma + \beta \cdot \delta}{\gamma + \beta \cdot D} \qquad \textbf{(21.4.9)}$$

and

$$SC_2 = \frac{D - \alpha \cdot \beta \cdot D - \alpha \cdot \gamma - \delta}{\gamma + \beta \cdot D}. \qquad \textbf{(21.4.10)}$$

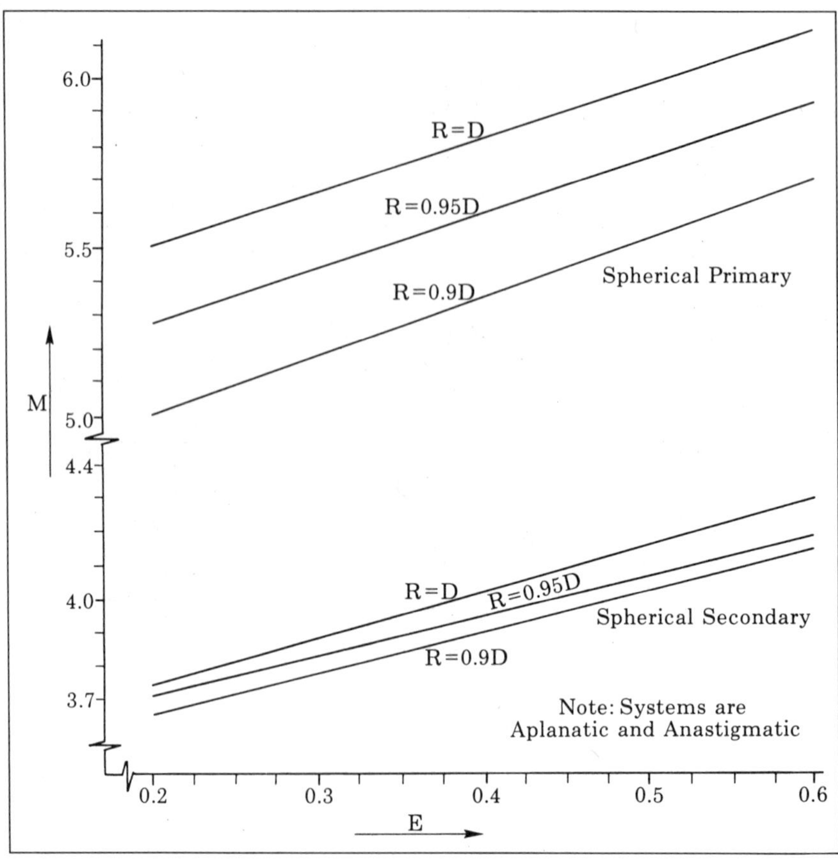

Fig. 21.6 *Parameters for Compact Schmidt-Cassegrain Telescopes.*

If, instead, we decide to aspherize the primary and leave the secondary spherical:

$$g = \frac{\delta + \alpha \cdot \gamma}{D} \tag{21.4.11}$$

and

$$SC_1 = \alpha \cdot \beta + \frac{\alpha \cdot \gamma + \delta}{D} - 1 . \tag{21.4.12}$$

Figs. 21.4 and 21.5 show the influence of the position of the corrector on the Schmidt power, g, and the asphericities, SC, of the mirrors for aplanatic systems. Fig. 21.4 shows data for an *f*/10 curved-field design with a mirror configuration similar to table 9.1. Fig. 21.5 shows data for an *f*/4 flat-field design with the mirror configuration given in table 9.3. Changes in the values of M and E change these graphs.

We have so far not taken correction of astigmatism into consideration. The systems characterized by the eqs. 21.4.7 through 21.4.12 are generally not corrected for astigmatism. However, Sigler has shown that within a family of aplanatic systems (i.e., systems in which spherical aberration and coma are already corrected), there are members for which astigmatism is corrected as well. For certain combinations of M and E, therefore, these systems can be made both aplanatic and anastigmatic.

Fig. 21.6 graphs values of M as the parameters E and R vary, for compact designs (R lies between 0.9D and 1.0D) for a spherical primary and aspheric secondary, and for a spherical secondary and aspheric primary.

For a value of $E = 0.4$ and $R = 0.95$, systems that are both aplanatic and anastigmatic have values of M equal to 5.6 and 4.0, for the cases of a spherical primary and spherical secondary, respectively.

If a compact Schmidt-Cassegrain that is both aplanatic and anastigmatic for combinations other than those indicated in fig. 21.6 is desired, then it is necessary to aspherize both mirrors. The power of the Schmidt corrector and the deformation factors of the mirrors are:

$$g = \frac{\delta \cdot \vartheta + \varepsilon \cdot \gamma}{D(\vartheta + D \cdot \gamma)} \tag{21.4.13}$$

$$SC_1 = \beta\left(\alpha + \frac{\varepsilon - \alpha \cdot \gamma \cdot D - \delta \cdot D - \alpha \cdot \vartheta}{\vartheta + \gamma \cdot D}\right) + \frac{\delta \cdot \vartheta + \varepsilon \cdot \gamma}{D(\vartheta + D \cdot \gamma)} - 1 \tag{21.4.14}$$

$$SC_2 = \frac{\varepsilon - \alpha \cdot \gamma \cdot D - \delta \cdot D - \alpha \cdot \vartheta}{\vartheta + \gamma \cdot D}. \tag{21.4.15}$$

21.5 Designing a Houghton-Cassegrain

The Houghton-Cassegrain offers an important advantage: all surfaces can be left spherical, because the Houghton corrector can correct both spherical aberration and coma for every position of the corrector. This is not the case for the Schmidt and Maksutov correctors, which eliminate spherical aberration, but correct coma for one position only, if both mirrors are spherical. The Houghton corrector can correct both because it consists of two elements. This also implies that the color aberration can be corrected.

If the elements are placed close together and have equal power of opposite signs, they can be made of the same kind of glass. Furthermore, if we make radii of curvature in pairs (i.e., $r_3 = -r_1$ and $r_4 = -r_2$), then the lenses can be tested against each other by means of interference. Although the literature advises against testing the surfaces of the two elements against each other, amateurs should be able to get away with this because they can retouch the surfaces locally, and because the concave surface can usually be independently tested by a simple test such as the Foucault or Ronchi test.

In chapter 13, we saw that the color aberration of the *f*/10 system was sufficiently low, but for the flat-field *f*/5.3 design, this was not the case. Color aberration can be reduced by making the elements from glasses having the same index of refraction at the design color, but different dispersions. Glasses with different indexes of refraction could be chosen, too, but the design optimization procedure would be very time-consuming.

Sigler derived the following equations for the Seidel coefficients, in the two spherical mirror configuration of the Houghton-Cassegrain (ref. 21.3), without the corrector:

Spherical aberration:

$$A = A_4 = 1 - \frac{(M+1)^2(M-1)(1-R)}{M^3} \tag{21.5.1}$$

Coma:

$$B = A_5 = \frac{2}{M^2} + \frac{(M+1)^2(M-1)\cdot R}{M^3} \tag{21.5.2}$$

Astigmatism:

$$C = A_6 = \frac{4(M-R)}{M^2(1-R)} - \frac{(M+1)^2(M-1)R^2}{M^3(1-R)}. \tag{21.5.3}$$

For the corrector, he defined a number of calculation constants:

$$A_1 = \frac{n+2}{n(n-1)^2} \tag{21.5.4}$$

$$A_2 = \frac{2(2n+1)}{(n-1)^2} \tag{21.5.5}$$

$$A_3 = \frac{2(n+1)}{n(n-1)}. \tag{21.5.6}$$

In these equations n is the refractive index of the corrector for which the system is designed.

Sigler then derived the equation for the Seidel coefficient for the spherical aberration of a complete Houghton-Cassegrain system:

$$A = \frac{2Q(2A_1 - A_2)}{L^3} + A_4. \tag{21.5.7}$$

Spherical aberration must always be corrected, so A must equal zero. In that case, the following equation is valid for coma:

$$B = \frac{-2Q\cdot A_3}{L^2} - D\cdot A_4 + A_5. \tag{21.5.8}$$

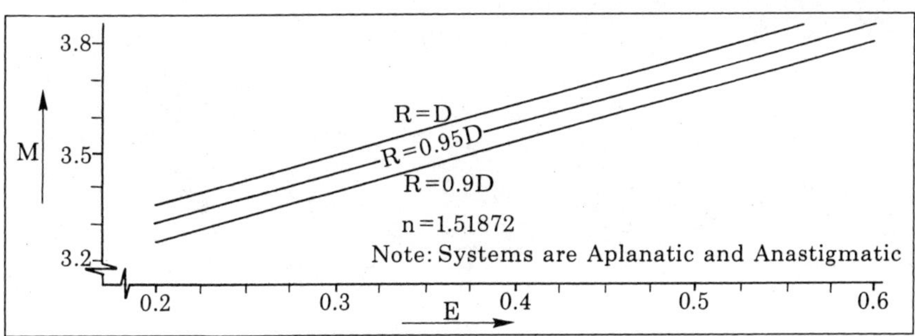

Fig. 21.7 *Parameters for Compact Houghton-Cassegrain Telescopes.*

When we have an aplanatic system, so both $A = 0$ and $B = 0$, we have the following equation for astigmatism:

$$C = D^2 \cdot A_4 - 2D \cdot A_5 + A_6 . \qquad \textbf{(21.5.9)}$$

In these equations, Q is the bending coefficient of both corrector lenses, that is, $(r_1+r_2)/(r_2-r_1)$; D is the distance from the corrector to the primary mirror expressed in units of the focal length of this mirror; L is the focal length of the corrector element, expressed in units of the focal length of the primary mirror, f_1.

From the equation for astigmatism, we find that astigmatism also can be corrected when the corrector is at a distance from the primary mirror given by the following equation:

$$D = \frac{A_5 \pm \sqrt{A_5^2 - (A_4 \cdot A_6)}}{A_4} . \qquad \textbf{(21.5.10)}$$

D is again expressed in units of the focal length of the primary mirror. This is a quadratic function with two possible solutions, one—a compact design—with the corrector close to the primary mirror, the other—a non-compact design—with the corrector far from the primary. For a compact system, then, depending on the distance of the focal plane from the front side of the primary and the position of the corrector, an aplanatic and anastigmatic design exists, for a certain secondary magnification. Fig. 21.7 shows the values of M for these combinations, as E varies, for values of R between 0.9 D and D. Here M is close to 3.6 for all values of R and E.

From the equations for spherical aberration and coma, we can derive the focal length of the corrector element and the bending coefficient of the corrector lenses, as follows:

$$L = \frac{(D \cdot A_4 - A_5)(2 \cdot A_1 - A_2)}{A_3 \cdot A_4} \qquad \textbf{(21.5.11)}$$

$$Q = \frac{(A_5 - D \cdot A_4) \cdot L^2}{2 \cdot A_3} . \qquad \textbf{(21.5.12)}$$

With these values, the radii of curvature of the first lens of the corrector can be calculated:

$$r_1 = \frac{2 \cdot f_1 \cdot L(n-1)}{Q+1} \qquad \textbf{(21.5.13)}$$

$$r_2 = \frac{2 \cdot f_1 \cdot L(n-1)}{Q-1} . \qquad \textbf{(21.5.14)}$$

Since the curvatures of the two lenses are equal in value but opposite in sign, it follows that the curvatures r_3 and r_4 of the second lens are $-r_1$ and $-r_2$ respectively.

21.6 Designing a Maksutov-Cassegrain

The design procedure for a Maksutov-Cassegrain differs from those for the catadioptric systems discussed above. While the Schmidt corrector and the Houghton corrector have, in third order, little power and consist of relatively thin lenses with moderate curvatures, a Maksutov corrector always has considerable power and relatively thick lenses with—and this is the most important difference—relatively strong curvatures. The Seidel theory has some difficulties with these optical surfaces. For this reason no direct derivations are made, as for the Schmidt and Houghton correctors. For the Maksutov-Cassegrain the Seidel coefficients for the corrector and the two-mirror systems are determined separately and added afterwards. The maximum admissible values of these Seidel sums were discussed in section 20.6.

Designing a Maksutov-Cassegrain is a trial-and-error procedure. The radii of curvature of a Maksutov corrector are determined by the shape, the axial thickness, the focal length of the primary, and the refractive index at the design color. After the Seidel sums are calculated, a certain corrector is defined and the system is analyzed. Then the shape and the position of the corrector are changed until the system meets the desired criteria. After the configuration of the two-mirror set-up has been determined, the Seidel coefficients (for a mirror part with two spherical mirrors, as in section 21.5) can be calculated:

$$A_{cass} = 1 - \frac{(M+1)^2(M-1)(1-R)}{M^3} \qquad \textbf{(21.6.1)}$$

$$B_{cass} = \frac{2}{M^2} + \frac{(M+1)^2(M-1)R}{M^3} \qquad \textbf{(21.6.2)}$$

$$C_{cass} = \frac{4(M-R)}{M^2(1-R)} - \frac{(M+1)^2(M-1)R^2}{M^3(1-R)} . \qquad \textbf{(21.6.3)}$$

Unfortunately the Seidel coefficients of the corrector are not simple expressions. According to Sigler (ref. 21.3):

$$A_{corr} = \frac{-32n^4h^3}{(n+1)(n^2-1)^2(Q^2-1)^3} \cdot \{(Q+1)^3 - Y[Yn^2(Q-1)-(n^2-1)(Q+1)] \cdot [Y \cdot n(Q-1)-(n-1)(Q+1)]^2\} \quad \textbf{(21.6.4)}$$

$$B_{corr} = \frac{-16n^2h^2}{(n+1)(n^2-1)(Q^2-1)^2} \cdot \left\{(Q+1)^2 - Y[Yn^2(Q-1)-(n^2-1)(Q+1)] \cdot [Yn(Q-1)-(n-1)(Q+1)] \cdot \left[1-\frac{2n^2}{(n^2-1)(Q+1)}\right]\right\} + D \cdot A_{corr} \quad \textbf{(21.6.5)}$$

$$C_{corr} = \frac{-8h}{(n+1)(Q_2-1)} \cdot \left\{(Q+1) - Y[Yn^2(Q-1)-(Q+1)(n^2-1)] \left[1-\frac{2n^2}{(n^2-1)(Q+1)}\right]^2\right\} + 2 \cdot D \cdot B_{corr} - D^2 \cdot A_{corr}. \quad \textbf{(21.6.6)}$$

In these formulae, n is the refractive index at the design color; h is the ratio of the focal length of primary to the axial corrector thickness; Q is the bending coefficient, that is, $(r_1+r_2) / (r_2-r_1)$; D is the distance from the corrector to the primary, expressed in units of the focal length of the primary; and Y is the ratio of both intercept heights in the corrector:

$$Y = 1 + \frac{2n}{(n+1)(Q-1)}. \quad \textbf{(21.6.7)}$$

Formulae 21.6.4 to 21.6.6 can be greatly simplified by defining the following auxiliary quantities:

$$n_1 = n + 1 \quad q_1 = Q + 1$$

$$n_2 = n^2 - 1 \quad q_2 = Q^2 - 1$$

$$n_3 = n - 1 \quad q_3 = Q - 1$$

$$Y = 1 - \frac{2n}{n_1 \cdot q_3} \tag{21.6.8}$$

$$Q_1 = Y \cdot n \cdot q_3 - n_3 \cdot q_1 \tag{21.6.9}$$

$$Q_2 = Y(Y \cdot n^2 \cdot q_3 - n_2 \cdot q_1) \tag{21.6.10}$$

$$Q_3 = \frac{-8h}{n_1 \cdot q_2} \tag{21.6.11}$$

$$Q_4 = 1 - \frac{2n^2}{n_2 \cdot q_1}\,.. \tag{21.6.12}$$

The resultant simplified equations for the corrector are:

$$A_{corr} = \frac{4 \cdot Q_3 \cdot h^2 \cdot n^4}{q_2^2 \cdot n_2^2} \cdot (q_1^3 - Q_2 \cdot Q_1^2) \tag{21.6.13}$$

$$B_{corr} = \frac{2 \cdot Q_3 \cdot h \cdot n^2}{q_2 \cdot n_2} \cdot (q_1^2 - Q_2 \cdot Q_1 \cdot Q_4) + D \cdot A_{corr} \tag{21.6.14}$$

$$C_{corr} = Q_3 \cdot (q_1 - Q_2 \cdot Q_4^2) + 2D \cdot B_{corr} - D^2 \cdot A_{corr}\,. \tag{21.6.15}$$

Adding the Seidel coefficients for the mirror part and for the corrector (see above), we get the Seidel coefficients for the complete Maksutov-Cassegrain system:

$$\sum A = A_{cass} + A_{corr} \tag{21.6.16}$$

$$\sum B = B_{cass} + B_{corr} \tag{21.6.17}$$

$$\sum C = C_{cass} + C_{corr}\,. \tag{21.6.18}$$

When these values meet the criteria, the corrector can be designed. First, the power of the corrector must be determined:

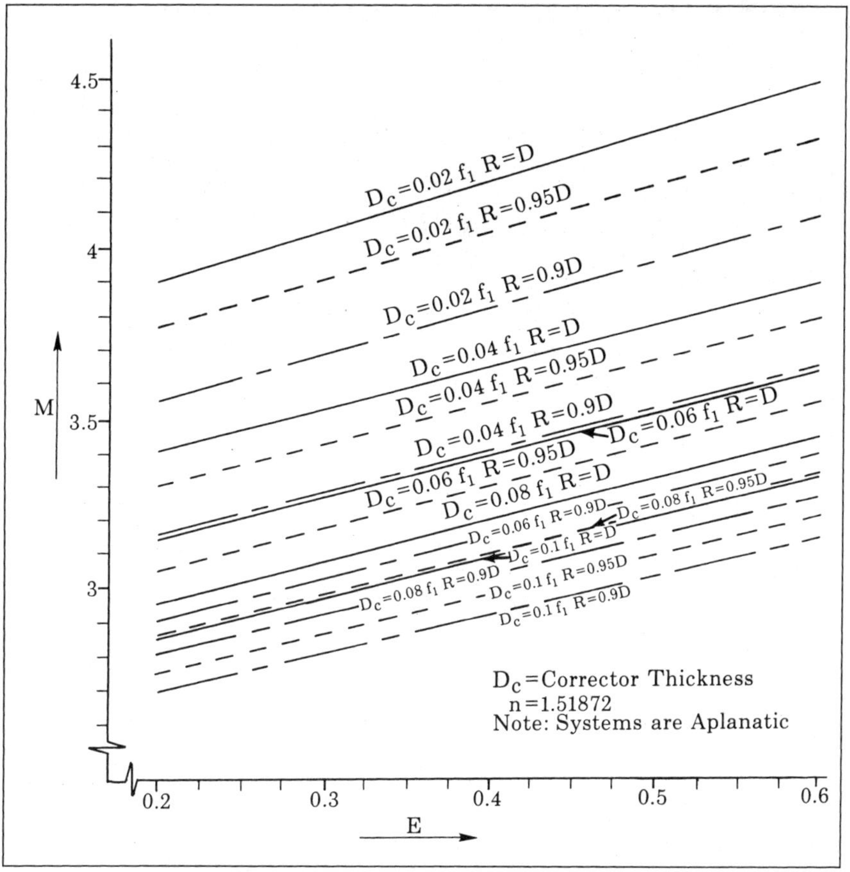

Fig. 21.8 *Parameters for Compact Maksutov-Cassegrain Telescopes.*

$$K = \frac{-4hn^2}{f_1 \cdot n_1 \cdot q_2}. \tag{21.6.19}$$

As discussed in section 10.3, to obtain a better color correction, the power of the corrector must be somewhat adjusted. The correction factor, K_c, is approximately 0.97. The curvatures of the corrector are then:

$$r_1 = \frac{2 \cdot n_3}{K \cdot K_c \cdot q_1} \tag{21.6.20}$$

$$r_2 = \frac{2 \cdot n_3}{K \cdot K_c \cdot q_3}. \tag{21.6.21}$$

When all surfaces are held spherical, coma can be corrected as well, but only under certain conditions. Fig. 21.8 shows how *M* varies relative to *E*, with various corrector thicknesses and distances, for aplanatic compact designs. As we saw in

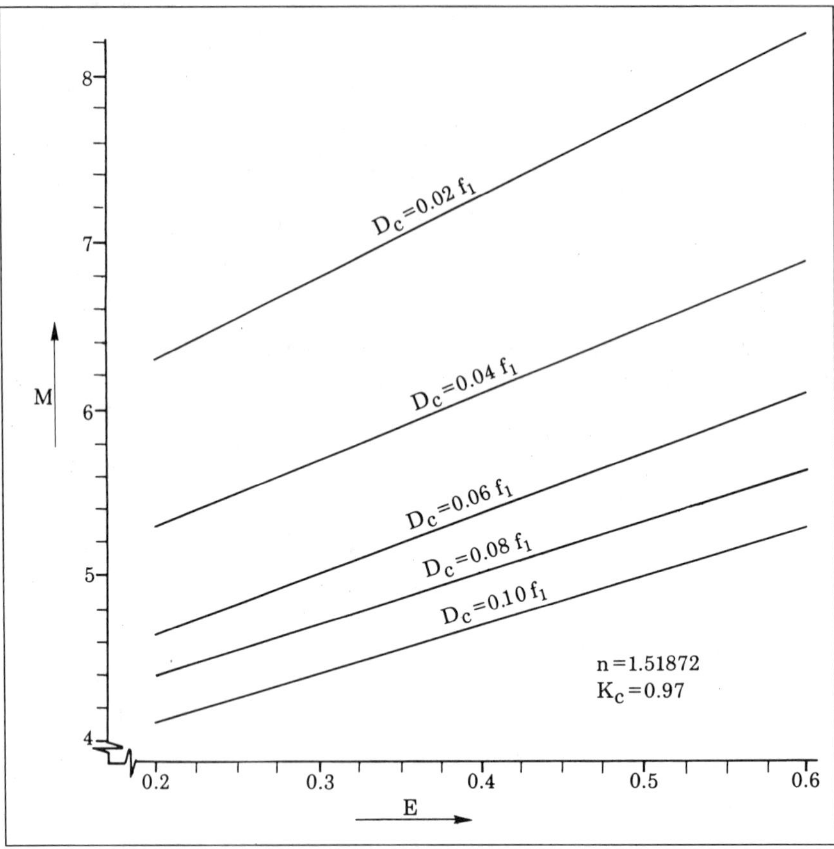

Fig. 21.9 *Aluminized-Spot Secondary Mirror Maksutov-Cassegrain Designs.*

chapter 10, the degree of spherical aberration in a Maksutov system depends rather strongly on the thickness of the corrector D_c. The value of M also depends on that thickness. For instance, for $D_c = 0.02\,f_1$, the value of M is in the range 3.5 to 4.0, while for $D_c = 0.1\,f_1$, M is approximately 2.7. For thin correctors the influence of D/R is larger than for thick correctors.

For a given configuration outside fig. 21.8, coma as well as astigmatism can be corrected by aspherizing one of the surfaces of the mirrors or by moving the corrector farther away from the primary (the latter method results in a non-compact design).

In designs with an aluminized-spot secondary mirror, the second curvature of the corrector must be equal to the curvature of the secondary mirror. This places an extra constraint on the design, and the designer is thus deprived of an important degree of freedom. With this system, as we saw in chapter 11, spherical aberration can be corrected, but coma and astigmatism remain present.

In principle spherical surfaces are used. From an analysis it appears that for

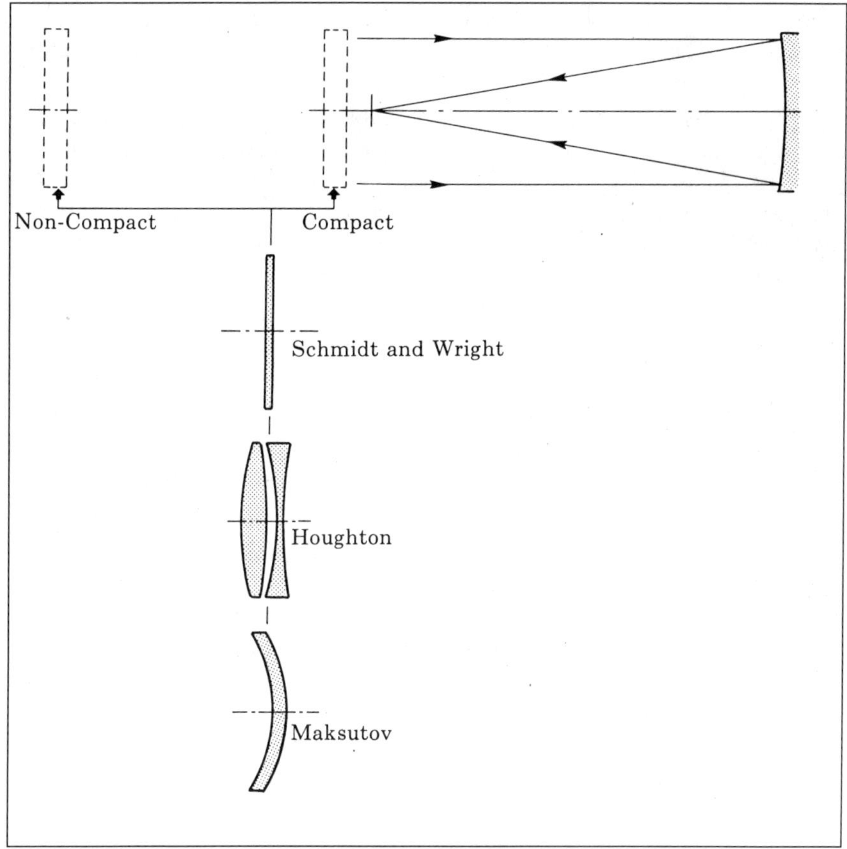

Fig. 21.10 *Compact and Non-Compact Single-Mirror Catadioptrics.*

any given combination of system parameters (the focal length of the primary, the distance of focus behind the primary, the axial thickness of the corrector, the design refractive index, and the extra color correction factor), there is only one secondary magnification for which spherical aberration can be corrected. (In systems with separate secondaries, this is not the case.)

The design proceeds as follows. First, a secondary magnification is chosen, and the system is optimized until spherical aberration is at a minimum. Note that the distance from the corrector to the primary mirror must be the same as the distance between the mirrors. The predesign is correct once the curvatures calculated for the secondary mirror and for the back side of the corrector are equal. If the mirror curvature is larger, the value chosen for the secondary magnification was too small; pick a new value and repeat the calculation.

Once the correct secondary magnification has been found, further optimization by skew-ray tracing is necessary. This is more complicated than for systems with a separate secondary because the curvature of the secondary mirror and the curvature of the back side of the corrector, which must be equal, are changing at the same time. Note that when the back side curvature is altered, the front side curvature must be recalculated.

Fig. 21.9 shows how the values of the secondary magnification vary depending on the back focus distance, with the corrector thickness, D_c, as an additional parameter. Fixed values are the design refractive index, 1.51872 (BK7 for 546.1 nm), and the color correction factor, 0.97. Both the degree of correction of spherical aberration and particularly the secondary magnification, M, depend strongly on the thickness of the corrector. For instance, for a typical value, $E = 0.4$, for corrector thicknesses D_c:

$$D_c = 0.02f_1 \quad : M = 7.3$$
$$D_c = 0.1f_1 \quad : M = 4.7$$

21.7 Designing Single-Mirror Catadioptrics (Astrocameras)

Various combinations of a single mirror and a corrector were treated in chapters 8 (Schmidt), 10 (Maksutov) and 13 (Wright and Houghton). These systems are normally used for photographic applications where the image lies between the corrector and the mirror. With these systems, it is possible to achieve good sharpness over a large, more or less curved field with a minimum number of optical components (fig. 21.10).

When a much larger central obstruction, compared with the image diameter, is allowed, it is possible to bring the image outside the tube by using a diagonal. In the following sections, the design procedure of these single-mirror catadioptrics will be discussed.

Seidel coefficients of a spherical mirror are:

Spherical Aberration	A=1
Coma	B=2
Astigmatism	C=4

In all cases, for a single-mirror catadioptric, spherical aberration must be corrected, that is, the Seidel coefficient for spherical aberration must be zero ($A = 0$). For an astrocamera, coma must be corrected as well, so $B = 0$.

21.8 Designing Schmidt and Wright Cameras

Schmidt and Wright cameras are aplanatic systems, that is, both spherical aberration and coma are corrected. The Schmidt camera, with its corrector at the center of curvature of a spherical mirror, is also anastigmatic. In a Wright camera the corrector is closer to the mirror, so the mirror must be aspheric in order to correct coma.

The Seidel coefficients for both cameras are:

$$A = 1 + SC - g \tag{21.8.1}$$

$$B = 2 - gD . \tag{21.8.2}$$

When the power of the corrector, g, and D are both 0, the corrector is a plane-parallel piece of glass in contact with the primary. To correct spherical aberration for this case, the mirror must be parabolic, that is, SC must be –1.

The following equations give the power of the corrector and its distance from the primary in units of the focal length of the primary for all spherical aberration- and coma-free systems:

$$SC = \frac{2}{D} - 1 \tag{21.8.3}$$

$$g = \frac{2}{D} . \tag{21.8.4}$$

Astigmatism can be corrected only when the corrector is located at the center of curvature of the primary, i.e., in the special case of the Schmidt camera, where $D = 2$. In the case of the Schmidt-Newtonian, with a spherical mirror and the corrector displaced from the center of the curvature, coma is present.

21.9 Designing a Houghton Camera

For the Houghton camera the mirror is always spherical. To design a Houghton camera, a method of calculation similar to that for the Schmidt and Wright cameras can be used. First, the position of the corrector must be given. Spherical aberration and coma can be corrected for every given position of the corrector. Astigmatism can only be corrected when the corrector is near the center of curvature of the primary.

Using the same symbols as we used as in section 21.5., the Seidel coefficients for the Houghton camera are:

$$A_{corr} = \frac{2}{L^3(n-1)^2}\left\{\frac{n+2}{2n} \cdot 4Q - 2Q(2n+1)\right\} \tag{21.9.1}$$

$$B_{corr} = \frac{-4Q(n+1)}{L^2(n-1)n} + D \cdot A_{corr} . \tag{21.9.2}$$

Calculate these auxiliary quantities:

$$A_1 = \frac{n+2}{n(n-1)^2} \tag{21.9.3}$$

$$A_2 = \frac{2(2n+1)}{(n-1)^2} \tag{21.9.4}$$

$$A_3 = \frac{2(n+1)}{n(n-1)}. \quad (21.9.5)$$

Then find the following:

$$L = \frac{(D-2)(2A_1 - A_2)}{A_3} \quad (21.9.6)$$

$$Q = \frac{(2-D) \cdot L^2}{2 \cdot A_3}. \quad (21.9.7)$$

The radii of curvature of the first lens are:

$$r_1 = \frac{2 \cdot f_1 \cdot L(n-1)}{Q+1} \quad (21.9.8)$$

$$r_2 = \frac{2 \cdot f_1 \cdot L(n-1)}{Q-1}. \quad (21.9.9)$$

Since the curvatures of the two lenses are equal but opposite, the radii of the second lens are $r_3 = -r_1$ and $r_4 = -r_2$.

21.10 Designing a Maksutov Camera

The mirror of the Maksutov camera is spherical, and the meniscus corrector lies nearer the mirror than the corrector in a Schmidt camera. Spherical aberration can be corrected for every position of the corrector. Coma can only be corrected for one corrector location; this depends on the thickness and refractive index of the corrector.

The method of derivation of the corrector shape is identical to that used for the Maksutov-Cassegrain (section 21.6). However, the magnitude of the aberrations that must be corrected in a single-mirror Maksutov telescope is different from that in a Makustov-Cassegrain. As we saw in section 21.7, for a single spherical mirror, the Seidel coefficients are $A = 1$, $B = 2$, and $C = 4$. The Seidel coefficient for the corrector should have the opposite sign. For example, a visual instrument should have $A_{corr} = -1$, and an astrocamera should have $B_{corr} = -2$.

As we showed in chapter 10, as the corrector thickness increases, it becomes less strongly curved, and the distance between the corrector and primary at which coma is corrected decreases. Furthermore, the residual aberrations also decrease. As we saw for a 200 mm *f*/3 Maksutov camera, a corrector thickness greater than 30 mm is impractical. However, the correction of astigmatism requires a thick corrector.

21.11 The Shape of the Schmidt Corrector

In sections 21.4 (Designing a Schmidt-Cassegrain) and 21.8 (Designing Schmidt and Wright Cameras) we employed third-order predesign methods, for which only the overall power of the Schmidt corrector was required. After the data of the predesign have been determined, the predesign must be optimized by means of ray tracing (section 21.12), and to do this, we need to determine the exact shape of the corrector.

The equations given below are valid for the Schmidt camera, the Schmidt-Cassegrain, and the Wright camera. The relative power factor g (defined in section 20.3.2) comes into play. In the Schmidt camera, it is 1.0 by definition; in a Wright camera this factor will be larger than one, and in most Schmidt-Cassegrain systems, smaller than one.

In section 8.3, we pointed out that the equation of the corrector surface may consist of two, three, or even more terms. For slow systems, two terms often suffice to describe the surface adequately, but for fast systems, the third term must also be taken into account.

The coefficients in the terms depend on the position of the neutral zone, the radius of curvature of the mirror, and the refractive index of the glass. The shape of the Schmidt profile must be chosen so that the optical path length from the entrance pupil to the focus is the same for all zones. The constants can be derived by using a paraxial ray and a second ray through the neutral zone.

We will use the following quantities:

h_n = radius of the neutral zone

D = diameter of the corrector (entrance pupil)

NZ = relative position of the neutral zone (= $2h_n/D$)

r = radius of curvature of the mirror

n = refractive index for the design color

z = deformation.

To reduce the complexity of the formulae, we introduce the following auxiliary quantities, based on the derivation given in ref. 21.4:

$$h_p = \frac{h_n}{r} \tag{21.11.1}$$

$$P_h = 2 \cdot h_p \cdot \sqrt{1 - h_p^2} \tag{21.11.2}$$

$$f = \frac{h_n}{P_h} \tag{21.11.3}$$

$$a = r\left(1 - \frac{1}{2\sqrt{1 - h_p^2}}\right) \tag{21.11.4}$$

$$b = r - a \tag{21.11.5}$$

$$c = r\sqrt{1 - h_p^2} \tag{21.11.6}$$

$$x = \frac{(r + a - b - c)}{1 - n}. \tag{21.11.7}$$

The two-term formula for the Schmidt shape given in section 8.3 applies when the neutral zone lies at 86.6% of the radius ($NZ = 0.866$), because the color aberration is then reduced to a minimum. For an arbitrary position of the neutral zone, the two-term equations are:

$$z = Ah^2 + Bh^4 \tag{21.11.8}$$

$$A = \frac{(NZ)^2 \cdot D^2}{8(n-1)r^3} \tag{21.11.9}$$

$$B = \frac{-1}{4(n-1)r^3}. \tag{21.11.10}$$

These equations are used in the ray-trace program that is an option with this book.

Often the maker of a Schmidt corrector wishes to minimize the volume of glass that must be removed. This is achieved when the edge thickness of the corrector is the same as the center thickness or the thickness at the edge of a hole made for the secondary support. For this case, we calculate the position of the neutral zone as follows:

$$NZ = \sqrt{2 \times \frac{H_2^2 + H_1^2}{D^2}}$$

where H_1 is the minimum radius, either zero for the center or a value for the radius of the hole, and H_2 is the radius of the edge of the corrector.

The three-term equations of surface are:

$$z = Ah^2 + Bh^4 + Ch^6 \tag{21.11.11}$$

$$A = \frac{f - \frac{r}{2}}{f \cdot r(n-1)} \tag{21.11.12}$$

$$B = \frac{3x - 2h_n^2 \cdot A}{h_n^4} \tag{21.11.13}$$

$$C = \frac{h_n^2 \cdot A - 2x}{h_n^6}. \tag{21.11.14}$$

When a corrector has a relative power of g, then the coefficients of the surface are:

$$A = A \cdot g; \; B = B \cdot g \text{ and } C = C \cdot g. \qquad \textbf{(21.11.15)}$$

Please note that the values of *A*, *B* and *C* are NOT programmed into our ray-trace program. To use three-term instead of two-term surface equations, the designer must calculate the values of *A*, *B*, and *C* separately and insert them into the higher-order surface equation as aspheric coefficients.

21.12 Optimization Techniques

Despite their sometimes complicated appearance, the equations given in the previous sections, indicating the values of the corrector shapes or asphericities of the mirrors, are all third-order approximations. The primary advantage to using these formulae is that the method of design is straightforward, and little design experience is required. Although the formulae will lead to solutions that are close to the final design, it is usually necessary to optimize the systems by means of an exact ray trace, minimizing residual aberrations as much as possible.

Only for two-mirror Cassegrains is a ray trace optimization not necessary, because the third-order equations given in section 21.2 give designs that are very close to the final design. Nevertheless, an optimization procedure similar to the method for the catadioptric systems, described below, can be used. Unlike the straightforward calculations done in predesign, final design optimization involves some trial and error. During this work, the designer will gain a sense of the influence of small changes of optical parameters on the residual aberrations.

As we pointed out in section 21.3, the design formulae for catadioptric systems ignore the optical power of the corrector; only its corrective action is taken into account. When a real corrector element is introduced, the focal position and the focal length of the mirror system are changed, and residual aberrations appear.

The ray-tracing program first prompts the user for the data obtained from the predesign program. Next, the designer makes minor changes in the optical parameters to restore the original focal position. This is done by moving the corrector and secondary mirror combination, or, in the case of two-mirror systems, by changing the radius of the secondary mirror.

Because changes in the power of the corrector will result in a shift of the focal plane until the final design has been obtained, it is useful to define some tolerance for the position of the focal plane.

Now the process of optimization begins. It consists of finding a particular combination of two or three parameters—the shape of the corrector, the position of the corrector, and (if there is an aspheric mirror surface) the degree of asphericity—for which axial and off-axis aberrations are smallest.

During optimization, the designer should keep certain ground rules in mind, rather than changing every parameter in sight. In Schmidt correctors, for example, the paraxial power depends on the position of the neutral zone. A Schmidt corrector with its neutral zone at 86.6% of the radius has the minimum color aberration

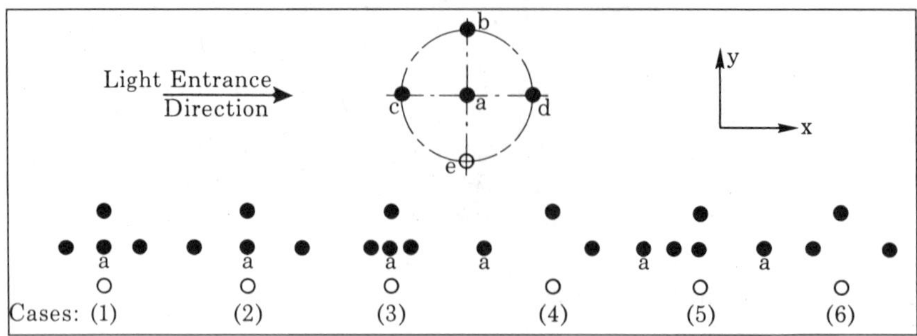

Fig. 21.11 *Optimization Procedure for an Optical System.*

possible. Other positions of the neutral zone may be chosen, of course, but they will result in greater color aberration. While optimizing the shape of a Maksutov corrector, the designer must remember to alter both corrector radii so that the corrector always satisfies the achromatization formula given in section 10.3.

Optimizing a system such as the Houghton is more difficult than optimizing a Schmidt or Maksutov system because there are more degrees of freedom. Start by changing the radii of curvature of both lenses in pairs at the same rate, and exhaust the possibilities before changing them independently. The distance between the lenses is another free parameter to test for its effect, but such changes should be made systematically. Only when color aberration cannot be suppressed sufficiently with lenses made of the same glass, should the designer resort to two different glasses with the same design index but different dispersion.

Spherical aberration must be eliminated first. Unless the axial spot diagram is smaller than the Airy disk, the system will never have value as an astronomical telescope. Next, trace an oblique beam consisting of five rays through the system, as shown in fig. 21.11. The tilt angle of this beam should be defined in the x-direction.

To trace this beam, use a square regular distribution with an intercept of 0.5 × D (see section 22.4.1). The principal ray will also be traced, and is shown as ray a. These rays will form a pattern in the focal plane that is symmetric with respect to the line c–d. The y-coordinates of the points c, a, and d will always be zero, and the y-coordinates of the points b and e will be of equal value but opposite sign. The coordinates of the five resulting intersection points help the designer decide what residual aberrations are present. Please refer to fig. 21.11 for the following cases.

In case 1,

$$x_c - x_a = x_a - x_d = y_a - y_b$$

and:

$$x_a = x_b .$$

At first sight, it appears that the image is simply out of focus, that is, that field curvature is present. However, this configuration may also indicate astigmatism if the intersecting plane happens to lie exactly midway between the tangential and sagittal focal surfaces. Consult fig. 4.12, which shows how curvature of field occurs.

If the intersection points remain on the same side of the principal ray as they do in the aperture pupil, then pure field curvature is present; if they are reversed, then astigmatism is present. If *b* and *e* are interchanged while *c* and *d* are not, then the sagittal focal surface is closer to the last optical surface. If *c* and *d* are interchanged, but *b* and *e* are not, then the tangential focal surface is closer to the last optical surface.

For cases 2 and 3:

$$x_c - x_a = x_a - x_d$$

and:

$$x_a = x_b \, .$$

This is pure astigmatism. These are similar to case 1, with the difference that the interception plane is not exactly midway between the tangential and sagittal focal surfaces. Refer to fig. 4.9, on astigmatism.

In case 4:

$$x_b - x_c = y_b - y_a$$

and *c* and *d* coincide.

This configuration indicates pure coma. When *c* and *d* do not coincide and/or the other conditions are not complied with, then curvature of field and/or astigmatism may also be present. Refer to fig. 4.7 for more about coma. Lastly, cases 5 and 6 demonstrate two of a very large number of possibilities that indicate a combination of aberrations is present.

The trial and error optimization procedure is continued until the optimum values of the power and position of the corrector and the asphericities of the mirrors, if any, are found. As more experience is gained, the process of optimization can be done in a shorter period of time.

21.13 Designing a Two-Element Achromatic Refractor Objective

21.13.1 Introduction

An achromatic refractor objective is somewhat more complicated to design than the previously discussed designs. Not only must the monochromatic aberrations be corrected, but also considerable attention must be devoted to correcting chromatic aberration.

For a given aperture and focal ratio, the designer has control over the kinds of glass used, the curvatures of the surfaces, the axial thicknesses of the lens elements, the width of the air gap, and whether the positive lens or negative lens will be placed in front. With these parameters, it is possible to correct longitudinal chromatic aberration, spherical aberration, coma, and spherochromatism in an achromatic doublet, as we saw in chapter 6. We also saw that astigmatism and curvature of field cannot be corrected in a doublet, but that over the 1- to 2-degree image angles normally used, these aberrations are not detrimental. Lateral color and distortion are very small, and need no separate correction.

It is possible to write an entirely automatic design program for an achromatic refractor objective, one that is small and simple enough to run on a personal computer. In such a program, the designer need only enter the aperture, desired focal ratio, and the kinds of glass available, and the program will design the system. The problem with such a program is that it does not give a beginner designer any insight into the effects of variations in various parameters on the image quality.

The method we describe allows the designer to control the computer and to remain a participant in the design process. Within seconds of changing a parameter, for example, increasing or decreasing a radius of curvature, the computer displays the results of the change. The designer rapidly gains a feel for the design, and an understanding of the sensitivity of each parameter of the design, while the tiresome calculations are done by the computer.

21.13.2 Doublet Design Procedure

The design of an achromatic refractor doublet is a trial and error process. The designer varies the available degrees of freedom until the aberrations have been reduced to an acceptable level.

It is important to work systematically, and to vary only one parameter at a time. In order to keep a good overview of the operations done, it is necessary to write down the main results on paper, or use a printer to keep a record of changes.

The design procedure includes four steps. These are:

1. Selecting the kinds of glass,
2. Correcting spherical aberration,
3. Correcting coma,
4. Minimizing the spherochromatism.

In practice, steps 2, 3, and 4 are carried out partly at the same time. For the sake of clarity, however, we will discuss them separately. If one or more of the above mentioned aberrations cannot be reduced to an acceptable level, the designer returns to the first step and selects another combination of glasses.

21.13.3 Achromatizing a Doublet Lens

Before discussing the factors that determine the color aberration of a doublet, we must first determine the axial longitudinal color aberration of a single positive lens. In fig. 21.12, we show the foci of a single lens in blue, green, and red light. We begin by asking: What is the difference, $(f_R - f_B)$, between the foci of a thin lens in red and blue light? (Note that subscript *R* means red, *B* means blue, and *G*

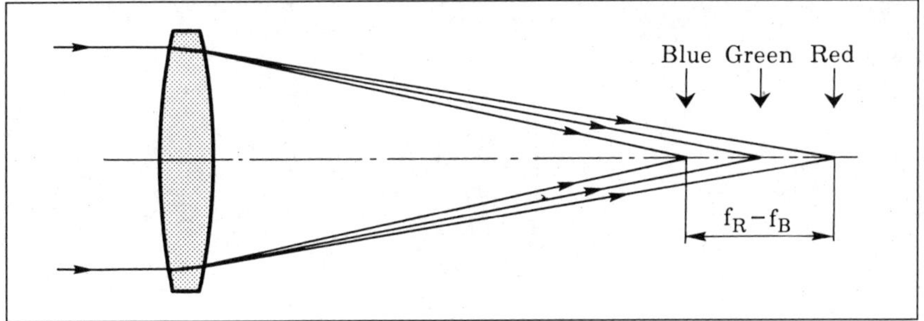

Fig. 21.12 *Longitudinal Chromatic Aberration for a Simple Lens.*

means green light.)

The following equation is valid for a thin lens:

$$\frac{1}{f} = (n-1)\left(\frac{1}{R_1} - \frac{1}{R_2}\right). \tag{21.13.1}$$

Thus:

$$\frac{1}{f_B} - \frac{1}{f_R} = (n_B - n_R)\left(\frac{1}{R_1} - \frac{1}{R_2}\right)$$

and therefore:

$$\frac{f_R - f_B}{f_B \cdot f_R} = (n_B - n_R)\left(\frac{1}{R_1} - \frac{1}{R_2}\right).$$

Now, since $(f_R - f_B)$ is small in comparison with f_G, we see that $f_B \cdot f_R \approx f_G^2$, so that:

$$\frac{f_R - f_B}{f_G^2} = (n_B - n_R)\left(\frac{1}{R_1} - \frac{1}{R_2}\right).$$

Then, because:

$$\frac{1}{f_G} = (n_G - 1)\left(\frac{1}{R_1} - \frac{1}{R_2}\right)$$

by substitution, we derive the following:

$$\frac{f_R - f_B}{f_G} = \frac{n_B - n_R}{n_G - 1}$$

which simplifies to:

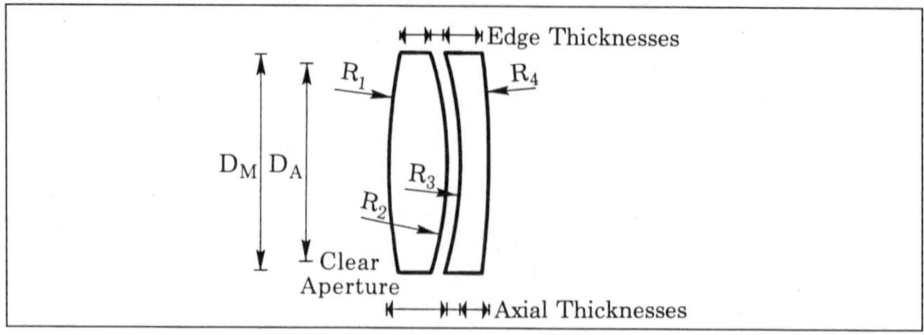

Fig. 21.13 *Dimensions Characterizing an Achromatic Doublet.*

$$f_R - f_B = \frac{f_G}{\dfrac{n_G - 1}{n_B - n_R}}.$$

The denominator in the term on the right is called the dispersion number, or Abbe number.

This means that the focal difference between red and blue for a single positive lens equals the focal distance divided by the dispersion number. This also implies that the axial color aberration will be small when the dispersion number is high.

The derivation of an expression for the color aberration of an achromatic doublet is somewhat more complicated, but is interesting because the expression yields insight into the factors that control this aberration. It is difficult, for instance, to see why a fluorite objective can have much better color correction than a doublet with conventional crown and flint glasses without understanding the derivation of these expressions. Fig. 21.13 shows the dimensions used to characterize an achromatic doublet. The quantities used in the derivation for color aberration are:

- R_1, R_2, R_3, and R_4 are the radii of curvature of the optical surfaces;
- n_{G1}, n_{B1}, n_{R1}, n_{G2}, n_{B2}, and n_{R2} are the refractive indices for the first (positive) and the second (negative) lenses in green, blue, and red light, respectively;
- f_{G1}, f_{B1}, f_{R1}, f_{G2}, f_{B2}, and f_{R2} are the focal lengths of the first (positive) and the second (negative) lenses in green, blue, and red light, respectively;
- $f_{G\,\text{comb}}$, $f_{B\,\text{comb}}$, and $f_{R\,\text{comb}}$ are the focal lengths for the combined system for green, blue, and red light, respectively;
- and V_1 and V_2 are the dispersion numbers for the first and second lenses.

Since the focal length of a combination of thin lenses close together is defined by the equation:

$$\frac{1}{f_{\text{comb}}} = \frac{1}{f_1} + \frac{1}{f_2} = (n_1 - 1)\left(\frac{1}{R_1} - \frac{1}{R_2}\right) + (n_2 - 1)\left(\frac{1}{R_3} - \frac{1}{R_4}\right)$$

the focal lengths of the doublet for blue and red light are:

$$\frac{1}{f_{B_{\text{comb}}}} = (n_{B_1} - 1)\left(\frac{1}{R_1} - \frac{1}{R_2}\right) + (n_{B_2} - 1)\left(\frac{1}{R_3} - \frac{1}{R_4}\right) \quad \textbf{(21.13.2)}$$

$$\frac{1}{f_{R_{\text{comb}}}} = (n_{R_1} - 1)\left(\frac{1}{R_1} - \frac{1}{R_2}\right) + (n_{R_2} - 1)\left(\frac{1}{R_3} - \frac{1}{R_4}\right). \quad \textbf{(21.13.3)}$$

To make an achromatic doublet for visual use, the focal lengths in red and blue light must be equal, and their reciprocals must also be equal. Combining eqs. 21.13.2 and 21.13.3:

$$(n_{B_1} - 1)\left(\frac{1}{R_1} - \frac{1}{R_2}\right) + (n_{B_2} - 1)\left(\frac{1}{R_3} - \frac{1}{R_4}\right)$$
$$= (n_{R_1} - 1)\left(\frac{1}{R_1} - \frac{1}{R_2}\right) + (n_{R_2} - 1)\left(\frac{1}{R_3} - \frac{1}{R_4}\right)$$

and simplifying:

$$\left(\frac{1}{R_1} - \frac{1}{R_2}\right)(n_{B_1} - n_{R_1}) + \left(\frac{1}{R_3} - \frac{1}{R_4}\right)(n_{B_2} - n_{R_2}) = 0. \quad \textbf{(21.13.4)}$$

Substituting dispersion numbers into eq. 21.13.4, we obtain:

$$\frac{1}{f_{G_1} \cdot V_1} + \frac{1}{f_{G_2} \cdot V_2} = 0$$

thus:

$$\frac{-f_{G_2}}{f_{G_1}} = \frac{V_1}{V_2}. \quad \textbf{(21.13.5)}$$

This means that the focal lengths of the components of an achromatic doublet are inversely proportional to their dispersion numbers. Consider the Fraunhofer doublet discussed in section 6.3, with a dispersion number for the positive element of 64.4, and for the negative element of 37.27. In order to obtain the same focal length for blue and red, the power of the positive element must be equal to (–64.4 / 37.27), or –1.73 times the power of the negative element. The powers of the individual elements are calculated from the total focal length of the system, $f_{G\,\text{comb}}$. Thus from eq. 21.13.5 and the expression for two thin lenses in contact, we find the powers of the elements in green light:

$$\frac{1}{f_{G_1}} = \frac{1}{f_{G_{comb}}} \cdot \left(\frac{V_1}{V_1 - V_2}\right) \tag{21.13.6}$$

$$\frac{1}{f_{G_2}} = \frac{1}{f_{G_{comb}}} \cdot \left(\frac{V_2}{V_2 - V_1}\right). \tag{21.13.7}$$

These are called the thin lens achromatization formulae. For example, given the 200 mm *f*/15 doublet described in section 6.3, we find these focal lengths for the front and rear elements:

$$\frac{1}{f_{G_1}} = \frac{1}{3000} \cdot \frac{64.4}{64.4 - 37.27} = \frac{1}{1264} \quad f_{G_1} = 1264\text{mm}$$

$$\frac{1}{f_{G_2}} = \frac{1}{3000} \cdot \frac{37.27}{37.27 - 64.4} = -\frac{1}{2184} \quad f_{G_2} = -2184\text{mm}.$$

We can also calculate the secondary spectrum of a doublet. This is the difference between the focal length of the doublet in green light and the focal length in red and blue, i.e., $(f_{RBcomb} - f_{Gcomb})$. We begin by deriving expressions for the powers of the individual elements:

$$\left.\begin{aligned} \frac{1}{f_{G_1}} &= \frac{1}{f_{G_{comb}}} \cdot \frac{V_1}{V_1 - V_2} \\ \frac{1}{f_{G_1}} &= (n_{G_1} - 1)\left(\frac{1}{R_1} - \frac{1}{R_2}\right) \\ V_1 &= \frac{n_{G_1} - 1}{n_{B_1} - n_{R_1}} \end{aligned}\right\}$$

$$\left(\frac{1}{R_1} - \frac{1}{R_2}\right) = \frac{1}{f_{G_{comb}}(V_1 - V_2)(n_{B_1} - n_{R_1})} \tag{21.13.8}$$

$$\left(\frac{1}{R_3} - \frac{1}{R_4}\right) = \frac{1}{f_{G_{comb}}(V_2 - V_1)(n_{B_2} - n_{R_2})}. \tag{21.13.9}$$

Next, we find the reciprocal focal lengths in green and in the combined red and blue light:

$$\frac{1}{f_{G_{comb}}} = \frac{1}{f_{G_1}} + \frac{1}{f_{G_2}} = (n_{G_1} - 1)\left(\frac{1}{R_1} - \frac{1}{R_2}\right) + (n_{G_2} - 1)\left(\frac{1}{R_3} - \frac{1}{R_4}\right)$$

$$\frac{1}{f_{RB_{\text{combo}}}} = \frac{1}{f_{RB_1}} + \frac{1}{f_{RB_2}} = (n_{B_1} - 1)\left(\frac{1}{R_1} - \frac{1}{R_2}\right) + (n_{B_2} - 1)\left(\frac{1}{R_3} - \frac{1}{R_4}\right).$$

From this we derive an expression for the reciprocal of the secondary spectrum, that is, the focus difference between green and the combined red and blue focal length:

$$\frac{1}{f_{RB_{\text{comb}}}} - \frac{1}{f_{G_{\text{comb}}}} = \frac{f_{G_{\text{comb}}} - f_{BR_{\text{comb}}}}{f_{G_{\text{comb}}} \cdot f_{BR_{\text{comb}}}} = \tag{21.13.10}$$

$$(n_{B_1} - n_{G_1})\left(\frac{1}{R_1} - \frac{1}{R_2}\right) + \quad (n_{B_2} - n_{G_2})\left(\frac{1}{R_3} - \frac{1}{R_4}\right).$$

Substituting eq. 21.13.8 and 21.13.9 into 21.13.10 and simplifying, we obtain a very important relationship that gives the amount of secondary spectrum of an achromatic doublet:

$$f_{G_{\text{comb}}} - f_{RB_{\text{comb}}} = \frac{\dfrac{n_{B_1} - n_{G_1}}{n_{B_1} - n_{R_1}} - \dfrac{n_{B_2} - n_{G_2}}{n_{B_2} - n_{R_2}}}{V_1 - V_2} \cdot f_{RB_{\text{comb}}} \tag{21.13.11}$$

The fractions in the numerator are the relative partial dispersions. By custom, we use the blue F-line (486.13 nm), the red C-line (656.27 nm), and the green e-line (546.07 nm) as reference wavelengths. The relative partial dispersion is expressed then as $P_{F,e}$, which is shorthand notation for:

$$P_{F,e} = \frac{n_F - n_e}{n_F - n_C}. \tag{21.13.12}$$

In the case of the classical Fraunhofer doublet given above, the partial dispersions are 0.4532 and 0.4689, and the Abbe V numbers are 64.40 and 37.27, for F and e light, respectively. Placing these values in eq. 21.13.11, we find the secondary spectrum to be:

$$\frac{\Delta f}{f} = \frac{0.4532 - 0.4689}{64.40 - 37.27} = -\frac{1}{1728}$$

or somewhat over 1/2000 of the focal length in green light.

For compactness, we rewrite eq. 21.13.11 as:

$$\Delta f = \frac{P_{F,e_1} - P_{F,e_2}}{V_1 - V_2} \cdot f. \tag{21.13.13}$$

Because of the importance of the values P and V in achromatizing doublets, optical glass manufacturers supply plots of these values, as a P–V diagram, for the optical glasses they supply.

Let us now explore how the designer can find suitable combinations of opti-

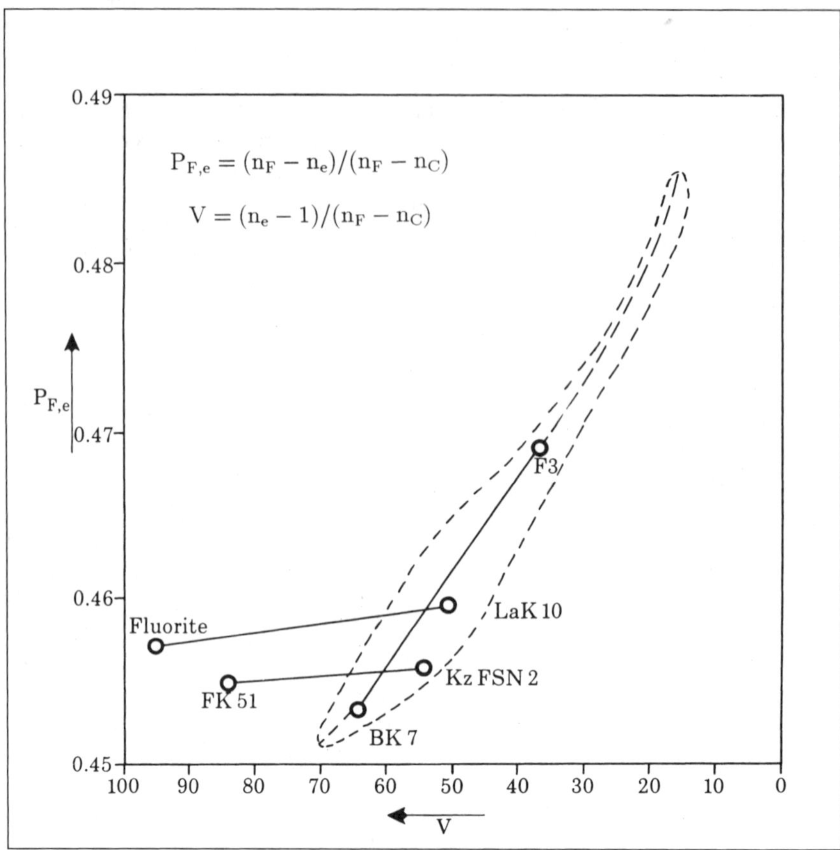

Fig. 21.14 *P-V Diagrams for Schott Glass Combinations for Some Achromatic Refractor Objectives.*

cal glasses for a doublet with the help of a *P–V* diagram. From eq. 21.13.13, it is evident that to minimize secondary spectrum, we must find a combination in which the difference between P_1 and P_2 is as small as possible and the difference between V_1 and V_2 is as large as possible.

Fig. 21.14 shows the $P_{F.e}$ versus *V* diagram for Schott glasses. The three lines connecting types of glass represent the three refractor objectives described in section 6.3, i.e., the Fraunhofer objective using BK7 and F3 glasses, the Apoklaas with FK51 and KzFSN2 glasses, and the Fluorite objective made with LaK10 and fluorite.

To minimize the secondary spectrum of the combination, the two glasses must be chosen so that the connecting line runs as much as possible horizontally, resulting in a small ΔP, and the distance between the points in the horizontal direction is as large as possible, yielding a large ΔV.

Notice that the line for the Fraunhofer objective is rather steeply inclined. The reason for this is that *P* and *V* are closely related. All "normal" glasses lying near a straight line in the P–V diagram are shown in the figure as a thick dotted line. For normal glasses it is not possible to find suitable combinations of glasses, that is, combinations for which the P-values are close together.

For certain of the modern glasses, however, the close relationship between P and V is broken to some extent. This is true especially for fluor-crown glasses that deviate quite strongly from the normal glasses. One very striking exception is fluorite, an artificial crystal. Both fluor-crown and fluorite lie far from the line of normal glasses, so that it is relatively easy to find a large ΔV for two materials.

Furthermore, it is rather easy to find a suitable negative element glass in combination with fluor-crown or fluorite, having approximately equal $P_{F,e}$-values. An additional advantage of fluor-crown and fluorite is their relatively low refractive indices. This facilitates finding a glass for the negative element with a higher refractive index. This is favorable because, as we saw in section 6.1, for easy correction of spherical aberration the negative element glass should be chosen to have a higher refractive index than the positive element.

This is absolutely necessary when a cemented doublet is being designed because spherical aberration cannot be corrected if the design indices are equal (assuming spherical lenses). For a cemented doublet, there are additional problems. Only a limited number of matched glass combinations exist for an aplanatic design (i.e., a design free of both spherical aberration and coma) with good color correction, and these can be found only by trial and error. This is a laborious process because spherochromatism remains to be corrected, and because the choice also depends on the focal ratio of the doublet. Therefore, it is not surprising that we often find cemented doublets that have not been fully corrected for coma.

The designer must decide at the outset whether he wants to design a normal achromat or an achromat with special (but expensive) glasses. As we have seen, with the normal glasses a secondary spectrum of 1/2000 f will be obtained. With fluor-crown glass, this can be diminished to 1/8000 f, and with fluorite to 1/16,000 f. The newer glasses also allow higher speed refractor objectives to be designed: with fluor-crown, *f*/10 is possible in a 200 mm refractor, and with fluorite *f*/8 is possible. This results in shorter tube lengths than conventional refractors.

Based on the foregoing discussion and on chapter 6, the following rules apply to the choice of glass:

1. The glass of the positive lens must have a higher Abbe number than the negative lens.
2. A large difference between the Abbe numbers is desirable because less strongly curved surfaces are necessary, allowing spherical aberration and coma to be more easily minimized.
3. The negative lens should preferably have a higher refractive index than the positive lens, especially when the lenses are cemented.

4. The difference of the relative partial dispersions of the two glasses should be as small as practical so that the secondary spectrum will be small.

As soon as a choice of glass has been made, the powers of the positive and negative elements can be computed using eqs. 21.13.6 and 21.13.7.

$$\text{First Lens: } \frac{1}{f_{G_1}} = \frac{1}{f_{G_{\text{comb}}}} \cdot \left(\frac{V_1}{V_1 - V_2}\right) \tag{21.13.14}$$

$$\text{Second Lens: } \frac{1}{f_{G_2}} = \frac{1}{f_{G_{\text{comb}}}} \cdot \left(\frac{V_2}{V_2 - V_1}\right). \tag{21.13.15}$$

These formulae give the powers of the two elements for which, to a first approximation, the foci for red and blue will coincide. Strictly speaking, these formulae are valid only for very thin lenses. Since they have finite thicknesses, the red and blue foci differ slightly. This can be corrected by giving the positive or the negative element a slightly higher power by means of an amplification factor. The determination of this factor is discussed later on.

The fact that lenses have finite thicknesses sometimes causes another effect: for certain combinations of glasses the final design can have unacceptably large residual spherical aberration, coma, or spherochromatism. If this happens, the designer must select another combination of glasses. The number of possible glass combinations is very large. From a selection of 30 suitable crown and 30 suitable flint glasses, there are 900 possible combinations.

A major concern of the professional designer, particularly when large series of optics must be made, is the cost of the glass and of manufacturing the lenses. In that case the designer will try to find a combination of glass types which allows the lowest cost consistent with an acceptable level of aberrations. For the amateur telescope maker, cost is often a less important factor because he will make only one objective. Therefore, the amateur can select more expensive glasses, and obtain a lower level of secondary spectrum.

There is still another difference between professional and amateur manufacturing. The amateur can remove a small amount of residual spherical aberration by aspherizing one of the surfaces. Because of the high costs of generating aspheric surfaces, the professional must nearly always specify spherical surfaces.

21.13.4 Correcting Spherical Aberration

Although we will discuss designing a Fraunhofer doublet (positive lens in front), the following method is also valid for the Steinheil form, with minor modifications. Fig. 21.13 indicates the various characteristics of the two-element achromatic doublet.

From the equations giving the power of the system as a function of the Abbe numbers and the radii:

$$\frac{1}{f_{G_1}} = \frac{1}{f_{G_{\text{comb}}}} \cdot \left(\frac{V_1}{V_1 - V_2}\right) \tag{21.13.16}$$

$$\frac{1}{f_{G_1}} = (n_{G_1} - 1)\left(\frac{1}{R_1} - \frac{1}{R_2}\right) \tag{21.13.17}$$

we find the powers of the two elements:

$$\left(\frac{1}{R_1} - \frac{1}{R_2}\right) = \frac{1}{f_{G_{\text{comb}}}(n_{G_1} - 1)} \cdot \left(\frac{V_1}{V_1 - V_2}\right) \tag{21.13.18}$$

$$\left(\frac{1}{R_3} - \frac{1}{R_4}\right) = \frac{1}{f_{G_{\text{comb}}}(n_{G_2} - 1)} \cdot \left(\frac{V_2}{V_2 - V_1}\right) \tag{21.13.19}$$

so that the second and fourth radii will be:

$$R_2 = \frac{1}{\frac{1}{R} - \frac{1}{f_{G_{\text{comb}}}(n_G - 1)} \cdot \left(\frac{V_1}{V_1 - V_2}\right)} \tag{21.13.20}$$

$$R_4 = \frac{1}{\frac{1}{R_3} - \frac{1}{f_{G_{\text{comb}}}(n_{G_2} - 1)} \cdot \left(\frac{V_2}{V_2 - V_1}\right)}. \tag{21.13.21}$$

For best results we will not use V_d as our dispersion number, but instead use:

$$V = \frac{n_e - 1}{n_F - n_C}.$$

This is because, for visual use, the minimum focal distance should be in green light at 555 nm wavelength, which corresponds to *e*-light (546.07 nm). Note that the peak spectral sensitivity of the eye shifts toward the blue in the dark-adapted eye (see fig. 6.7).

For calculating the longitudinal spherical aberration in *e*-light, the following data are necessary:

- the diameters of the elements D,
- the four radii of curvature R_1, R_2, R_3, and R_4,
- the axial lens thicknesses,
- the axial width of the air gap, and
- the refractive indices for both glasses for *e*-light.

For a cemented doublet, the width of the air gap is zero, and $R_2 = R_3$.

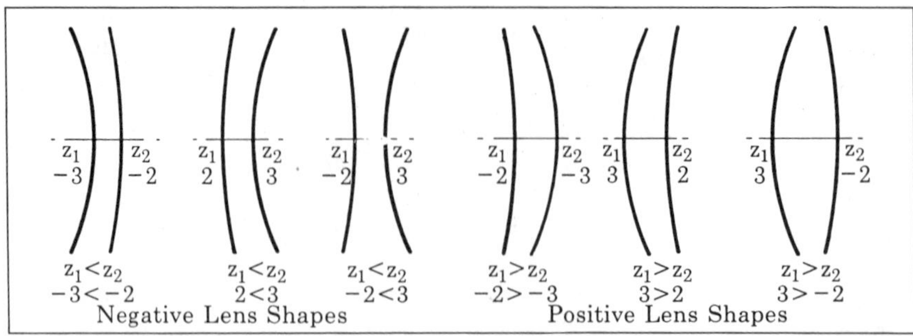

Fig. 21.15 *Lens Shape Recognition Procedure.*

It is customary to set the minimum edge thickness for the positive lens and the minimum axial thickness for the negative lens at the beginning of the design procedure. For a 200 mm lens, these will be around 15 mm. In a positive lens, the center thickness is always larger than the edge thickness. The center thickness of the positive shaped lens is calculated from the edge thickness and the sagittas of both surfaces. In computer aided design it is essential to use a procedure allowing the computer to recognize whether a lens is positive or negative based only on its curvatures, without the need to calculate its power. Lens-shape recognition is important not only for the glass lenses, but also for the shape of the airspace between them. Applying the following method will allow you to avoid the embarrassment of designing a lens with a negative edge or center thickness.

First, calculate the sagittas of both surfaces, taking the signs of the curvatures into account:

$$Z = \frac{D^2}{4R\left(1 + \sqrt{1 - \left(\frac{D}{2R}\right)^2}\right)}. \tag{21.13.22}$$

When the curvature is positive (convex to the left), then Z is positive. When the curvature is negative (concave to the left), then Z is negative. For a flat surface, $R = \infty$, so $Z = 0$.

Fig. 21.15 shows all possible cases of curvature in a lens; the first three are negative, the others positive. When the signs are taken into account, then:

- $Z_1 < Z_2$ for all negative-shape lenses,
- $Z_1 > Z_2$ for all positive-shape lenses.

When a lens is negative, its center thickness is set equal to the minimum value selected. When a lens is positive, then the center thickness is calculated from:

$$T_{\text{center}} = T_{\text{edge}} + Z_1 - Z_2. \tag{21.13.23}$$

Do not forget to use the correct signs for Z_1 and Z_2.

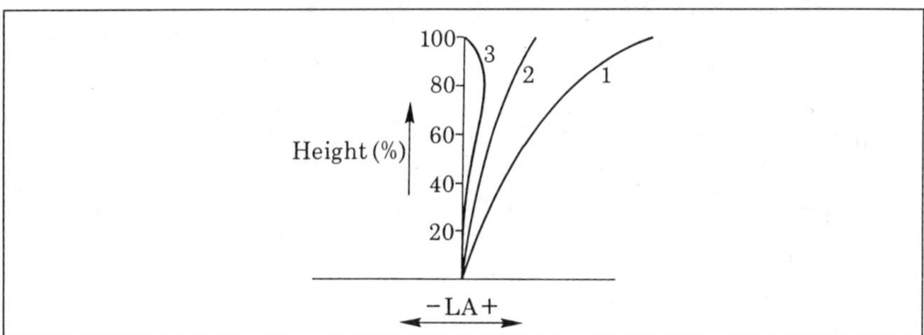

Fig. 21.16 *LA-Curves Encountered in the Design of an Achromatic Doublet.*

When the center thickness of the negative lens is set equal to the edge thickness of the positive lens, the calculation is simplified. In that particular case, the center thickness can be calculated with the following formula, whether it is a positive or negative lens:

$$T_c = T_1 - ((Z_1 > Z_2) \cdot (Z_1 - Z_2)). \qquad \textbf{(21.13.24)}$$

$(Z_1 > Z_2)$ is a Boolean expression which is evaluated as −1, if true. If the expression is false, it is evaluated as zero.

In an air-spaced doublet, the air gap between the lenses is typically set to 1 or 2 millimeters. If the shape of the air lens is positive, then the edge distance must be checked. If this is zero or negative, then the axial distance must be increased. This can also be done automatically using the procedure above.

We now determine the longitudinal spherical aberration, then proceed to minimize it by an iterative process. The initial values of the radii of curvatures are not very critical, but it is useful to study some existing designs before starting. The designs given in table 6.1. make good starting points. On the basis of these designs, we might assume an initial R_1 equal to 2/3 of the focal length of the objective. The value of R_2 follows from eq. 21.13.20. Since the starting value of R_3 is assumed to be the same as R_2, we next compute R_4 from eq. 21.13.21. Following this, we compute the longitudinal spherical aberration for various zones. In the first trial, we generally find a large amount of spherical aberration, for example, curve 1 in fig. 21.16. In order to reduce this aberration, the lenses must be "bent," that is, the radii must be changed without changing the net power of the lens.

It may be that the LA cannot be reduced sufficiently as long as the values of R_2 and R_3 are held the same; making them different is an effective tool in reducing spherical aberration. Suppose we set R_3 equal to $1.01R_2$ for a Fraunhofer design, or R_3 equal to $0.99R_2$ for a Steinheil type, then recalculate the *LA*.

Eventually an *LA* such as curve 3 in fig. 21.16 will be obtained for one particular combination of R_1 and R_3 / R_2. This must then be checked to see whether the *LA* has been reduced sufficiently, and to decide whether the shape of the *LA* curve is optimum. This point is not easy to judge.

Fig. 21.17 graphs four kinds of *LA* curves; of course, one can also get curves

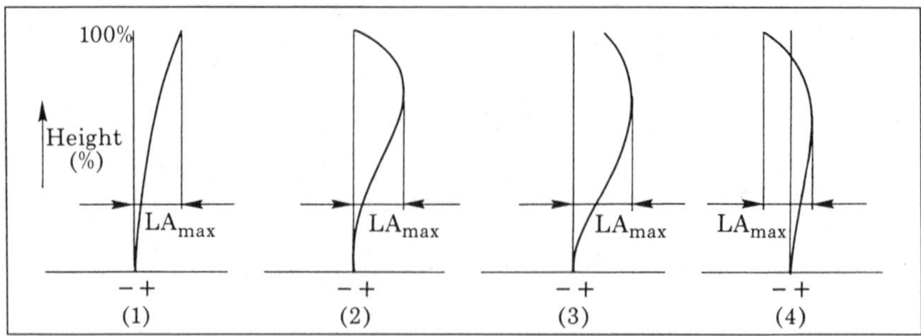

Fig. 21.17 *Types of LA-Curves.*

that are mirror images of these. The maximum deviation, LA_{max}, is the same in all four cases. Which is best? In the literature we often see *LA* curves having something similar to shape 2, the case for which the edge ray has the same focal distance as the paraxial ray. This *LA* shape offers no guarantee that the transverse spherical aberration (that is, the spread at best focus) is as small as it can be. Quite often, a curve of type 3 has a smaller transverse spherical aberration. At this point, the magnitude of the transverse spherical aberration should be calculated using the $H' \tan U'$-method, or by carrying out a complete ray-trace analysis, resulting in a spot diagram. Note that the transverse aberration should not exceed the diameter of the Airy disk for green light.

However, rather than carry out the full calculation of transverse aberrations, it is often easier to apply a simple rule-of-thumb guideline and postpone a complete ray-trace analysis until the design has been finalized. We recommend the following maximum allowable values of *LA* between the rim ray and the paraxial ray:

f/20	2.0 mm
f/15	1.2 mm
f/10	0.6 mm
f/8	0.4 mm
f/5	0.16 mm.

Although these criteria are independent of the aperture, they depend somewhat on the shape of the *LA*-curve.

21.13.5 Correcting Coma

When a design has been found for which spherical aberration is sufficiently small, we must still check that coma is low enough. Often this will not be the case because spherical aberration and coma occur independently of each other. This means that we must find other lens shapes for which both spherical aberration and coma are simultaneously corrected. Although it is possible to test for coma by calculating a complete off-axis spot diagram, during optimization it is easier to calculate the offense against the sine condition, or *OSC*, in which the magnitude of

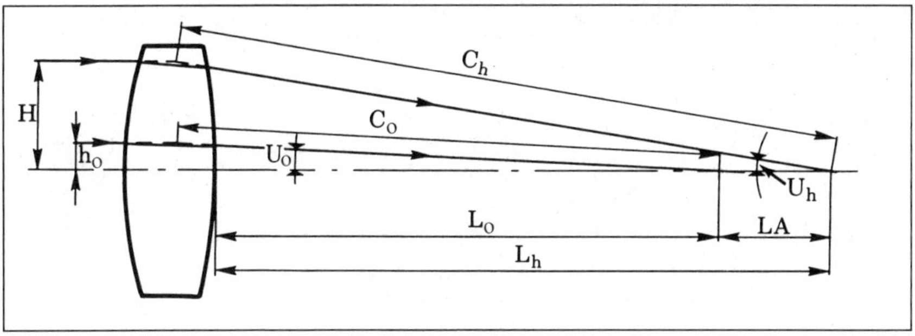

Fig. 21.18 *The Sine Condition and OSC.*

the coma present is expressed as a single number. The principle of the *OSC* was explained in section 4.2.2, and is shown in fig. 4.8.

Fig. 21.18 shows an optical system that does not meet the sine condition, and also suffers from spherical aberration. The lengths of lines C_h and C_o are:

$$C_h = \frac{H}{\sin U_h} \tag{21.13.25}$$

$$C_o = \frac{h_o}{\sin U_o}. \tag{21.13.26}$$

In these equations, h_o is the entrance height of the paraxial ray, and H belongs to a zonal ray. When the spherical aberration, *LA*, is zero, then the offense against the sine condition is:

$$OSC = \left(\frac{C_h}{C_o} - 1\right) = \left(\frac{H}{\sin U_h} \cdot \frac{\sin U_o}{h_o} - 1\right). \tag{21.13.27}$$

If the *LA* is not zero, a correction factor, L_o / L_h, is introduced, and the *OSC* is:

$$OSC = \left(\frac{\sin U_o}{\sin U_h} \cdot \frac{H}{h_o} \cdot \frac{L_o}{L_h} - 1\right). \tag{21.13.28}$$

Of course the same zones must be used to calculate the *OSC* as are used for the *LA* calculations. In the optical literature, a maximum allowable value of the *OSC* of 0.0025 is sometimes quoted. If stringent demands are made with respect to freedom from coma, however, the *OSC* must be reduced to 0.001 or less.

Although we have treated *LA* and *OSC* separately here, in `LENSDES`, our semi-automatic design program, the values of *LA* and *OSC* are calculated and displayed simultaneously on the monitor screen for various zones. Because the designer can monitor both aberrations during all stages of the design process, optimization proceeds smoothly.

If spherical aberration has been sufficiently corrected but the value of the *OSC* is still too large, change the radius of curvature of the first lens, R_1. From the

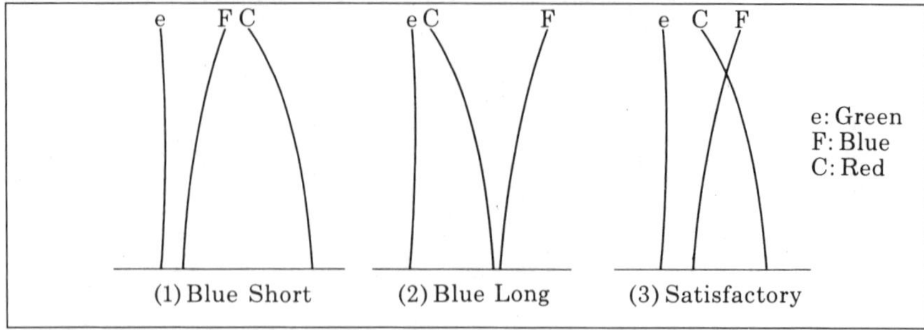

Fig. 21.19 *LA-Curves for an Achromatic Doublet.*

new values of R_1 and R_3 / R_2 the whole procedure of the *LA* and *OSC* correction, as described above, is carried out again, and repeated until *LA* and *OSC* have been reduced to an acceptable level. Once this situation has been attained, the lens is corrected for *LA* and coma in green light. This does not necessarily mean that chromatic aberration for red and blue light have yet been corrected sufficiently. This is discussed next.

21.13.6 Reducing Spherochromatism

Spherochromatism is the change of spherical aberration with wavelength. As we saw in chapter 6, an achromatic doublet that is corrected for green light will be undercorrected for red and overcorrected for blue.

Fig. 21.19 graphs a number of different possibilities. In case 1, the lens has too short a focal length in blue (F) light with respect to red; in case 2 the blue focal length is too long. Normally refractor objectives intended for visual use are corrected so that the *C* and *F* curves intersect somewhere between 50 and 100% of the semidiameter, as in case 3. The optimum position for the point of intersection of the two curves depends on the shape of the *C* and *F* curves, but it should be chosen so that the diameters of the spot diagrams for blue and red light are roughly equally large when green is at best focus.

After bending the two lenses so that *LA* and *OSC* are corrected, you will often discover that the red light and blue light curves are situated as in case 2. The reason for this is that the powers of the lenses are based on eqs. 21.13.20 and 21.13.21, which are valid only for thin lenses. Now you must move the *F* curve with respect to the C curve. This is accomplished by giving the first lens a slightly stronger power (i.e., shorter focal length), resulting in the more favorable situation shown in case 3. The "lens power" factor, *LP*, of the first lens in a Fraunhofer type doublet normally lies between 1.00 and 1.01, while that of the negative first lens of a Steinheil objective lies between 1.01 and 1.02.

With the introduction of this factor, the focal length of a Fraunhofer lens will be slightly reduced, while that of a Steinheil type will become somewhat longer. These effects are normally negligible, but if it is necessary, the original focal length can be achieved by rescaling. After the lens power factor has been introduced, the whole procedure for correcting spherical aberration and coma, and checking the position of the *F* and *C* curves must be repeated.

It is recommended that you introduce the lens power factor early in the design process, as soon as *LA* has been reduced to a reasonable value. This saves considerable time, since the intersection height of the blue and red light curves will hardly change during subsequent bendings because the color correction of a thin doublet is only slightly dependent on the state of the bending.

In summary, we have seen that for a given set of glasses, the design process works by altering only three parameters,

- the curvature ratio R_3 / R_2, which mainly influences the *LA*;
- the radius of curvature of the first surface, R_1, which mainly influences the *OSC* value; and
- the lens power factor of the front lens, LP, which influences the relative positions of the *LA* curves for red and blue light.

The designer works by manipulating these three variables until a satisfactory design solution has been achieved.

Once the designer has gained some experience, it is advisable to begin a new design with suitable starting values of R_3 / R_2 and *LP*. For a Fraunhofer objective, we suggest:

$$\frac{R_3}{R_2} = 1.02 \text{ and } LP = 1.005 \ .$$

For a Steinheil design:

$$\frac{R_3}{R_2} = 0.98 \text{ and } LP = 1.015 \ .$$

While the value of *LP* seldom deviates far from 1.0 for close elements, the value of R_3 / R_2 can deviate rather strongly in the case of extreme glass combinations.

Although the iterative design method looks cumbersome at first sight, experience shows that a final design can be obtained within twenty or thirty computer trials. Because the computer needs only a few seconds per trial, a final design can usually be achieved within an hour of starting.

We recommend that the aspiring lens designer maintain a good overview of these results by using a matrix with the variables R_1 and R_3 / R_2 as the axes. As the *LA* and *OSC* values are found, note them on the matrix. The direction toward each optimum combination is then obvious, and can be quickly found. As the designer proceeds, a second matrix with a finer grid can be used to locate the optimum more precisely.

The final design should be checked by a complete axial and off-axis ray trace for three colors.

The procedure described above is valid for a broken doublet with different inner radii. For a cemented doublet it may be impossible to obtain a low *OSC* value in combination with a low *LA* and good color correction. The designer may have to choose between accepting a lens that is not fully corrected for coma or seeking other combinations of glass.

21.14 Other Degrees of Freedom

If the designer cannot reduce the aberrations for a particular combination of glasses, a number of degrees of freedom remain. These are the thicknesses of the elements and the width of the air gap, if there is one. The air space is most useful in controlling spherical aberration. Spherochromatism can be reduced by making the air gap larger. This approach leads to a design called the Clark objective.

Another reason for selecting a larger lens separation is that the designer can achieve an aplanatic system with equal R_3 and R_2. This simplifies fabrication because the surfaces can be tested against each other by interference. If the ratio R_3 / R_2 approaches unity as the separation of the elements is increased, then this goal can be achieved.

Unfortunately, a large lens distance has some disadvantages: it is very sensitive to decentering, and lateral color may occur. For this reason, the Clark objective is not often used.

The designer may also opt for a Steinheil objective rather than a Fraunhofer. However, only in exceptional cases can a better design be achieved with the Steinheil form.

21.15 An Alternate Method of Designing a Doublet

While the iterative method of designing an achromatic doublet is reasonably fast with a computer, an alternative exists. To achromatize the system and bring the blue and red color to a single focus, the designer can use a special application of the optical path difference (*OPD*) method proposed by Conrady, and called the *D–d* method. Kingslake has used this method also, and has worked out an elegant graphical procedure for minimizing spherical aberration and coma. This method is detailed in ref. 21.4.

21.16 Designing a Three-Element Apochromatic Refractor Objective

As we have seen in previous sections, a doublet refractor objective corrected for visual use brings red (*C*-light) and blue (*F*-light) to a common focus, while the shortest focus lies in the green (*e*-light). We then say that such an objective has been corrected for two colors (fig. 6.6). Violet light still comes to focus well behind the *C* and *F* foci, so that the violet image is blurred.

The color aberration curve of a doublet (shown in fig. 6.6) can be made considerably flatter by using special glasses or fluorite. But a less expensive alterna-

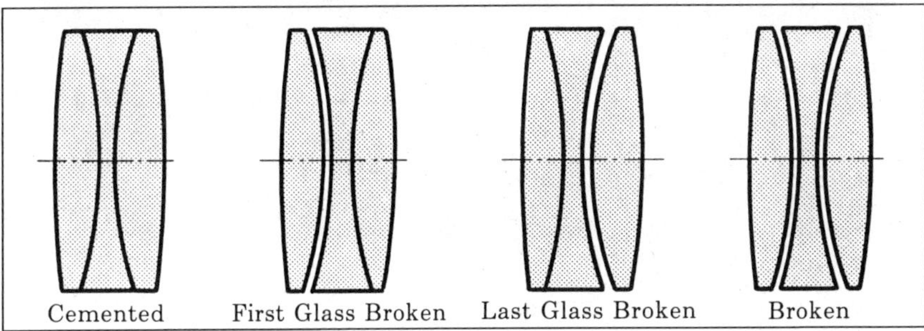

Fig. 21.20 *Types of Triplet Refractor Objectives.*

tive can be achieved when three different kinds of glass are used. Then it is possible to bring three colors, and with some combinations of glass, four colors, into common focus.

Such an objective may be suitable for photographic applications as well as visual use. Additionally, triplet refractors can be designed to work at focal ratios between *f*/6 and *f*/10, as compared with the *f*/15 focal ratio of a conventional doublet. A triplet corrected for three colors is called an apochromat, whereas a triplet corrected for four colors, according to a proposal by Herzberger, is termed a superachromat. Some triplets corrected for five colors have been designed. In this section we discuss the design of an apochromat and a superachromat; we assume that the reader is already familiar with the material we have presented on doublet objectives.

21.16.1 Choosing Glass for a Triplet

Although the sequence of the three elements of a triplet can be freely chosen, normally the system is constructed of:

- a positive biconvex element
- a negative biconcave element
- a positive biconvex element.

Such a system is shown in fig. 21.20. We will also apply this sequence in the equations.

For a good color correction, the proper glass is of the utmost importance. Finding a suitable combination is considerably more laborious and takes more time than for a doublet. In 1959, Herzberger (ref. 21.5) formulated the conditions these glasses must satisfy for systems corrected for both three and four colors. The choice is closely connected with the behavior of the dispersion of the three glasses with respect to each other. We define this dispersion with respect to green. For the blue partial dispersion:

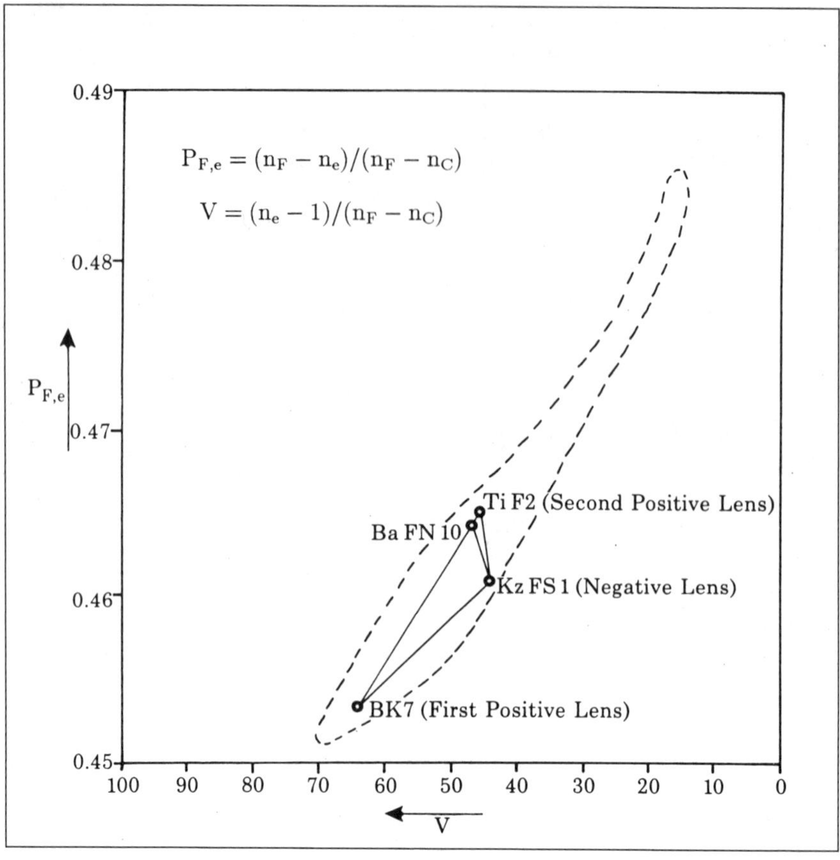

Fig. 21.21 *Two Glass Combinations for Triplet Refractor Objectives.*

$$P_{F,e} = \frac{n_F - n_e}{n_F - n_C}.$$

For the red:

$$P_{e,C} = \frac{n_e - n_C}{n_F - n_C}.$$

And for violet:

$$P_{g,e} = \frac{n_g - n_e}{n_F - n_C}.$$

These three quantities define the dispersion behavior for the full range from red to violet.

In the same way we did in fig. 21.14 for the doublet, we plot a number of glasses in the $P_{F,e} - V$ diagram. A combination of glasses suitable for a triplet will

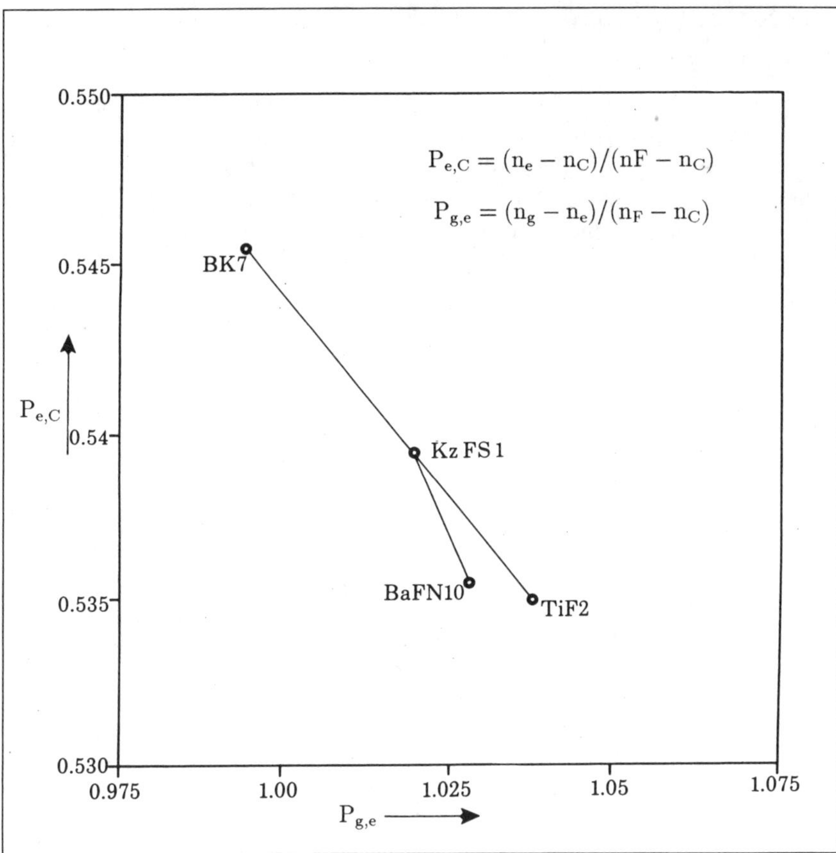

Fig. 21.22 P – P *Plots of Two Glass Combinations for Triplet Refractor Objectives.*

form a wide triangle in this particular diagram. The larger the area of this triangle is, the easier the design will be. This also facilitates the design of *f*/6 to *f*/10 triplets with low spherochromatism. Fig. 21.21 shows two of many possible combinations. The first consists of BK7, KzFS1, and BaFN10 (517642, 613443, and 670471, respectively). This combination is the same one used in the Christen triplet described in section 13.5. For the second combination, the BaFN10, a relatively cheap glass, is replaced by the considerably more expensive glass TiF2 (533460). In general a suitable combination of glasses for a triplet consists of:

- a glass with a relatively high V-number and relatively low P-number, such as a crown with an Abbe number between 60 and 70;
- a glass with a relatively low V-value and high P-value such as one of the KzFS glasses; and

- a glass with a low V-value and a P-value even higher than that of the second glass, usually a dense flint.

As we mentioned earlier, the first element is biconvex, the second biconcave, and the third element biconvex. Under no circumstances can these glasses lie on a straight line in the *P* – *V* diagram. If this were the case, the elements would have to have very high powers, which would hamper the correction of the monochromatic aberrations. In principle, there are numerous possible combinations of glass for an apochromat, that is, for an objective to be corrected for three colors, such as red, green and blue.

When we require that the lens bring four colors to a common focus (red, green, blue, and violet, for example), the three glasses must satisfy an additional condition (see figs. 21.21 and 21.22). The partial dispersion on the vertical axis is the red side value; the violet side partial dispersion is plotted on the horizontal axis. It can be shown mathematically that when four colors are to be corrected, the three glasses in the *P* – *P* diagram will lie on a straight line (ref. 21.5). This is the case shown for the combination that includes TiF2. The glass combination with BaFN10 glass does not satisfy the straight line condition; with this combination only three colors can be brought into a common focus.

Strictly speaking, these conditions are valid only for three thin lenses placed close together. It should be immediately obvious that the number of suitable combinations for a four-color triplet is lower than for a three-color triplet. Despite this, Herzberger found several hundred glass combinations for which four-color triplets could be designed.

Because these glasses are in many cases exotic, and therefore expensive, the choice of glass for a triplet is often based on economic considerations. Exotic glasses are also often unstable, or can be etched by weak acids or water. Before making a final choice, check the price of the glass, its chemical and physical properties, and the cost of grinding it.

21.16.2 The Powers of the Elements

Once a suitable glass combination has been found, the powers of the three elements are determined, then the radii of curvature calculated. To calculate the focal lengths, f_1, f_2, and f_3, we need three independent equations. The first equation concerns the combined focal length of the system, f:

$$\frac{1}{f_1} + \frac{1}{f_2} + \frac{1}{f_3} = \frac{1}{f}. \qquad \textbf{(21.16.1)}$$

The second condition to be fulfilled is that red and blue light must be brought to a common focus. As shown for a doublet in eq. 21.13.5, a triplet with the same *C* and *F* focus satisfies the following equation:

$$\frac{1}{f_1 \cdot V_1} + \frac{1}{f_2 \cdot V_2} + \frac{1}{f_3 \cdot V_3} = 0. \qquad \textbf{(21.16.2)}$$

The third condition to satisfy is that green light must have the same focus as the combined red-blue focus. In a way similar to that shown in eq. 21.13.10, for a triplet:

$$\frac{P_1}{f_1 \cdot V_1} + \frac{P_2}{f_2 \cdot V_2} + \frac{P_3}{f_3 \cdot V_3} = 0\,. \tag{21.16.3}$$

Solving these three equations for the powers of the three elements, we find:

$$\frac{1}{f_1} = \frac{1}{f} \cdot \frac{(P_2 - P_3) \cdot V_1}{V_1(P_2 - P_3) + V_2(P_3 - P_1) + V_3(P_1 - P_2)} \tag{21.16.4}$$

$$\frac{1}{f_2} = \frac{1}{f} \cdot \frac{(P_3 - P_1) \cdot V_2}{V_1(P_2 - P_3) + V_2(P_3 - P_1) + V_3(P_1 - P_2)} \tag{21.16.5}$$

$$\frac{1}{f_3} = \frac{1}{f} \cdot \frac{(P_1 - P_2) \cdot V_3}{V_1(P_2 - P_3) + V_2(P_3 - P_1) + V_3(P_1 - P_2)}\,. \tag{21.16.6}$$

For P-values, we use the blue side $P_{F,e}$ quantities, and for the V-values:

$$V = \frac{n_e - 1}{n_F - n_C}\,. \tag{21.16.7}$$

21.16.3 Designing a Triplet

We will first discuss the design procedure for a broken triplet, and follow that with the design of a cemented triplet. In designing a broken doublet, we saw that after finding the powers of both lenses, we chose a trial value for R_1 and calculated R_2 from it. R_3 follows from an initial choice of R_3 / R_2, and from R_3 we determine the value for R_4. Analysis of the *LA* and *OSC* in green light follows. For a given set of spacing, lens powers, and lens thicknesses, only one combination of R_1 and R_3 / R_2 exists for an aplanatic doublet. The system is then fixed because all available degrees of freedom have been used.

In a broken triplet, the number of degrees of freedom is larger than it is in a doublet because we can change R_5 / R_4 as well as R_1 and R_3/R_2. We also have two lens power factors, LP_1 and LP_2, for the first and the second element with respect to the third element. In the design of a broken triplet, we use the variables R_1, R_3/R_2 and R_5/R_4 to correct spherical aberration and coma, while LP_1 and LP_2 are used to minimize the influence of spherochromatism.

We start with a matrix displaying the values of R_3 / R_2 and R_5 / R_4 on the two axes for one starting value of R_1. Through trial and error, we determine what particular combination of both variables gives the smallest *LA* and *OSC*. We repeat the procedure for other values of R_1 until the best values of *LA* and *OSC* are obtained. In effect, we are constructing a three-dimensional matrix with R_1 as the third variable.

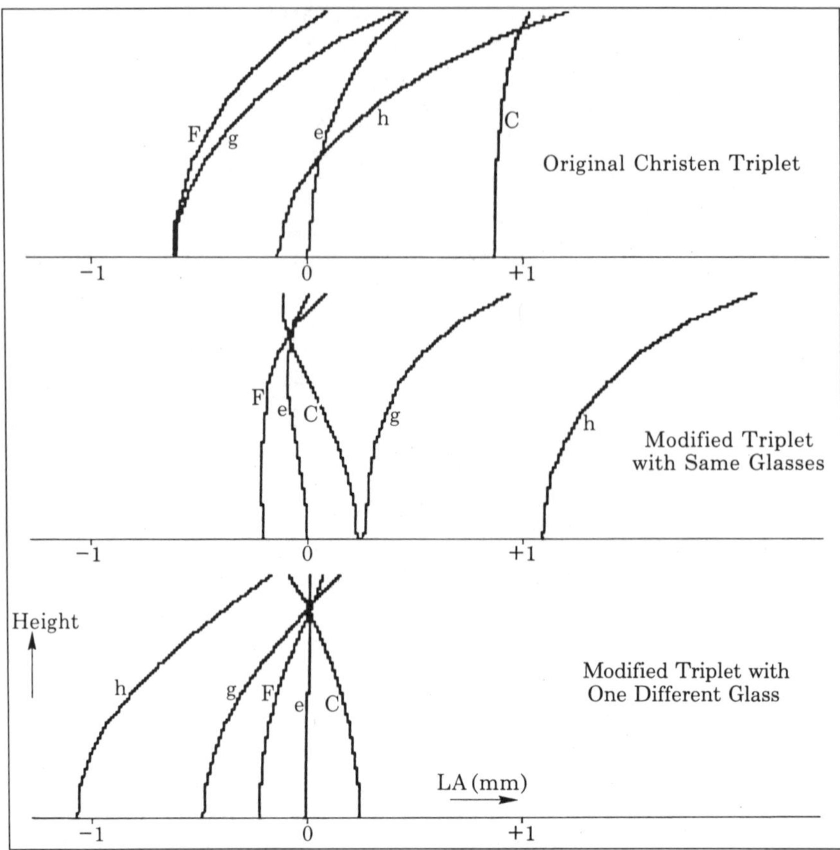

Fig. 21.23 *LA Curves for Three 200 mm f/10 Triplet Refractor Objectives.*

As is the case for doublets, triplets have spherochromatism because the lenses have finite thickness. During the first trials, the three or four desired colors generally do not focus at a common point on the *LA*-curves. As soon as the value of *LA* has been sufficiently decreased, we vary slightly the powers of the first and second lens with respect to the third lens. After several trials, the colors are made to coincide and intersect somewhere between the 70% and 100% zone. After correcting spherochromatism, we continue to optimize *LA* and *OSC*. In a broken triplet the final values of LP_1 and LP_2 are normally very close to one.

More difficult is the design of a system in which the inner radii are equal (i.e., $R_3 = R_2$ and $R_5 = R_4$). A design of this type is called a cemented triplet, though the optical surfaces need not be cemented. Instead, they can be filled with an immersion oil, or left with a small air space. Triplets can also be partly broken, as shown in fig. 21.20. For a fully cemented triplet we have only one degree of freedom once the powers of the three elements are fixed, and that is R_1. All other radii depend on the chosen value of R_1. Often it is not possible to correct *LA* and *OSC*

Table 21.1

Three 200mm *f*/10 Triplet Designs

(All dimensions in mm)

	Example 1	Example 2	Example 3
R_1 (radius of curvature)	2080	2402.1	1825
T_1 (axial thickness or distance)	22.5	25.7151	30.1584
M_1 (medium—six-digit code)	517642	517642	517642
R_2	–507.5	–498.7921	–449.641
T_2	12.5	1	1
M_2	613443	Air	Air
R_3	476.25	–499.2909	–434.5556
T_3	22.5	12.5	12
M_3	670471	613443	613443
R_4	–2080	413.9897	501.4963
T_4	1998.9	1	1
M_4	Air	Air	Air
R_5		418.9989	462.7206
T_5		28.5082	32.4128
M_5		670471	533460
R_6		–1868.7758	–973.0116
T_6		1980.66	1989.86
M_6		Air	Air

sufficiently with R_1 alone; and we must vary LP_1 and LP_2 to achieve a good design.

Contrary to the situation with the broken triplet, the values of LP_1 and LP_2 needed to achieve an aplanatic design can deviate quite strongly from 1.0, so the designer should keep a strict eye on the color correction. Often it is necessary to compromise between good color correction and good off-axis correction.

21.16.4 Examples of Triplets

To obtain a better sense of the correctability of various triplets, we compare three 200 mm *f*/10 designs. *LA* curves for each lens in *C*, *e*, *F*, *g*, and *h* light are shown in fig. 21.23. Design data for the three triplet designs appear in table 21.1.

Example 1: The immersion triplet designed by Roland Christen consists of BK7, KzFS1, and BaFN10 glasses. From the spot diagrams shown in fig. 13.13, we see that this objective has about the same chromatic performance for visual use as the two-element Apoklaas with expensive glass types (see fig. 6.10), but that the triplet's violet blur is much larger. Because the triplet is cemented or oil-immersed, coma is uncorrected.

Example 2: A broken triplet designed by the authors to be constructed from the same glasses as the Christen triplet. Note that the *C*, *e*, and *F* lines intersect at one zone. This system is a true apochromat.

Example 3: A broken triplet with the same first two glasses as the previous examples, but with third glass the more expensive TiF2. In this case, four colors (C, e, f, and g) have been brought to a common focus in one zone. This system can be called a superachromat and is much better for photographic use.

The results obtained in examples 2 and 3 are in accord with fig. 21.22. In example 2, it appears the glasses do not lie on a straight line in the *P–P* diagram, so four colors cannot be corrected.

21.17 Thick Optical Elements

As we saw in section 3.4, a reflecting prism behaves as a plane-parallel plate. We also saw that some effects will occur when a plane-parallel glass plate is inserted into a non-parallel light beam. The most striking effect is a longitudinal displacement of the image. Oblique beams are displaced to the side. In section 3.4, we gave a first-order approximation formula for the longitudinal displacement:

$$S = \frac{n-1}{n} \cdot t \,. \qquad \textbf{(21.17.1)}$$

However, the shift increases slightly more than this as the angle increases, causing overcorrected spherical aberration. Eq. 21.17.1 implies that the image displacement is dependent on the wavelength, so that for blue light the displacement will be larger than it is for red. This produces longitudinal chromatic aberration.

When a beam strikes a flat plate at an angle, as shown in fig. 3.11, it is shifted to the side. The magnitude of the shift is approximately:

$$l = \frac{n-1}{n} \cdot t \cdot \alpha \ (\alpha \text{ in radians}). \qquad \textbf{(21.17.2)}$$

For oblique, non-parallel beams, the plane-parallel plate introduces coma, astigmatism, and lateral color which increases as the focal ratio of the beam decreases, so fast systems will suffer more from the aberrations than slow systems. Refs. 21.6 and 21.7 give formulae with which the magnitudes of the aberrations can be calculated. (These formulae are not treated here because the reader can investigate the magnitudes of the aberrations rather easily with the computer program that is an option with this book.) Whenever an instrument is used in combination with a prism, it is necessary to carry out a ray-trace analysis of the system with the prism.

The spherical aberration and chromatic aberration caused by a prism can be corrected by placing a weak positive lens near the prism. This lens must be designed in such a way that it just corrects the aberrations introduced by the prism.

Chapter 22

How to Use the Telescope Design Programs

22.1 Capabilities

We have written three computer programs to be used in conjunction with this book. The programs are suitable for use with any IBM PC compatible and most work-alikes. The IBM executable file version of these programs is available from the publisher for a nominal charge. (An order form may be found in the back of this book.)

`TDESIGN` is a powerful telescope design program. It allows the user to select a design type—any Cassegrain or catadioptric form treated in chapter 21—and produces a pre-design based on third-order aberration theory.

`LENSDES` allows the user to design high-performance doublet and triplet lenses. These are completed designs, and do not need further optimization.

`RAYTRACE` is a fast and powerful ray tracing program. It can trace up to 2,500 rays through axially symmetric, tilted, or decentered systems with flat, spherical, conic, or aspherically deformed optical surfaces, with or without vignetting. The output of `RAYTRACE` can be in graphical or tabular form.

Each of these programs is present as a compiled executable file with the extension `.EXE`. The executable file runs directly from DOS. The graphics in the programs require a CGA, EGA or Hercules graphics card in the computer; printing screen graphics requires loading the memory-resident program `GRAPHICS.COM` before starting the design programs.

We have worked hard to make these programs easy to use. Once you have become familiar with them, you should be able to run them without consulting the documentation. We recommend, however, that you read this chapter carefully so that you are aware of the capabilities and limitations of the programs, and the meaning of all of the commands.

The distribution diskette also contains `FIRST`, a program to modify the default parameters used by `RAYTRACE`; `INT.REC`, a file of startup parameters created by `FIRST`; and `README.TXT`, a text file that updates this documentation to the current release of the program.

You may run `FIRST` to set the ratio between the height and width of the monitor screen and of a graphics printer; `FIRST` also controls form feed at the end of printouts, sets the default path for design files, and permits display of exit angles for tabular longitudinal traces (useful in eyepiece design). The values entered in the `FIRST` program are saved in the file `INT.REC`, which is read by `RAYTRACE` if it is present in the current directory when `RAYTRACE` is started; if `INT.REC` is not in the current directory, default values are used.

Finally, on the program distribution diskette you will find files with extensions `.RLD`, `.TDS`, and `.DES`. These files are optical design files. Files with the extension `.RLD` have been created by `LENSDES`, the doublet and triplet design program, and may be retrieved by this program as design examples. Files with the `.TDS` extension have been created by `TDESIGN`, the telescope design program, and may be retrieved by this program only as design examples. The `.DES` files have been saved by `RAYTRACE`, the optical ray-trace program. `RAYTRACE` can also retrieve `.RLD` and `.TDS` files created by `LENSDES` and `TDESIGN` to optimize a design or carry out an optical analysis.

You will find it is easier to become familiar with the programs if you load and work with several "canned" designs before striking out on your own.

The programs have been extensively tested against a large number of optical systems, and they give consistent and accurate results. However, large computer programs are seldom entirely free of bugs. The authors and publisher are interested in hearing about any errors you find. Please be specific in describing the problem. Tell us what you did and exactly what happened. If you were working on a telescope design file, please make a copy of the file for us. Mail your comments and bug reports to Willmann-Bell, Inc. P.O. Box 35025, Richmond, VA 23235.

22.2 Designing Telescopes with `TDESIGN`

`TDESIGN` allows you to select any Cassegrain or catadioptric system shown in the telescope design tree (fig. 22.1) and produces a pre-design based on third-order aberration theory to your specifications. If you want to design lenses for refractors, see section 22.3 for a description of `LENSDES`. After developing a predesign with `TDESIGN`, you will use `RAYTRACE` to optimize it.

All of the telescope designs shown in the design tree are examined in this book. In the section that follows we give a brief summary of each design's characteristics, advantages, and disadvantages to help you decide which designs are most interesting.

22.2.1 Designs Available with `TDESIGN`

Cassegrain Telescopes The shortest possible configuration for a certain focal length is achieved with the Cassegrain-like telescopes. The disadvantages of two-mirror systems are the relatively large obstruction caused by the secondary mirror and the need for a spider or window. Because Cassegrain systems are normally not closed, air turbulence and atmospheric deterioration of the mirrors can occur. Another factor is the relatively strong field curvature, especially when the diameter of the secondary must be held small. Field curvature is closely connected with the

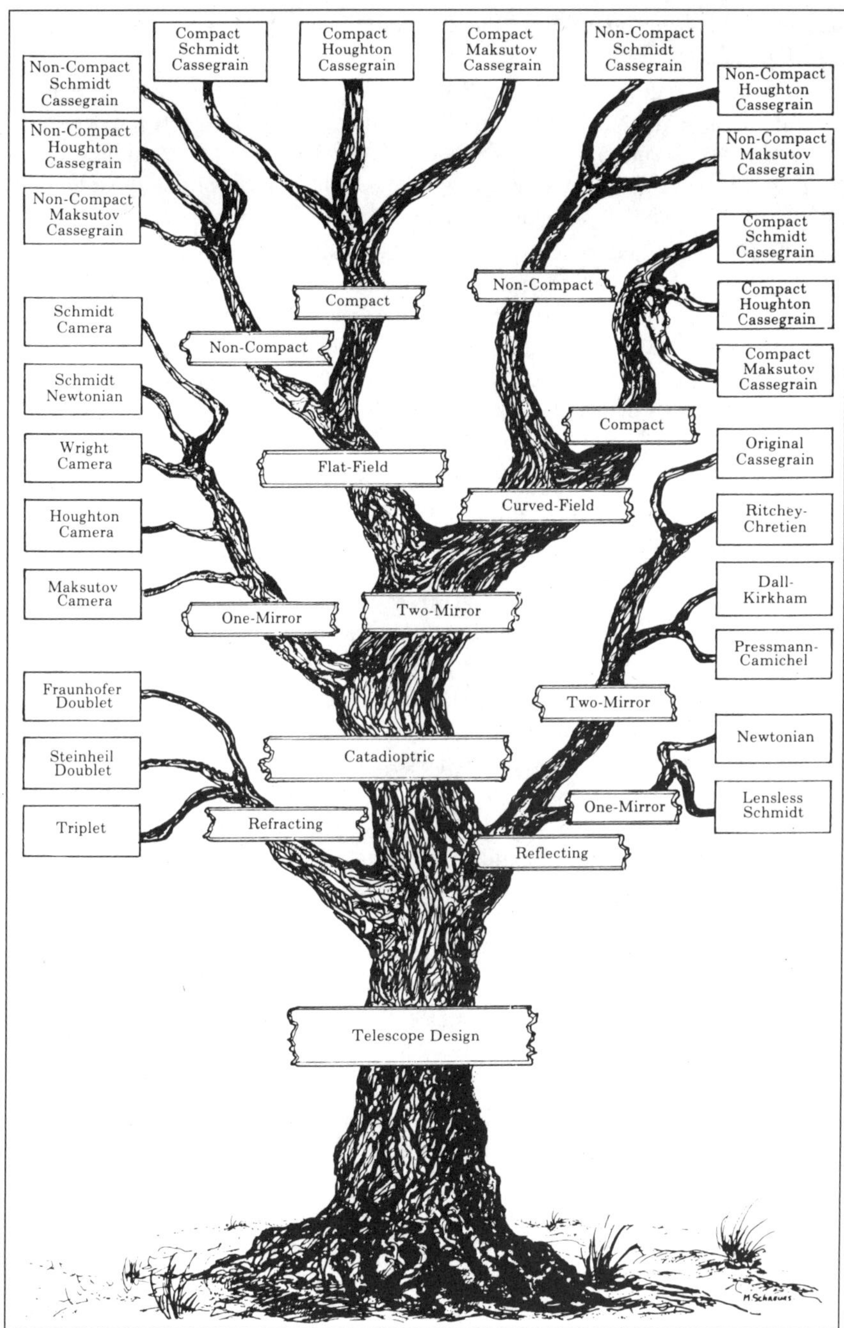

Fig. 22.1 *The Telescope Design Tree.*

curvature of the secondary mirror and, therefore, with the value of the secondary magnification.

Of the Cassegrain designs, the Ritchey-Chrétien is the most difficult to manufacture because the two hyperbolic mirrors must be exactly matched in shape. It offers the best image quality possible with a two-mirror configuration, and is aplanatic. This type is usually built for photographic applications. Although it has astigmatism, off-axis star images are round when curved film is used.

The Dall-Kirkham is easier to build because the secondary mirror can be kept spherical, but the instrument has strong coma, so the usable field is small. The original Cassegrain lies between the Ritchey-Chrétien and the Dall-Kirkham with respect to both difficulty of fabrication and image quality. A virtually unknown type of Cassegrain is the Pressmann-Camichel, with a spherical primary; it has coma even stronger than that of the Dall-Kirkham, resulting in a very small usable field.

One-Mirror Catadioptrics Consisting of a spherical or aspheric mirror with a corrector placed in front of it, this group provides good off-axis performance. The corrector closes the tube, protecting the mirror from the weather. In the compact configuration, the spider can be omitted because the secondary mirror or film holder can be attached to the corrector. In one-mirror catadioptrics, however, only concentric systems can be entirely free from astigmatism.

The best-known single-mirror catadioptric is the Schmidt camera. A concentric configuration with the corrector placed in the center of curvature of the spherical mirror, the Schmidt system is both aplanatic and anastigmatic. A derivative system, the Schmidt-Newtonian, also has a spherical mirror. In it, the corrector is placed much closer to the mirror, resulting in a compact system. Because the principle of concentricity has been abandoned, the system suffers from both coma and astigmatism. For a compact Schmidt system to be free of coma, the mirror must be ellipsoidal, and the corrector must have considerably more power than in a noncompact system.

The Wright camera has its corrector at the paraxial focal plane, and offers the photographer a flat focal surface. Another one-mirror catadioptric is the Houghton system; with the two-lens Houghton corrector, a compact system with all spherical surfaces is possible.

An aplanatic Maksutov configuration has a greater tube length than the Houghton, but is shorter than the Schmidt camera. In its original form, the Maksutov camera is not compact in most cases, so the film holder must be supported by a spider. The length of the tube depends on the thickness of the corrector.

Two-Mirror Catadioptrics This, the largest group shown in the design tree, can be designed with a flat or a curved field, in a compact or a non-compact configuration, and with a variety of types of correctors. In most cases, the focal plane is outside the system, behind the primary mirror, for good accessibility.

In flat-field designs, the radii of curvature of the primary and secondary mirrors are approximately equal. For the image to lie behind the primary mirror, a relatively large secondary mirror, typically 50% to 60% of the diameter of the entrance pupil, is needed. Therefore, these flat-field instruments are not well suited for visual applications.

In designs that allow a curved field, the secondary mirror can be more strongly curved and smaller than in flat-field designs, typically only 30% to 40% of the entrance pupil. Curved film or a field flattener must be used when these instruments are employed for photography.

In comparison with the compact design, the non-compact design two-mirror catadioptric suffers from greater tube length, a secondary mirror that must be supported by a spider, and, for the same degree of light drop-off at the edge of the field, larger primary and secondary mirrors. However, for flat-field designs, the baffle system required can be smaller or even omitted.

When the designer chooses between a non-compact or a compact-built system, he must take into account not only these points, but also the possibilities for correcting aberrations. Of course, all types are corrected for spherical aberration. For the Schmidt-Cassegrain, the correction of coma in a compact design requires at least one aspheric mirror, while in a non-compact design both mirrors can be left spherical for one particular position of the corrector. In the Houghton-Cassegrain, in both the compact and non-compact designs, an aplanatic system can be achieved with two spherical mirrors and a corrector doublet. With Maksutov-Cassegrains, however, the correction of coma depends on the thickness of the corrector with respect to the focal length of the primary. For each thickness ratio, there is a single secondary magnification that yields an aplanatic design.

An interesting alternative is a Maksutov system with an aluminized spot secondary on the back of the corrector. This system, however, is not aplanatic, and no satisfactory flat-field designs are possible. (We have so far considered only aplanatic systems.)

For simultaneous correction of coma and astigmatism, the situation is more complicated. The reader is referred to chapter 21 for a detailed description. Depending on the type of corrector and the configuration, the designer must find one specific position for the corrector and two aspheric mirrors if he is to obtain a system that is both aplanatic and anastigmatic.

With the exception of a few non-compact flat-field designs, all Cassegrain-derived telescopes require a baffle tube system. Depending on the width of the field, baffling may cause a considerable light drop-off toward the edge of the field. This is a decided disadvantage compared to the one-mirror catadioptrics, which do not require such baffles.

22.2.2 Using `TDESIGN`

`TDESIGN` produces preliminary optical designs that must be further optimized, or "predesigns." This is because the formulae used in the program are based on third order Seidel theory (see chapter 21), which is an approximation. In most cases, predesigns are quite accurate—within 5% of the final values and often much better.

Designing begins with choosing what sort of system to design. We stress that different designs require different design procedures. Systems with Schmidt and Houghton correctors are easy to design. Construction data are displayed on the screen immediately because these predesigns result from straightforward calculations.

For systems using Maksutov correctors, however, the designer must proceed by trial and error until a satisfactory predesign has been obtained. We recommend learning to use the program by designing systems similar to those described in this book.

After preliminary construction data have been produced by `TDESIGN`, you can save design data as a disk file. This design file is then loaded into, and optimized by skew-ray tracing with, `RAYTRACE` (described in section 22.4).

Starting `TDESIGN`: Load the program by typing `TDESIGN`. As soon as the program loads, the main menu appears on the screen:

```
C=Create new design  R=Retrieve file  H=New file path  ESC=Exit
```

Creating a new design is the primary function of `TDESIGN`. Selecting `C` will send you through a series of menus and requests for design parameters. You will see only the menus and enter only the information required for the design you are creating. If you feel a bit lost, refer to the telescope design tree shown in fig. 22.1; it shows all the design options `TDESIGN` offers.

The `Retrieve` option allows you to display a file on the disk. This is fully described at the end of the section. Pressing the `ESC` key stops the program and returns you to `DOS`.

Creating a Design: Begin by entering a design title. Then select a design class from the Telescope System Selection menu:

```
1. Two-Mirror Cassegrain
2. One-Mirror Catadioptric
3. Two-Mirror Catadioptric
```

Telescope System Selection: `Two-Mirror Cassegrains` For two-mirror Cassegrains, you must first specify what criteria you wish to use in specifying the system. There are three options: the focal length of the primary mirror and the system focal length, or the maximum axial obstruction, or the maximum mirror separation. You enter your choice of one specification set at the Design-Limiting Specifications menu:

```
1. Set Primary and System Focal Length
2. Set Maximum Axial Obstruction
3. Set Maximum Mirror Separation
```

Telescope Specification Entry: Once you have selected which set of limiting factors you wish to work from, `TDESIGN` requests further information based on your choice.

Regardless of your selection, it asks the clear aperture of the telescope and the position of the focal surface with respect to the primary. Depending on your selection, however, `TDESIGN` will ask for the focal length of the primary mirror and the system focal length, or the maximum axial obstruction, or the maximum mirror separation.

`TDESIGN` will accept any reasonable input values, but rejects impossible conditions. For example, you cannot specify a system focal length that is less than the focal length of the primary.

If you have chosen the axial obstruction as the design-limiting factor, you must remember that `TDESIGN` does not take off-axis vignetting or the optical power of the corrector into account when it makes a predesign. If it is important to keep the secondary as small as possible, the designer must make a reasonable allowance for these effects. We recommend an initial maximum obstruction that is 80% of the maximum value you can accept; if you want the obstruction smaller than 50%, then specify a maximum axial obstruction of 40%.

The maximum permissible distance between the mirrors is usually chosen for a practical reason, for example, that the instrument must fit into an existing tube or will be housed in a small observatory. `TDESIGN` requires that the distance between the mirrors be smaller than the focal length of the primary mirror.

You next select the optical design type of the telescope:

```
1. Classical Cassegrain
2. Dall-Kirkham Cassegrain
3. Ritchey-Chrétien Cassegrain
4. Pressmann-Camichel
```

Within a few seconds of selecting the design type, `TDESIGN` will display the construction parameters of the predesign.

Telescope System Selection: `One-Mirror Catadioptrics` There are three types of one-mirror catadioptric to choose from, based on the type of corrector:

```
1. Schmidt Design
2. Houghton Design
3. Maksutov Design
```

Your choice influences the questions and design parameters you will need to supply later on in the design process. First, however, `TDESIGN` needs to know the telescope's specifications: clear aperture, primary and system focal length, primary to corrector distance for some designs, corrector refractive index and thickness.

Although `TDESIGN` will not accept impossible values, it does not signal an error if rays pass twice through the corrector, since this might be the designer's intent. For one-mirror catadioptrics, the corrector-to-primary distance must be greater than half the focal length. This distance is always positive. For the Schmidt- and the Houghton-derived systems, the design will be aplanatic, that is, simultaneously corrected for spherical aberration and coma. If the designer deliberately places the Schmidt corrector inside the center of curvature of the primary, `TDESIGN` will design a Wright system. For Schmidt-Newtonians, the program places a normal Schmidt camera corrector at the desired position, but of course coma is not corrected.

For Schmidt systems, `TDESIGN` asks the thickness of the corrector and the refractive index of the glass at the design wavelength. `TDESIGN` accepts index values between 1.4 and 2.1.

For Houghton systems, you must enter the refractive index and the primary-to-corrector separation. The program sets the edge thickness of the positive element and the center thickness of the negative element to 1/40th of the aperture, and spaces the lenses so that they are no closer than 0.15 mm. All Houghton designs produced by `TDESIGN` are aplanatic, that is, corrected for spherical aberration and coma.

Unlike the systems described above, Maksutov systems must be designed by trial and error. First, you must enter the refractive index for the design color and thickness. The refractive index must lie between 1.4 and 2.1, and the thickness must be larger than 1/20th of the clear aperture. Next, an initial bending coefficient (a value describing the shape of the corrector) is entered; this value can be arbitrary, a number between 20 and 80. The larger the thickness of the corrector with respect to the focal length of the primary, the smaller this factor can be. The program calculates the Seidel coefficients of the corrector and the mirror separately and subsequently adds them.

`TDESIGN` analyzes the system and prints the Seidel coefficients, then asks:

```
Optimize or Final Design (O/F)?
```

If you select `O`, the following question is asked:

```
Optimize Shape or Position corrector (S/P)?
```

Begin a design by selecting `S` to alter the shape of the corrector. Vary the value of `S` until spherical aberration reaches an absolute value less than 0.01. During this procedure, coma and astigmatism will change, too.

When you have reduced spherical aberration to an acceptable value, select `P` and change the position of the corrector. Changing the position influences coma and astigmatism, but does not influence spherical aberration. After a number of changes of `P`, you will obtain a shape and position that minimize both spherical aberration and coma.

Now select `F` to do a final design. `TDESIGN` now allows you to introduce an extra factor, `Kc`, for achromatizing the corrector, and completes the design. After a few seconds, you will see the construction data on the screen.

Telescope System Selection: `Two-Mirror Catadioptrics` This immediately takes you to the Possible Catadioptric Systems menu, which asks what type of corrector you want your system to have:

```
1. Schmidt Design
```

```
2. Houghton Design
3. Maksutov Design
```

Your choice influences the questions and design parameters the program will need later in the design process.

You will next see the Possible Focal Plane Design Choices menu:

```
1. Curved-Field Design
2. Flat-Field Design
```

Typing `1` selects a two-mirror catadioptric Cassegrain telescope with a curved field, intended mainly for visual use. When `2` is selected, `TDESIGN` designs a two-mirror catadioptric astrocamera, which will have, in principle, a flat field. `TDESIGN` requires somewhat different specifications for the two types of designs; we discuss the peculiarities of flat-field designs below.

Design Limiting Specifications: You must next specify what criteria you wish to use in specifying the system. There are three options:

```
1. Set Primary and System Focal Length
2. Set Maximum Axial Obstruction
3. Set Maximum Mirror Separation
```

Enter your choice of specification. This choice, and the selection of a curved- or flat-field design, affects the specification questions `TDESIGN` asks later.

Flat- versus Curved-Field Systems: Unlike those in curved-field designs, the radii of curvature of the two mirrors in flat-field designs are approximately the same. In the flat-field specifications, `TDESIGN` therefore asks not for the focal length of the primary mirror, but for the focal length of the system. You may specify the ratio of the curvatures of the two mirrors, in the `F-sec/F-prim Ratio` menu, as a value between 0.8 and 1.1. Your choice depends on the type of corrector you have chosen and its optical power, and is a matter of your design experience. Once this is done, `TDESIGN` computes the diameter of the central obstruction. If this is too large, you must reduce the back focus distance, select a new curvature ratio to flatten the field, then calculate a new central obstruction.

`TDESIGN` does not check whether the rays pass the corrector twice, but it will not accept impossible values. For two-mirror catadioptrics, the program only accepts primary-to-secondary values that are less than 0.9 times the corrector-to-primary distance.

At this point, `TDESIGN` branches into three major telescope types: Schmidt, Houghton, and Maksutov.

Schmidt Designs: In designing a Schmidt-Cassegrain, select one of the following options:

```
1. Spherical Mirrors
2. One Aspheric Mirror
3. Two Aspheric Mirrors
```

If both mirrors are spherical, TDESIGN automatically places the corrector so that the system will be free of both coma and spherical aberration, i.e., aplanatic. At this point, the program computes and displays construction data for the system. Although this is not a compact design, the designer can move the corrector closer to the primary, but then, of course, the design is no longer coma-free.

When one mirror is chosen to be aspheric, the designer is also asked to specify whether the primary or secondary mirror is to be aspheric. The corrector can be placed in any position; TDESIGN prompts the designer for the position desired. The resulting system is always aplanatic.

When two aspheric mirrors are chosen, it is possible to find an aplanatic and anastigmatic compact Schmidt-Cassegrain design for every mirror configuration. A few seconds after the desired distance between the corrector and the primary mirror has been entered, the construction data of the Schmidt-Cassegrain predesign are displayed.

Houghton Designs: For Houghton correctors, astigmatism is corrected for some positions of the corrector. The program will ask whether you want an anastigmatic design. If you do, two possible corrector positions will be displayed after a few seconds. If an aplanatic (but not anastigmatic) system is desired, you must enter your choice of distance from the corrector to the primary mirror.

Maksutov Systems: For Maksutov correctors, the design is an iterative process exactly like that described above for the one-mirror catadioptric Maksutov camera. Please refer to the design instructions found there.

Displaying the Results: When all necessary data for a system have been entered, TDESIGN displays the design on the screen in tabular form. The following menu then appears at the bottom of the screen:

```
P=Print; S=Save design; B=Begin again; ESC=Exit
```

The P command prints a hard copy of the results.

The S command stores the design on disk; the computer will ask for a file name. The extension .TDS need not be keyed in; it is added automatically by the computer. The maximum allowable filename length is eight characters, and the program will not accept longer names.

The B command takes you back to the opening menu, to start all over again.

If you press the ESC key, TDESIGN returns you to DOS.

Recalling a File: Pressing R in the initial menu recalls a telescope design file. A list of all .TDS files saved on the selected diskette or directory will appear on the screen. Type the desired filename without the .TDS extension. After the file has loaded, the design is displayed on the screen. A file retrieved in this way cannot be edited, redesigned, or optimized, but the user can view the contents of the file. Any .TDS file can be loaded into RAYTRACE for detailed evaluation.

22.3 Lens Design with LENSDES

LENSDES is a design program for doublet and triplet refractor objectives. The designs it produces do not require further optimization, though you may enjoy checking your latest efforts using RAYTRACE.

22.3.1 Designing Lenses

The design of a refractor objective used to be a time-consuming task, but with the aid of LENSDES, you can complete a design in a few minutes, providing you initially select the proper kinds of glass. Design time for a doublet is typically five to fifteen minutes; for a triplet, half an hour to two hours. The design proceeds rapidly because the analysis of intermediate results takes only a few seconds.

In this context, it is interesting to consider J. H. Wyld's well-known treatise on lens design (see *Amateur Telescope Making, Book III*, p. 581 or in the new ediditon published by Willmann-Bell: *Amateur Telescope Making Book 2*, p.139). Wyld comprehensively describes how to design and ray-trace an achromatic doublet with goniometric and logarithmic tables, and describes the perseverance required of the designer to carry out this painstaking job.

When a programmable pocket calculator is used, there is a significant decrease in design time—to several evenings. With LENSDES, the authors found they could redesign the Apoklaas or the fluorite objective shown in table 6.1 in a very short time. Naturally, the better the original choice of glasses and the more experienced the designer, the shorter the design time will be.

It is possible, using a fully automatic program, to decrease the design time even further. However, the involvement of the designer is reduced to nothing, and he does not gain a "feel" for the sensitivity of the design to varying parameters. For that reason we have left the program semi-automatic. This means that the designer must participate actively in the various steps of the design process even though all of the calculations are carried out by the computer. The greatest advantage of this hands-on process is that the designer obtains a good idea of the tolerances of the system parameters.

The program is used in almost the same way to design both doublets and triplets. Of course, because a triplet contains an additional lens, more data must be input and more choices must be made during optimizations.

New designs are generally based on the standard refractive indices given in the catalog of the glass manufacturer. When the glass is delivered, the manufacturer will also supply the measured refractive indices of the blanks. The final design must be re-optimized based on the actual indices since these often deviate slightly from the indices in the catalog. The reader is advised to review sections 21.13 through 21.16 before starting a design in order to obtain some necessary background.

22.3.2 Using LENSDES

Load the program by typing LENSDES at the DOS prompt. As soon as the program loads, the "short" menu appears on the screen:

```
Enter program choice (L for list).
```

Type `L`. The following menu of commands will appear:

```
Create lens design
Display lens design
Print lens design and results
One-color lens analysis
Three-color lens analysis
Graph spherical aberration
Zone change for analysis
Retrieve lens design file
Save lens design file
File path ( )
Modify lens radii
Indices modifications
Airgap and thickness modification
New title
Here is design reference
List of commands
ESC=exit design session.
```

All of the commands except `ESC` are entered by pressing the key of the first letter of the command. Only some of the commands—`C`, `Z`, `R`, `F`, `H`, and `L`—operate when the program has just been loaded and before a lens design has been entered into the computer. All other commands require a design to be in the computer and will not operate without one.

`Create lens design` To start a new design, enter `C`. The program will first ask for the name (limited to 40 characters) of the new design. The design procedure then begins.

`Display lens design` The current lens design and design parameters can be displayed with the `D` command. If you are starting a new design, this command will operate only after the design parameters have been entered and `LENSDES` has made a predesign.

`Print lens design and results` The `P` command tells the printer to print a hard copy of the design and an analysis of it.

`One-color lens analysis` `O` causes `LENSDES` to make a one-color analysis of the current lens design at the first of the indices listed for each surface. The focal length, focal ratio, spherical aberration, and OSC are also calculated.

`Three-color lens analysis` `T` initiates a three-color analysis of the current lens design, and shows the results in tabular form on the computer screen.

`Graph spherical aberration` `G` shows a graph of spherical aberration in three colors; thus spherochromatism is shown. The program must run an analysis of all colors that have not already been analyzed. If neither the `T` nor the `O` command was given before the `G` command, all three colors will be analyzed and tabulated before the graph is drawn.

Zone change for analysis LENSDES traces five zones. The 1% and 100% zones are always traced; the remaining three zones are initially set to 50%, 70%, and 90%. Using the Z command, these can be changed to any integer values.

Retrieve lens design file The R command retrieves a design from the disk. LENSDES first shows all .RLD (Refractor Lens Design) files. To load the contents of a file, enter the filename without the .RLD extension. After the file has loaded, the design can be analyzed and optimized. LENSDES allows you to specify new indices (such as melt indices) for the optical glasses, or to move the lenses with respect to each other (thereby changing the airgap) using the I and A commands.

Save lens design file S saves a design on a diskette. After the command has been given, the computer will list all the .RLD files. Enter any filename (up to eight characters) without an .RLD extension. If a longer filename is entered, the program will not accept it. If the filename you enter is the same as that of an already existing file, that file will be overwritten.

File path (..) This shows the current file path to the design file and permits changing the path. The current path appears between parentheses.

Modify lens radii Use M for optimizing a design. When this key is struck, LENSDES asks for a new first radius, a new ratio R3/R2, and a new power ratio Lens1/Lens2, if you are designing a doublet. If a triplet is being designed, LENSDES asks for the ratios R5/R4 and power ratios Lens1/Lens3 and Lens2/Lens3 as well. To leave a value unchanged, simply press the ENTER key. For cemented and immersion lenses, R3/R2 and R5/R4 are held at unity.

Indices modifications I allows you to change the refractive indices of the glasses. This is extremely useful for re-optimizing a design when actual melt indices are known.

Airgap and thickness modification You can alter the airgap and/or the lens thickness with the A command. This is particularly useful for doing a tolerance analysis about these parameters. The degree of correction depends somewhat on the thickness of the lenses and very strongly on the airgaps.

New title With the N command, the current design can be given a new name.

Here is design reference The H command summons a comprehensive help screen. It will remind you of the best and most efficient approaches to optimizing a design.

List of commands L shows the list of commands on the screen. These commands are active from any point in LENSDES, even without the command list.

ESC = exit design session Press the ESC key to quit the program and return to DOS. LENSDES does not save the current design automatically; always check that you have saved your work before quitting.

22.3.3 Doublet Design with LENSDES

The method used and background information on its application are given in section 21.13. Before beginning your design session, choose glass types for the elements and have available the refractive indices for the three colors of interest.

Load the program by typing `LENSDES`. Begin the design procedure with the `Create` command. After you have entered a name for the design, the computer will ask whether you wish to design a doublet or a triplet. It will then ask:

```
Enter lens diameter?
```

Only positive values are accepted, preferably lenses equal to or larger than 100 units. To design a smaller lens, scale it down afterwards—the focal ratio will remain the same, of course!

After the lens diameter has been entered, the computer will ask:

```
Enter focal length of the lens?
```

Both positive and negative focal lengths will be accepted. The absolute value of the focal length should be at least 5 times its diameter. It is not necessary to indicate whether a Fraunhofer or a Steinheil configuration is chosen, because this is calculated by `LENSDES` with the aid of the Abbe numbers of the two glasses: the positive lens will have the higher Abbe number and the negative lens the lower.

After the focal length has been entered, the computer will ask for the refractive indices. The refractive indices should be given for one lens at a time. The first index `(N1)` is the design index of the color for which the design is optimized for spherical aberration and offense against sine condition. The computer next asks for the red side index (longer wave length) and the blue side index (shorter wave length). Then it asks for the refractive indices of the second lens in the same way; these indices must be given for the same wavelengths as for the first lens.

After this the computer asks:

```
Enter thickness at edge of positive lens?
```

and:

```
Enter thickness at center of negative lens?
```

The program assumes the edge thickness you enter is for a 4% oversize blank. This value should not be too small because thin edges quickly chip. The center thickness should not be too small either, or the lens will be liable to break. Furthermore, overly thin elements bend during grinding and polishing, and are difficult to figure. Lastly, thin lenses bend during use, causing aberrations.

After the edge and center thickness have been entered, the computer asks:

```
Enter airgap?
```

The airgap is the distance between the lenses on the optical axis. This distance should be made reasonably large, i.e., larger than about 3 millimeters for a 200 mm aperture. After the system has been ground and polished, any residual spherical aberration can be corrected by moving the lenses with respect to each other. For cemented or oil immersion doublets, however, the airgap should be set to zero.

After these data have been entered, the computer makes a predesign. The predesign has a first lens curvature of 2/3 of the given focal length, and other optimization parameters are preset. The program automatically calculates the other radii of curvature and the thickness of the positive lens, and checks the airgap. It also checks that the distance between the edges of the lenses is equal to or greater than zero. If it is not, the program increases the distance between the lenses until the edge distance is 0.025 mm. The predesign is then printed on the screen, and the program is ready to do analyses or to optimize the design manually.

To explore the aberrations of the design, enter `T` to initiate a three-color analysis. After a few seconds, you will see the longitudinal spherical aberrations and OSC values for all three colors. If you want to see the same information in graphic form, enter a `G` for a graph of the spherochromatism.

If the design is not correct, you may now modify it by entering the `M` command. This allows you to vary three parameters:

- the radius of curvature of the first surface (`R1`)
- the ratio of the radii (`R3/R2`)
- the ratio of the powers (`Lens1/Lens2`)

Changing `R1` has the greatest effect on the OSC. Changing `R3/R2` influences the LA, and changing `Lens1/Lens2` influences the spherochromatism. Of course, when any one parameter is altered, it influences all of the aberrations. Whenever you are optimizing a design, it is important to make notes or print out successive designs and aberrations.

Begin an optimization by changing the power ratio, ignoring the LA and OSC. Look first to see in which zone the red and blue curves intersect, and whether the design is achromatic. For airspaced designs, slight changes in the power ratio are usually needed, so keep `R1` and `R3/R2` constant by pressing the `ENTER` key, and alter `Lens1/Lens2`.

If the design is achromatic—that is, if the red and blue curves intersect somewhere between 50 and 90% of the semidiameter—vary the ratio of the radii while keeping the other parameters constant to suppress LA.

Once the spherical aberration has been reduced to an acceptable level, change `R1` to reduce the OSC. As you alter the parameters, you will gain insight into which changes improve the design. When you have completed the first set of optimizations, note how much the position of the achromatic zone has moved and how much spherical aberration has been introduced, then reoptimize the design. After several trial and error runs, you will be able to create a good design easily.

During the optimizations, the computer constantly checks that the inside curvatures do not become too strong. Very exotic designs with curves like Christmas tree balls are generally not practical. Whenever the radius of curvature becomes smaller than the diameter of the lens, the program generates an error message and returns to the menu.

After you have purchased glass for the lenses, you should reoptimize the design for the measured melt indices. At that time, start the program with the `R` command, enter the filename (without the `.RLD` extension), and enter the new true values using the `I` command.

After this change has been made, the computer makes a new predesign. Enter the original design parameters and check the correction with the new indices. Because actual melt indices are always close to the catalogue indices, re-optimization takes only a short time.

A final note about doublet design: if you are new to lens design, alter only one parameter at a time. The more design experience you gain, the faster you can achieve a good design.

22.3.4 Triplet Design with LENSDES

Using `LENSDES` for a triplet is nearly the same as using it for a doublet. But there are some important differences.

Naturally, the program will ask for the refractive indices of three lenses rather than two. The three lenses must be three different glasses. It will also ask for the edge thickness of the second positive lens and the second air gap.

When designing a triplet, the choice of a proper glass combination is very important, so much so that it is the choice of glass, and not the computer design process, that is the most time-consuming part of making a new design. Recommendations for proper glass choices for both three-color and four-color triplets are given in section 21.16.1.

With a triplet, the number of design parameters is higher than with a doublet. The additional parameters are the curvature ratio `R5/R4` and the lens power ratio `Lens2/Lens3`. Because there are more ways to achieve very low spherical aberration and OSC with apochromatic color correction, designing a triplet is considerably more complex than designing a doublet.

The optimization procedure is begun by altering both lens power ratios to achieve good color correction. Generally, it is possible to keep the power ratios close to unity. Each design starts with a fairly short `R1` (the default value is 3/8 of the focal length). The curvature ratios `R3/R2` and `R5/R4` are altered to reduce spherical aberration and OSC. That completed, make a new design with an `R1` about 7/8 of the focal length. Once the second design has been done, you can hazard a guess for a suitable value of `R1`.

Designing cemented and immersion objectives is difficult because the airgaps are fixed at zero and both curvature ratios are fixed at unity. Only the lens powers and `R1` are free parameters. It may take a long time to find a good design; making a good glass choice is essential to success.

We have found that some users of `LENSDES` do not succeed in designing a triplet on the first try. If you do not start with a good combination of glasses you may not reach an acceptable design. If you have trouble, read section 21.16 again to be sure you have selected suitable glass types.

22.3.5 Rescaling Doublet and Triplet Designs

When you have completed the design of a telescope objective with LENSDES, you may notice that the effective focal length of the design is no longer the value you specified initially. This is because the routine the program uses to calculate the powers of the original lens elements does not take into account their thicknesses, the airgaps, or whether the positive or negative element is first.

When the airgaps are small and the lenses thin, the final focal length will be quite close to the original focal length specification. When the airgaps exceed 0.25% of the focal length and the lenses are thick, relatively large deviations may occur. If this difference is not acceptable, the design must be rescaled.

There are two ways to rescale. The easiest method is to multiply the radii of curvature, thicknesses, and airgaps by the ratio between the desired focal length and the focal length of the lens design. Because the aberrations will change a little and the zone of spherochromatism will shift, it is necessary to check and possibly correct the design. If the new aberrations are unacceptable, then you must use the second rescaling method.

Create a new design giving a desired focal length scaled proportionately from the previous design. For example, if the focal length in the first design came out 0.92 of the desired focal length, begin the design with a focal length the reciprocal of 0.92, or 1.087 times the focal length you want, and enter proportionally scaled values for the radii, thicknesses, and airgaps. (You may wish to retain the original lens thicknesses and airgaps for mechanical reasons, but the resulting design will, of course, then depart slightly from the desired focal length, but by a much smaller amount.) Optimize the new design; then check its focal length. It may be necessary to repeat this process several times if the focal length must exactly match some value.

22.4 The Telescope Optics Ray-Tracing Program

RAYTRACE is a powerful ray-trace program suitable for examining axially symmetric, tilted, or decentered optical systems with flat, spherical, conic, and aspheric surfaces as far as these surfaces are rotationally symmetric, using the skew ray-tracing methods described in chapter 20. The program can present the results as a table or as a graph.

22.4.1 Using RAYTRACE

Load the program by typing RAYTRACE from the DOS prompt.

Key Commands

After the program has loaded, the following appears at the bottom of the screen:

```
Enter the program choice (L for list).
```

Entering L will display the Main Menu:

```
Create new design
```

```
Display design parameters
Toggle to scale for MONITOR/printer
Print screen
X = examine design
Alter analysis results
File path ( )
Retrieve design
Save design
Modify design radii/distances
Indices modification
Vignetting/decentering modifications
New design title
Here is design reference
ESC = Exit program
```

The keyboard is ready to receive a command as soon as the previous command has been completed; it is not necessary, once you know all of the commands, to use the `L` command to display the list of possible commands.

`Create new design` The `C` command allows you to start a new design. Designs are limited to a maximum of 35 surfaces.

`Display design parameters` After a design has been entered, it may be modified, retrieved, or displayed. The display allows you to re-examine a system you are optimizing.

`Toggle to scale for MONITOR/printer` This command allows you to set the height and width proportions for the monitor and printer if their scales differ. This is needed for spot diagrams only; otherwise, scale doesn't matter. (If the default scales are correct, no adjustments are needed. If they are not right, run `FIRST` to set them correctly for your system.) When the menu reads:

```
Toggle to scale for MONITOR/printer
```

the height to width ratio is correct for the monitor, but the program may give a distorted diagram on a printout. When the menu reads:

```
Toggle to scale for monitor/PRINTER
```

the height to width ratio is correct for the printer, but the program may give a distorted diagram on the monitor.

`Print screen` With this command, a printout of the current screen is made. This may be a tabular or graphic output of the design parameters.

`X = examine design` This command initiates an analysis of the performance of the current design. The analyses possible are described below.

`Alter analysis results` After a design has been analyzed, the `A` command allows you to refocus, convert the output from tabular to graphic, or convert it from graphic to tabular.

File path ( ) You can change the drive or directory on which design files will be stored by entering the desired path, i.e., B:\WORK or C:.

Retrieve design A design which has been saved on disk can be retrieved and loaded into the memory of the computer. The full filename, including the extension, must be entered.

Save design The current design can be saved as a disk file. This file may be retrieved later for further work. All files saved will automatically be given a .DES extension.

Modify design radii/distances The current design can be modified. This includes altering radii, shapes, thicknesses, and distances between the elements. Refractive indices cannot be changed with the M command.

Indices modification The I command allows you to change the refractive indices of the elements.

Vignetting/decentering modifications If a design contains vignetting or decentering parameters, these can be modified with the V command. When a design has been entered or retrieved without vignetting parameters, these may be added to the design with this command.

New design title Once a design has been modified, it is no longer the same design. The N command allows you to change the name of the current design. The filename does not depend on the title.

Here is design reference The H command calls up a summary of the method for entering data and of the analyses possible in RAYTRACE. If you need an extended description of RAYTRACE, please read on.

ESC = Exit At any point in the program, hitting the ESC key permits the user to quit the program and return to DOS.

Loading and Saving Design Files

There are two ways to load a design into RAYTRACE: by retrieving a file using the R command, and by creating a file using the C command, then entering each of the parameters manually. Once you have entered a design, you may save it as a disk file for later reference.

Retrieving a file The designer will often wish to ray-trace a design created with the TDESIGN or LENSDES. RAYTRACE can load files with extensions .RLD, .TDS, and .DES. .RLD files are those saved by LENSDES (.RLD means Refractor Lens Design). The TDS files are those saved by TDESIGN (.TDS means Telescope Design). .DES files (DESign) are those that have been saved by RAYTRACE in a previous operation.

To retrieve a file, enter the R command; then, at the prompt, enter the filename and its extension:

```
BERRY022.RLD
RUMAKDES.TDS
DIANE001.DES
```

Alternatively, typing `L` and `Enter` to exit from this procedure. If you request a file that does not exist on the disk or in the directory, the program will give an error message and prompt for program choice again.

When a design is retrieved, all parameters are displayed on the screen. Accept the file by pressing `RETURN`; the program will then proceed as if you had just created a new design.

Creating a New Design

To create a new design, use the `C` command. The first series of prompts is for general information about the system, and includes questions that affect the types of analysis you will be able to perform.

Pay strict attention to entering data into `RAYTRACE` correctly. Especially critical is complete adherence to all sign conventions. Especially confusing, therefore especially important to check, is whether you have correctly entered the sign of the refractive index. If light is reflected, thus traveling from right to left, the refractive index must be preceded by a minus sign, and all subsequent surfaces, until a second reflector occurs, must also use a minus sign with the refractive index.

`RAYTRACE` begins by asking you for a name to use with the design. This is not a filename, but a label for your information and reference. The name should be 60 characters or less in length. The program next asks:

```
Enter entrance pupil diameter = ?
```

Enter the diameter of the entering bundle of light. This diameter need not be that of the aperture stop. If you do not know the diameter of the entrance pupil, you can determine it with the "Trace a ray" option discussed later. If you enter `0` (zero) or press `ENTER`, the program will ask you for the diameter of the entrance pupil with every analysis. This is very useful when the diameter of the aperture stop is initially unknown.

Next, the program asks:

```
Enter number of surfaces
(include stops & pupils) = ?
```

If you do not plan to make vignetting calculations, it is not necessary to include vignetting diameters such as annular diaphragms, obstructions, and holes. However, you must include the entrance pupil position if it is not at the first optical surface. For example, a Newtonian is entered as one surface, and a lensless Schmidt as two surfaces. For an eyepiece, include an extra surface positioned at the fictitious entrance pupil as described in section 16.4.

To examine the vignetting of a two-mirror Cassegrain without a baffle system, enter four obstructing surfaces: the secondary mirror blocking incoming light, the outside rim of the primary, the outside rim of the secondary, and the hole in the primary. Vignetting calculations are discussed in greater detail below.

Next, enter the number of refractive indices (to a maximum of five). When a design is traced for only one color, or when a design contains only mirrors, enter `1`. If, however, a design must be calculated in more colors, enter the number of colors to a maximum of five. Note that the program does not know what wavelengths have been chosen, so you must enter the refractive indices for each type of glass in the same sequence.

You must next enter the back focal length. This need not be the last optical surface, but may refer to any position you choose as a reference. Enter this position relative to the last optical surface. It causes no problem to enter the back focal length as `0` or press `ENTER`.

```
Enter vignetting parameters (Y/Enter = N)
```

If you do not wish to make vignetting calculations, then type `ENTER` in response. If you enter `Y`, the program will ask the diameters of the optical components during the entry of the optical data. When `N` is entered, the program does not ask for these diameters. Note that `RAYTRACE` does not check whether the optical elements touch each other or if they have negative edge thicknesses.

If the system is axially symmetric, then answer `N` (press `ENTER`) to the question:

```
Enter non-centered parameters (Y/Enter=N) ?
```

If you answer `Y`, the program will ask for the decentering parameters at every surface as you enter the optical data. When `N` or `ENTER` is pressed, the program does not ask for these parameters.

Note that if the focal surface is not perpendicular to the optical axis of the last optical surface or vignetting dummy surface you enter, then the focal surface must be entered as a separate optical surface.

`RAYTRACE` next asks for the optical parameters of the system. The screen clears and the entrance pupil and number of surfaces are printed at the top:

```
                    CREATE NEW DESIGN

new design
    Entrance pupil: 200             Surfaces: 3


Surface 1:
Surface code:S=Spherical C=conic  F=Flat M=Schmidt Corr.
             P=Pupil    B=Baffle T=Stop 1-5=#Aspheric Coeffs.

    Enter Surface code =
```

Then the following appears:

```
Surface 1: Enter radius of curvature = ?
```

Input the radius of curvature. Pay close attention to the sign convention. *If the center of curvature is right of the vertex of the surface, the sign should be positive; if the center of curvature is left of the vertex of the surface, the sign should be negative.* If the surface is a flat, a stop, an obstruction, or a diaphragm, enter `0` (zero).

```
Enter distance to the next surface = ?
```

Again, you must adhere strictly to the sign convention. *If the direction to the next surface is to the right, the sign of the distance must be positive; if the direction is to the left, the sign must be negative.*

```
Enter N(1) = ?
```

Enter the first refractive index. Once again, the sign convention is important. *If a ray continues its direction, the sign of the index is the same as the sign of the previous surface. (The initial index is 1.000, for air.) If the ray encounters a stop, the index is the same as for the previous surface. However, when a ray is reflected, the sign of the index changes.*

If you are tracing more than one wavelength, the program will now prompt:

```
N(2) = ?
```

until all of the refractive indices (up to five) have been entered. The surface codes are listed in the prompt to help you remember them.

Simple surfaces: `S` means the surface is spherical, `F` means flat, `B` means the surface is a baffle, and `T` means a stop. If you are running vignetting calculations, the program will ask for vignetting parameters; if the system is decentered, the program will ask for non-centered parameters. After you respond, it will continue to the next surface.

Conic surfaces: `C` means the surface is a conic section. The computer asks for conic surface deformation. If the surface is a paraboloid enter –1; if it's a sphere enter 0. If the surface is a prolate ellipsoid, enter a value between –1 and 0. A Dall-Kirkham primary might have a conic deformation of –0.65355. If the surface is an oblate ellipsoid, the value will be greater than 0, for example, 2.12 for a Wright primary. If the surface is a hyperboloid, the conic deformation must be less than –1; a value of –1.0123 would be typical of a Ritchey-Chrétien primary. After the deformation has been entered, the computer continues.

Schmidt surfaces: `M` means the surface is an aspheric Schmidt surface. The program expects a second-order surface and calculates the higher-order coefficients automatically after asking for the relevant parameters:

```
Enter relative power (+ for back, - for front) = 0.0000)?
Enter neutral zone position (0 = center, 1 = edge) = ?
Enter design index = 0.000000)?
Enter primary radius = ?


Enter distance to next surface = ?
   Enter N( 1 ) = ?
```

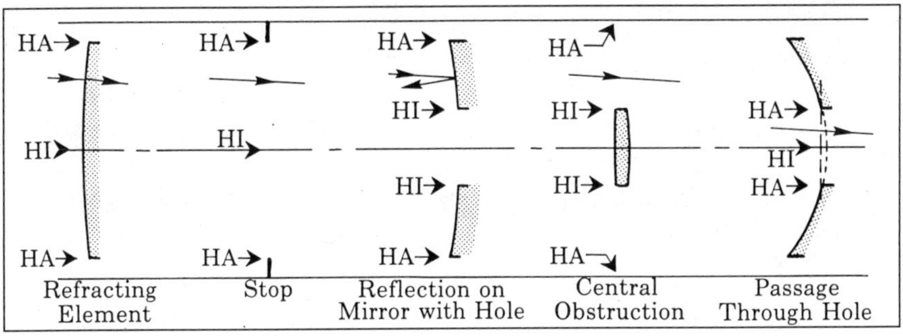

Fig. 22.2 *Vignetting Calculations Showing Maximum and Minimum Diameters of Optical Surfaces.*

The relative power of the Schmidt surface is that which is necessary to correct the spherical aberration of a primary mirror of the desired radius as specified in the fourth prompt. Full correction has an absolute value of 1; partial correction, a value between 1 and zero, or no correction; over-correction, absolute values greater than 1. If the surface is a back side corrector (glass-air), then the sign of the power will be positive; if it is a front side corrector (air-glass), the sign will be negative.

The position of the neutral zone indicates where the thickness of the corrector is at a minimum. This position is given as a relation to the diameter of the entrance pupil, not to that of the optical component. In a Compact Schmidt-Cassegrain telescope, for example, the diameter of the entrance pupil and the diameter of the corrector are the same; but for a Flat-Field Schmidt-Cassegrain Camera with internal stop, the diameter of the corrector is larger than that of the entrance pupil. The best color correction is given at a value of 0.866. The minimum amount of glass to remove depends on the oversized corrector diameter. This is explained more fully in section 21.11.

The primary radius is that of the primary for which the corrector is designed. Normally this radius is the same as the radius of curvature of the primary in a Schmidt Camera or a Schmidt-Cassegrain Telescope. The sign of the radius should match that of the primary.

The design index is the refractive index of the corrector glass at the wavelength for which the corrector is designed. If the instrument is intended for visual observation, give the index for yellow-green light (546.1 nm); if it is for a photographic instrument, enter an index for blue light (486.1 nm).

Higher-Order Surfaces: When the surface is an arbitrary higher-order surface, it is described as a function of the intercept height from the axis, h:

$$z = A_1 \cdot h^2 + A_2 \cdot h^4 + A_3 \cdot h^6 + A_4 \cdot h^8 + A_5 \cdot h^{10} . \qquad \textbf{(22.4.1)}$$

Enter the number of aspheric coefficients, to a maximum of five, including the intermediate zeros you wish to include in the surface description. After the number has been entered, the program will prompt you for each successive surface coefficient.

Vignetting and Non-Centered System Parameters: If you have told the program not to ask for vignetting parameters or non-centered parameters, the computer will clear the screen, print the entered data, and continue with the next surface.

However, if the program expects vignetting parameters, it will continue, using the notation shown in fig. 22.2:

```
Enter field stop or maximum diameter = ?
```

Enter the diameter of the optical component or of an annular diaphragm. If a secondary mirror or a baffle tube creates an obstruction at this position, then enter the internal diameter of the tube. The computer assumes that optical elements and stops are round.

```
Enter hole or obstruction minimum diameter = ?
```

Enter the diameter of a hole in a mirror or of any obstruction such as the external diameter of a baffle tube. When rays pass through a hole in a mirror, this should be treated as a diaphragm.

If you told the program to expect non-centered parameters, it will ask for these parameters:

```
Enter axial surface shift = ?
```

The sign of the axial shift depends on the direction the surface has been shifted. *If the surface is moved to the right, the sign is positive; if to the left, the sign is negative*. RAYTRACE assumes that the *rest of the optical system is shifted together with this shift,* so if only one surface in a design is shifted, you must enter the same value with the opposite sign as the shift at the next surface.

```
Enter surface decentering in x-direction = ?
```

RAYTRACE allows decentering of components only in the x–axis. Decentering in one direction usually gives adequate information for testing design sensitivity. The program assumes that the rest of the optical system is centered with respect to the decentered surface, so if only one surface is decentered, enter the same value with opposite sign for the decentering of the next surface.

```
Enter sine of surface tilt angle = ?
```

The sign of the sine of the angle is positive when the surface is tilted clockwise, and negative when it is tilted counterclockwise.

RAYTRACE allows components to be tilted in only one direction. As with decentering, if tilt were allowed in both directions, results would be difficult to interpret. Moreover, tilted telescope designs with components decentered and tilted in two directions are very rare.

Note that a surface tilt influences the distance between the surfaces and introduces decentering as described in section 20.9. RAYTRACE *assumes that the rest of the telescope design lies on the optical axis of the tilted surface,* so if only one surface is tilted, you must enter the next surface so that the system is centered with respect to the other surfaces.

When the last surface of a design is decentered, RAYTRACE assumes that the focal surface is perpendicular to the optical axis of the last optical surface. If the focal plane is tilted, as it is in a Schiefspiegler, then the focal surface must be entered as an extra surface like a tilted diaphragm.

After these data have been entered for each surface, RAYTRACE clears the screen and continues to the next surface.

After data have been entered for the last surface, the computer calculates the effective focal length, the back focal length, and the effective focal ratio. The program next displays these values and all optical data, and then waits for a command. (This is the same point as that at which the program continues after a design has been retrieved from disk.)

To save the newly entered data to disk, use the S command. RAYTRACE will list all .RLD, .TDS, and .DES files and ask the name of the file to be saved. The filename may be up to eight characters long; do not add the extension .DES, since RAYTRACE does this automatically. To avoid saving a file, type L and press ENTER. After the program has saved the file, the computer returns to the command menu.

Examining an Optical System

The X command tells RAYTRACE to examine the optical properties of the system currently loaded into memory. Each analysis begins with a summary screen:

```
04-08-1988        SELECTION OF DESIGN ANALYSIS        11:55:36
gamma
Entrance pupil: 200.00 Focal ratio: 15.5 Effective focal length: 3096.6
                                          Back focal length: 3076.795
       [0] Trace a single ray
       [1] Longitudinal tabular
       [2] Longitudinal graphical
       [3] Transverse tabular
       [4] Transverse graphical
       [5] Sagittal tabular
       [6] Sagittal graphical
       [7] Skew tabular
       [8] Skew graphical
```

The entrance pupil is the diameter entered at the beginning of the program or through a modification of the design using the M command. The effective focal length is calculated assuming all optical components are positioned as in a centered system. When you are investigating a decentered design, there may be a small discrepancy in the results.

The focal ratio is the effective focal length divided by the diameter of the entrance pupil. The back focal length is the distance from the vertex of the last surface to the paraxial focus. These values are all calculated using the first set of refractive indices.

If more than one index per surface is entered, the following prompt appears:

```
Enter number of color to trace?
```

The number you enter must be equal to or less than the number of indices per surface.

After this, `RAYTRACE` continues with a menu of possible types of analysis:

```
(0) Trace a single ray
(1) Longitudinal tabular
(2) Longitudinal graphical
(3) Transverse tabular
(4) Transverse graphical
(5) Sagittal tabular
(6) Sagittal graphical
(7) Skew tabular
(8) Skew graphical
```

Enter the number of the selected analysis, or type `ESC` to exit.

Options `0` through `6` perform simple one-ray or one-plane analyses of systems. They are powerful tools for exploring the surface-by-surface passage of one or more rays through the optical system.

Option `0` is used to trace a ray through the optical system. This option is useful mainly for determining the size of the optical components and for checking the data entered for a tilted-component telescope design. A ray in the meridional plane can be defined by its object parameters: for an object at infinite distance, the angle and the zone; for an object at finite distance, the distance, the height, and the zone. An intercepting plane can also be defined—usually the position of the focal surface—when entering the distance from the last surface to the focus (BFD).

Options `1` and `2` are used to make longitudinal analyses. With these options, the longitudinal spherical aberration can be determined. These analyses cannot be carried out for tilted-component telescope designs, because with the shifting and tilting of optical components, the optical axis is no longer definable. For these designs, options `3` and `4` are applicable.

Options `3` and `4` are used to make transverse analyses. With these options, the minimum meridional blur can be determined. When an off-axis analysis is done, the position of the best focus in the tangential plane, or best tangential focus, can be determined.

Options `5` and `6` are also used to make transverse analyses. The difference from options `3` and `4` is that the calculations are for the sagittal, rather than the tangential, plane. In this case, when an off-axis analysis is done, the position of the best focus in the sagittal plane, or best sagittal focus, can be determined.

Spot Diagrams from the Skew-Ray Trace: Options `7` and `8` are the most powerful and most difficult menu choices to use. When you select them, you must select a series of ray-trace parameters, beginning with the distribution of rays on the entrance pupil:

```
SELECT INTERCEPT DISTRIBUTION AT THE ENTRANCE PUPIL

(1) Regular square distribution
(2) Regular concentric distribution
```

Enter the number of the selected analysis.

Under the head:

```
SELECTION OF STARTING LIGHT RAYS
```

you will be prompted for the diameter of the entrance pupil. If the system has an internal stop, the diameter of the entering bundle may be different from the stop diameter (for an example, see fig. 19.10). In this case, the diameter of the entering bundle must first be determined by ray-tracing, and the correct value subsequently filled in. If you enter `0`, `RAYTRACE` will ask the diameter with every analysis.

The entrance pupil intercept determines the number of zones for which the ray-tracing analysis is carried out. An intercept distance of 20 mm on an aperture of 200 mm will trace 11 zones, 5 above the optical axis, 5 below the optical axis, plus the principal ray. Note that all spot diagrams are calculated for the full aperture pupil; only in this way can non-rotationally symmetric systems be analyzed in all directions.

The distance from last surface to focus, or back focal distance, is requested next. The BFD is the axial distance from the last surface to a chosen reference intercepting plane. Begin by entering the distance to the paraxial focal plane (this is the back focal length shown on the summary screen), and search for a better focal position in subsequent analyses. The spot diagram will lie on this plane of focus.

The object distance is, in most cases, infinite. Enter `0` to specify this. If the value is finite, the actual distance from the object to the aperture stop must be given as a positive value in the same units as used for the design. Note that when you examine a design at a finite distance, the first surface you have specified must act as the entrance pupil.

For infinite object distances, you next enter the sine of incidence angle in the X-plane. This is the sine of the incidence angle of an oblique bundle of rays with the axis in the X-Z plane. For small angles, the sine can be taken the same as the angle expressed in radians. For a bundle entering parallel to the optical axis, this value is, of course, `0`.

The sine of incidence angle in the Y-plane is precisely the same as above, but for the angle with respect to the optical axis in the Y-direction. Ray tracing in the Y-direction is not necessary for rotationally symmetric systems because the system has the same characteristics and aberrations for all directions. Only tilted and decentered systems, i.e., systems which are not rotationally symmetric, require separate analysis in the Y-direction.

For finite object distances, `RAYTRACE` will calculate the values of X-sine and Y-sine from the values the user supplies for the X displacement (or height), of the object above or below the optical axis in the X-Z plane; and the Y displacement (or width), of the object in the Y-Z plane. The ray-trace calculations are carried out for rays in a diverging beam originating at a point source at these X,Y coordinates. They can be used, for example, to calculate measurement data for a Foucault test, given the position of the light source.

If you have specified that you wish a graphic analysis, `RAYTRACE` continues by asking:

```
Enter X-distance from optical axis = ?
```

For the graphic presentations of the meridional, sagittal, and skew-ray tracing, the value of the off-axis distance of the image point is the product of the angle of incidence and the focal length. When the X-sine of incidence is 0.01 and the focal length 2000 mm, for example, the X-distance from the axis should be 2000 × 0.01, or 20 mm. If the system suffers from distortion, this image may be lower or higher. This datum can also be obtained from the tabular display of ray-trace data.

```
Enter Y-distance from optical axis = ?
```

This is same as for the X-distance. If the object is on the Y-axis, that is, if either the Y-sine or the Y-axis displacement is zero, this question will be omitted.

```
Enter width of plotting screen = ?
```

The value chosen depends on the size of the aberrations you expect to see. If you enter 0.5 mm, the full width of the plotting screen will represent 0.5 mm. For scale, `RAYTRACE` shows a bar 0.025 mm long on the screen, so that a quick estimate of the aberration is possible. If the plotting screen width is too small, the aberration curve or the spot diagram may not fit within the screen, and the analysis will have to be repeated with a larger plotting screen.

If the plotting screen is too small, you may also miss an off-axis image entirely, especially if you estimate the off-axis distance incorrectly. In that case, you will not see a spot diagram or aberration curve on the screen because it falls outside the chosen position of the plotting screen.

If this happens, you will need to increase the size of the plotting screen until you locate the aberration curve or spot diagram. Read its X-distance from the axis; then make a new plot centered on the curve or spot diagram, using an appropriate plotting screen width.

Displaying the System Analysis: Once all necessary data are in the computer, `RAYTRACE` carries out its optical calculations. Results may be presented in tabular or graphic form. (Graphics output is available only if the computer has CGA graphics.) You may print graphic results (provided you have installed `GRAPHICS.COM` in memory before starting `RAYTRACE`); modify the display; or modify

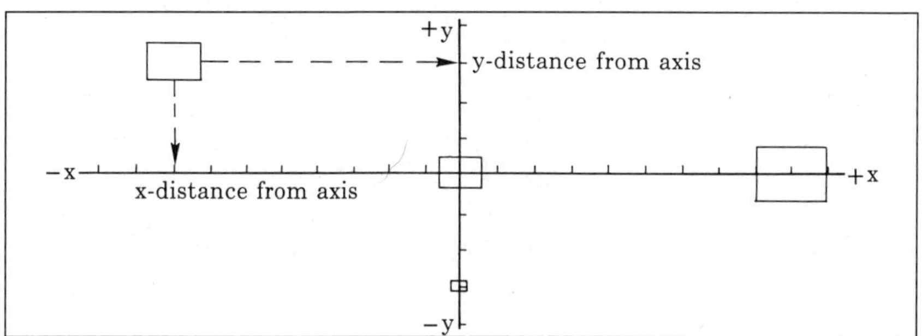

Fig. 22.3 *Sizes and Positions of the Plotting Screen.*

the optical system or direction of the incoming rays, and run another analysis.

The longitudinal, meridional, and sagittal aberration graphs appear on a system of coordinate axes. The display of LA (longitudinal aberration), because it is a symmetrical aberration, uses only the upper half of the coordinate axis system, unless vignetting data for the design are present and a vignetting analysis is requested. The graphics-mode skew-ray trace presents a spot diagram.

Modifying a Design: The `M` command allows you to change the optical parameters of a design. During the modification, the data of the design will appear in the same sequence as you entered them. If it is not necessary to modify a value, simply press the `ENTER` key.

Altering the Display: `RAYTRACE` stores the numerical results in memory, so you may display the results (using the A command) in as many ways as the program allows. Only results for the rays that have actually been traced can be altered, so if the chosen starting light rays have to be altered, a new analysis must be done.

After `A` has been pressed, the screen clears and `RAYTRACE` offers you the chance to switch from graphics to tabular, or tabular to graphics, mode. You may, of course, remain in the present mode.

If the presentation of the results is not fully satisfactory, you can change the location and size of the plotting screen. If you do not wish to alter a value, hit the `ENTER` key. This capability is extremely useful for centering or enlarging an aberration figure on the screen. Fig. 22.3 shows some possibilities for the apparent size of the plotting screen and its position with respect to the optical axis.

You may also focus the optical system by changing the intercept plane. Normally analyses are begun with the intercept plane at the paraxial focus. When aberrations are present, however, the image blur is usually not smallest there. By moving this plane—"refocusing"—you will determine the optimum position for the intercepting surface.

Finally, you may re-specify the focal surface curvature. If you know the radius of curvature, you can select and display the off-axis spot diagram lying on this surface. If the radius of curvature is not known, you must determine the optimum position of an off-axis interception point by refocusing, then calculate by hand the radius of curvature from the off-axis distance and the focal shift, using the sagitta formula.

22.4.2 Vignetting Calculations

The principles behind the vignetting analysis are explained in section 19.4.

The calculation can be carried out with either the tabular or the graphic tracing method, and the result of the calculation is shown on the screen as, for instance, "19 of 81 rays are vignetted." Tabular data are labeled with a V to designate vignetted rays; graphical rays are not plotted if they are vignetted.

Much more data must be input for an analysis of vignetting than for a normal ray-tracing analysis. All limitations of the optical elements, stops, and central obstructions must be included (see fig. 22.2). Each stop or central obstruction is treated as an optical surface with zero power and the same refractive index as the medium in which it is placed.

When tracing a bundle of rays through a mirror with a hole, you may either calculate the position of the edge of the hole by hand and enter the surface as a plane, or treat the surface as one with the same radius of curvature and position as the mirror and the same refractive index as the adjacent medium. In the latter case, it is not necessary to calculate the position of the edge of the hole with respect to the vertex of the curvature.

A vignetting analysis is rather slow because every vignetting surface is treated as an optical surface. For an adequate understanding of vignetting, three or four off-axis points must be checked (see fig. 19.12). In spot diagrams, the effect of vignetting on the light distribution can be seen nicely.

22.4.3 Tilted and Decentered Surfaces

The optical calculations for tilted and decentered surfaces are described in section 20.9. In RAYTRACE, the decentering parameters are entered and the system is calculated as a centered system. Axial shifts are very easily handled: the axial distance of the particular surface is increased or decreased according to the shift. Tilted surfaces are more difficult, but the notation used in RAYTRACE is made clear in section 20.9. For a decentered system, the focal surface must be treated as a specific optical element with the same refractive index as the surrounding medium.

Tilted-Component Telescopes: When calculating TCTs, the greatest difficulties occur when the focal plane is tilted; that is, the position of smallest spot diagram is ahead of the focal plane on one side of the axis, but behind it on the other side. Find the tilt angle of the true focal plane, phi, with respect to the reference plane, from:

$$\phi = \arctan\frac{|df+| + |df-|}{|h+| + |h-|} \tag{22.4.2}$$

where $|df+|$ and $|df-|$ are the absolute values of the focal deviations, and $|h+|$ and $|h-|$ are the absolute values of the off-main-axis distances. This formula is valid when the reference plane is perpendicular to the main ray.

Because the focal plane is tilted, the shape of the spot diagram will be slightly distorted if the display plane is not tilted. If a distortion error of 2% is acceptable, the maximum permissible tilt angle is arccos(0.98), or 11.5 degrees. For larger angles, the reference plane should be tilted to avoid any distortion due to tilt.

22.4.4 Notes on Vignetting Computations

It is possible when ray-tracing a system that a ray will miss a surface or that the iteration process to find the intercept of a higher-order aspheric surface will fail. It is also possible that a ray traveling from a dense medium to a less-dense medium will be internally reflected. `RAYTRACE` checks for these eventualities and works as follows.

1. **For Trace a Single Ray Calculations:** The program will stop the printout at the surface previous to the one which the ray misses or in which it is reflected internally.
2. **For Vignetting Calculations:** The program adds all rays which miss a surface or are internally reflected to the total number of vignetted rays.
3. **For Non-Vignetting Calculations:** The program counts the number of rays that miss a surface, are reflected internally, or, if an aspheric surface, for which the iteration process fails. `RAYTRACE` does not record how the ray was lost, and does not continue to trace a lost ray. After the analysis is complete, the program prints the number of "vignetted" rays.

22.4.5 Data Input Exercises

We recommend that you key in the following seven system designs treated in this book as examples. By keying in the specifications from the tables in the book you will be able to achieve a working understanding of the data input procedure.

1. 200 mm *f*/4 Newtonian parabolic mirror
2. 200 mm *f*/8 Ritchey-Chrétien (table 7.4)
3. 200 mm *f*/10 Apoklaas doublet (table 6.1)
4. 200 mm *f*/10 compact Schmidt-Cassegrain (table 9.1)
5. 200 mm *f*/10 compact Schmidt-Cassegrain vignetting calculations (section 19.4)
6. 200 mm *f*/10 Apoklaas with second lens shifted 1 mm with respect to the first (table 6.1)
7. 200 mm *f*/14.7 Trischiefspiegler (table 12.1)

We recommend that you also run the program with these examples, repeating the calculations that the authors themselves have made, and compare the resulting spot diagrams with those shown in this book.

22.5 Optimizing Predesigns from TDESIGN

TDESIGN creates third-order approximations of telescope designs. As noted in chapter 21, these designs, especially those for catadioptrics, need optimization by skew-ray tracing. When rays are traced, they will show deviations from the predesign, especially in the focal length of the telescope and position of the focal surface (BFL). The size of the deviation depends on the power of the corrector used. Two-mirror designs do not normally need optimization, but when strong curves are present, it is best to ray-trace the system in order to gain a thorough understanding of the design and its performance.

Iteration is always necessary in the design of a two-mirror catadioptric, to choose the right secondary magnification factor to make an anastigmatic or aplanatic design with the desired telescope focal length and back focal length behind the front of the primary mirror.

The telescope systems presented below were made using the curves for anastigmatic and aplanatic designs presented in chapter 21. These examples show quite dramatically the influence of the optical power of the corrector on the performance of each system.

Using RAYTRACE to optimize a design is a lot like optimizing a lens using LENSDES. The designer enters each change by hand and follows how much the aberrations change. (Use the M command to modify the optical system parameters in RAYTRACE.) This process is effective because the results TDESIGN produces are so close to final that manual optimization takes only a little time. More importantly, however, the process of manual optimization gives the designer a feeling for how critical the optical components are, and a sense of the tolerances.

If you notice that small changes during optimization produce large changes in optical performance, a good tolerance analysis is more than worthwhile. Of course, for any critical application, running a manual optimization does not obviate the necessity of running a full tolerance analysis.

22.5.1 The Wright Design

The Wright camera is just about the simplest case possible: an oblate ellipsoid with a Schmidt corrector at half the radius. Starting with the following inputs:

Clear Aperture	200 mm
Focal Length	750 mm
Position Corrector	750 mm

we obtain the following as a result of skew-ray tracing:

Focal Length	752.50 mm (+0.3%)
Back Focal Length	–747.49 mm (–0.3%)

These deviations cannot be reduced unless the focal length of the primary

Wright Camera		
	Predesign	Optimized Design
Clear aperture	200	200
Focal length	750	752.5
Geometric focal ratio	3.75	3.76
Primary radius	–1500.00	–1500
Primary aspheric deformation	1.000000	1.0000
Schmidt corrector (n = 1.51872) strength	2.000000	2.0000
Schmidt corrector thickness	5.00	5.00
Primary corrector distance	750.00	750.00
Back focal length (behind the primary)	–750.00	–747.49

mirror or the position of the neutral zone on the corrector is changed. The latter is currently 0.866 for the best color correction, and should not be modified.

The overall optical performance is good. The diameter of the axial spot diagram is 0.0015 mm, small enough that it needs no optimization. The spot diagram 22.5 mm off axis is 0.045 mm diameter and round because astigmatism is present. It is not possible to reduce coma in this design even though it is noticeable. In any case, when so little coma is present, it is not necessary to reduce it.

Wright cameras seldom need optimization.

22.5.2 The Schmidt-Cassegrain Telescope

The goal is an anastigmatic 200 mm *f*/10 Schmidt-Cassegrain design. With a few iterations of `TDESIGN`, we obtained a predesign for a telescope with a clear aperture of 200 mm and a focal length of 2000 mm, having a primary focal length of 350 mm, a back focal length of 175 mm (i.e., E=0.5), an aspheric secondary with a magnification of 5.75, and the corrector at 1/0.95 times the mirror separation.

When we ran a skew-ray trace, we found:

```
Focal length of telescope              1763.2 mm (-12%)
Back focal length (behind primary)   111.39 mm (-36%)
```

These values deviate strongly from those in the predesign. This difference is due to the optical power of the Schmidt corrector. The position of the neutral zone is especially important; if the neutral zone were at the optical axis, deviations would be absent but the color correction of the telescope would be poor.

There are two ways to optimize these parameters: change the curvature of the secondary mirror or change the distance between the mirrors and the corrector-to-primary distance. If the distance between the mirrors is fixed, or if the obstruction is fixed, then the curvature must be changed. However, if the radius of curvature is fixed, as it would be in a flat-field design, then the distances must be changed. When this is done, the distance between the mirrors and the corrector-to-primary distance must be changed proportionally.

We altered the radius of curvature of the secondary mirror:

```
R(4)    (TDESIGN)      -189.56
R(4)    (RAYTRACE)     -175.56
```

This now gave:

```
Focal length             2031.7 mm (+1.6%)
Back focal length           169.8 mm (-3%)
```

These departures are sufficiently close to the design goals to be worth optimizing again.

However, spherical aberration is –2.16 mm. By changing the power of the Schmidt corrector in several iterations, we obtained:

```
G (DESIGN)       0.823664
G (RAYTRACE)     0.841664
```

This system produces an axial spot diagram of 0.010 mm diameter. At 20 mm from the optical axis, the size on a curved focal surface is about 0.008 mm. No coma can be seen in the spot diagram, and the design is free of astigmatism.

This time we will alter the distances between the two mirrors and between the corrector and the primary. A few iterations give us the following parameters:

```
Mirror separation (TDESIGN)        -271.81 mm
Mirror separation (RAYTRACE)       -269.56 mm

Corrector-primary (TDESIGN)          286.1 mm
Corrector-primary (RAYTRACE)        283.85 mm
```

This gives new parameters:

```
Focal length            1998.9 mm (+0.05%)
Back focal length       177.6 mm (+1%)
```

These deviations are sufficiently small. The spherical aberration is now –1.4 mm.

A few iterations of changing the power of the Schmidt corrector yields:

```
G (TDESIGN)     0.823664
G (RAYTRACE)    0.835664
```

The axial spot diagram is 0.010 mm, and 20 mm from the optical axis, the size on a curved focal surface is about 0.006 mm. No comatic shape is visible in the spot diagram.

It is possible, of course, that when the design specifications are changed,

Schmidt-Cassegrain Telescope		
	Predesign	Optimized Design
Clear aperture	200	200
Focal length	2000	1998.9
Minimum secondary diameter	44.68	45.02
Geometric focal ratio	10.00	9.99
Photographic focal ratio	10.33	10.26
Primary radius	–700.00	–700.00
Primary aspheric deformation	0.000000	0.0000
Secondary radius	–189.56	–189.56
Secondary aspheric deformation	–0.622782	–0.622782
Primary to secondary distance	–271.81	–269.56
Schmidt corrector (n=1.51872) strength	0.823664	0.835664
Schmidt corrector thickness	5.00	5.00
Primary to corrector distance	286.50	283.85
Secondary to focal plane distance	446.81	447.16
Back focal length (behind the primary)	175.00	177.6

some coma will be introduced. If the design contains aspheric mirrors, then not only the power of the Schmidt corrector but also the aspheric deformation of the mirror must be changed. If the system has no aspheric mirrors, then the distance from corrector to primary must be changed.

Take care: although changing the Schmidt corrector is the fastest way to optimize a system, changing the mirror deformations very quickly introduces a large amount of coma.

22.5.3 The Houghton Camera

We started this design with a clear aperture of 200 mm and a focal length of 750 mm, with a corrector-to-primary distance of 750 mm and a design index of 1.51872 (BK7 at 546.1 nm). Other data were less important and did not influence the optimizations.

Results of skew-ray tracing with RAYTRACE were:

```
Focal length        +754.1 mm (=0.55%)
Back focal length   -748.5 mm (-0.2%)
```

These deviations are comparable with those of the Wright system. They cannot be optimized unless the curvature of the primary mirror is changed.

The optical performance is quite good; the axial spot diagram is 0.0016 mm in diameter, and 0.022 mm in diameter 22.5 mm from the axis. The best off-axis performance is obtained on a curved focal surface. The spot diagram on the optical axis is small, so it is not necessary to optimize. The size of the spot diagram on the

Houghton Camera		
	Predesign	Optimized Design
Clear aperture	200	200
Focal length	750	754.1
Geometric focal ratio	3.75	3.77
Primary radius	–1500.00	–1500
Houghton corrector (n = 1.51872)		
First radius	1185.40	1185.4
Thickness	11.43	11.43
Second radius	–2264.41	–2264.41
Air space	2.17	2.17
Third radius	–1185.40	–1185.4
Thickness	5.00	5.00
Fourth radius	2264.41	2264.41
Primary to corrector distance	750.00	750
Back focal length (behind the primary)	–750.00	–748.5

curved focal surface does not need optimization either.

Houghton camera designs seldom need further optimization. If spherical aberration and/or coma occur, first change `R(1)` and `R(3)`, keeping `R(1)=-R(3)`. Once spherical aberration is minimized, check the coma. If coma is present, change `R(2)` and `R(4)`, keeping `R(2)=-R(4)`. This, of course, introduces some spherical aberration which can be minimized by changing `R(1)` and `R(3)` again. After several iterations, you may begin to see some kind of relationship between the front and back curvatures. Once you know this relationship, you can complete the optimization much faster.

22.5.4 The Houghton-Cassegrain Telescope

Our goal was an anastigmatic 200 mm *f*/12.5 Houghton-Cassegrain telescope. Again after some iterations in `TDESIGN`, we settled on a design with a clear aperture of 200 mm and a telescope focal length of 2500 mm. The focal length of the primary was 715 mm (so the secondary magnification was 3.5), and the back focal length was 250 mm (E=0.35). The corrector distance came out to 1/0.95 of the mirror separation, and the design index for BK7 at 546.1 nm wavelength was 1.51872.

The skew-ray trace gave us:

```
Focal length              2465 mm (-1.4%)
Back focal length       229.21 mm (-8.3%)
```

As you can see, these results do not deviate very much from the design desired.

A look at the optical performance shows that the size of the axial spot dia-

Houghton-Cassegrain Telescope		
	Predesign	Optimized Design
Clear aperture	200	200
Focal length	2500	2465
Minimum secondary diameter	60.03	59.32
Geometric focal ratio	12.50	12.33
Photographic focal ratio	13.10	12.91
Primary radius	−1430.00	−1430.00
Secondary radius	−601.15	−601.15
Primary to secondary distance	−500.39	−500.39
Houghton corrector (n = 1.51872)		
First radius	1072.84	1072.84
Thickness	13.47	13.47
Second radius	−1317.28	−1317.28
Air space	1.02	1.02
Third radius	−1072.84	−1072.84
Thickness	5.00	5.00
Fourth radius	1317.28	1317.28
Primary to corrector distance	526.50	526.50
Secondary to focal plane distance	750.39	729.6
Back focal length (behind the primary)	250.00	229.21

gram is 0.0055 mm, which is more than sufficient. At a distance of 20 mm from the optical axis, the size of the spot diagram is 0.007 mm, which is also very good. The smallest off-axis spot diagram was obtained for a curved focal surface.

If we wish to reduce the differences between our design goal and the initial design, we can do the same as with the Schmidt-Cassegrain design. Using the original corrector from `TDESIGN` gives spot sizes comparable with the values above, which do not need further improvement. If there is any residual spherical aberration or coma, follow the iterative procedure described in section 22.5.3 above.

22.5.5 The Maksutov Camera

The Maksutov camera has the strongest corrector of any catadioptric design. Because of this, the deviations of the first-order values are larger than in any of the other systems. Therefore even `TDESIGN`, using third-order methods, requires an iterative optimization procedure.

Our design goal is a 200 mm clear aperture *f*/3.75 camera. To correct coma, we needed a corrector distance of 1008 mm for a 20 mm thick corrector designed for green light (546.1 nm) and a color correction factor of 0.98. Using BK7 glass, the design index is 1.51872.

Maksutov Camera		
	Predesign	Optimized Design
Clear aperture	200	200
Focal length	750	727.2
Geometric focal ratio	3.75	3.64
Primary radius	–1500.00	–1500.00
Maksutov corrector (n = 1.51872, shape factor = .98)		
First radius	–285.71	–285.71
Thickness	20.00	20.00
Second radius	–297.27	–297.27
Corrector to primary distance	1008.80	1008.80
Back focal length (behind the primary)	–750.00	–765.75

From skew-ray tracing, we see:

```
Focal length            727.2 mm (-3%)
Back focal length    765.75 mm (2.1%)
```

These values are close to the design values. They can be changed only by altering the thickness of the corrector or the focal length of the primary. Optical performance is excellent; the spot diagram on axis is 0.005 mm, 0.006 mm at 21.8 mm from the axis on a curved focal surface. `TDESIGN` has done a fine job!

If you use a corrector with a different thickness, some residual spherical aberration and/or coma may be introduced. Minimize spherical aberration by changing `R(1)` and `R(2)`. To keep the color correction optimal, these changes should be the same for both surfaces:

`R(1)`new = `R(1)`old + `dR`

`R(2)`new = `R(2)`old × (`R(1)`new/`R(1)`old)

Once the spherical aberration has been minimized, check for coma. If coma is present, the position of the corrector should be altered until coma has been eliminated. After this is done, check the color correction by ray-tracing two other colors of interest, red and blue, for example. If the color correction is not optimal, the first or second radius should be recalculated as discussed in chapter 10. This will influence spherical aberration and coma, so the new design must again be re-optimized.

22.5.6 The Maksutov-Cassegrain Telescope

The Maksutov-Cassegrain, because of the optical power of the secondary mirror, has an even stronger corrector than the Maksutov camera. We set out to design a 200 mm *f*/15 telescope. The focal length of the primary was 910 mm, giving a secondary magnification of 3.5. The back focal length was 230 mm (E=0.25). We chose a corrector made of BK7 (design index 1.51872 at 546.1 nm) with a thickness of 36.4 mm (0.04 Fprim) and a color correction factor of 0.98. The corrector distance was chosen to give us a coma-corrected system.

The skew-ray trace gave the following rather depressing output:

```
Focal length telescope   3508 mm (+16.9%)
Back focal length        477.7 mm (207%)
```

These results depart strongly from the data originally entered in `TDESIGN`. As in the Schmidt-Cassegrain design, the designer can alter the radius of curvature of the secondary mirror or change both the distance between the mirrors and the distance from the corrector to the primary mirror.

In this example, we decided to change the radius of curvature of the secondary mirror. We felt that by reducing the back focal length, the secondary would become flatter and thus also give us a longer focal length.

After some iterations, we obtained:

```
R(4) (TDESIGN)          -761.68 mm
R(4) (RAYTRACE)            -836.68
```

This gives new parameters of:

```
Focal length telescope          2775 mm (-7.5%)
Back focal length              242.3 mm (+5.3%)
```

If you want a design that more exactly matches the desired design parameters, the curvatures and the distance between the mirrors would have to be changed. (For flat-field designs, it would also be necessary to change the primary mirror because the ratio of Rsec/Rprim should be kept constant. After an analysis of the curvature of field, a new design can be made with `TDESIGN` if the focal surface is not flat. Optimizing flat-field designs is more time-consuming than optimizing curved field designs.)

Back to the design described here. With the corrector unchanged, there was residual spherical aberration. After a few iterations, the radii of curvature of the corrector became 3 mm stronger. Thus,

```
(TDESIGN)    R(1)=-425.21 mm   R(2)=-446.25 mm
(RAYTRACE)   R(1)=-422.31 mm   R(2)=-443.2 mm
```

The optical performance is very good. The axial diameter of the spot diagram is 0.003 mm, and the spot diameter 19.5 mm off-axis is only 0.005 mm. The best off-axis images lie on a curved focal surface. Neither coma nor astigmatism is present.

Maksutov-Cassegrain Telescope		
	Predesign	Optimized Design
Clear aperture	200	200
Focal length	3000	2775
Minimum secondary diameter	58.31	64.06
Geometric focal ratio	15.00	13.88
Photographic focal ratio	15.68	14.65
Primary radius	−1820.00	−1820.00
Secondary radius	−761.68	−836.68
Primary to secondary distance	−644.68	−644.68
Maksutov corrector (n = 1.51872, shape factor = .98)		
First radius	−425.21	−422.31
Thickness	36.40	36.40
Second radius	−446.25	−443.21
Corrector to primary distance	679.00	679.00
Secondary to focal plane distance	874.68	877.04
Back focal length (behind the primary)	230.00	242.3

22.5.7 Automatic Optimizations

It would, of course, be possible to write an automatic optimization program. We intentionally did not write such a program because then optimizing a design would not provide the user with hands-on design experience.

Besides, hand optimization is more satisfying and fun to do. You decide for yourself to what extent to optimize a design, what course of action you will follow, and what results you will be content with. The higher your goal, the more time optimization takes.

In optimizing your first telescope design, you may have trouble finding your way, and you could be disappointed if you take a wrong direction. We recommend working through the designs above before attempting totally new designs. When you are ready to work with your own designs, you will be experienced, capable, and ready to go.

Appendix A:

Schott Optical Glass Specifications

6-digit number	Glass Type	Cost vs BK 7	Glass density	n_r	n_C	n_d	n_e	n_F	n_g	n_h
434954	Fluorite	?	?	1.43171	1.43249	1.43388	1.43496	1.43704	1.43949	1.44151
437907	FK 54	19.83	3.18	1.43467	1.43552	1.43700	1.43815	1.44034	1.44291	1.44501
458678	Quartz	?	?	1.45515	1.45637	1.45846	1.46008	1.46313	1.46670	1.46962
465658	FK 3	2.97	2.27	1.46107	1.46232	1.46450	1.46619	1.46939	1.47315	1.47626
471673	FK 1	?	2.31	1.46728	1.46853	1.47069	1.47236	1.47552	1.47924	1.48230
479587	TiK 1	?	2.39	1.47479	1.47621	1.47869	1.48063	1.48436	1.48880	1.49249
486818	FK 52	18.58	3.64	1.48320	1.48424	1.48605	1.48747	1.49018	1.49337	1.49601
487704	FK 5	2.69	2.45	1.48410	1.48535	1.48749	1.48914	1.49227	1.49593	1.49894
487845	FK 51	13.54	3.73	1.48379	148480	1.48656	1.48794	1.49056	1.49365	1.49619
498651	BK 3	?	2.37	1.49457	1.49594	1.49831	1.50014	1.50360	1.50767	1.51101
498670	BK 10	3.09	2.39	1.49419	1.49552	1.49782	1.49960	1.50296	1.50690	1.51014
500614	K 11	?	2.50	1.49621	1.49764	1.50013	1.50207	1.50578	1.51019	1.51385
501564	K 10	1.93	2.52	1.49713	1.49867	1.50137	1.50349	1.50756	1.51243	1.51649
504669	PK 1	?	2.44	1.50011	1.50146	1.50378	1.50558	1.50898	1.51298	1.51627
505596	K 51	?	2.47	1.50109	1.50258	1.50518	1.50720	1.51106	1.51564	1.51945
508612	ZK N7	1.34	2.49	1.50445	1.50592	1.50847	1.51045	1.51423	1.51869	1.52238
510635	BK 1	1.28	2.46	1.50621	1.50763	1.51009	1.51201	1.51566	1.51998	1.52355
511510	TiF 1	?	2.47	1.50648	1.50818	1.51118	1.51356	1.51820	1.52384	1.52866
511604	K 7	1.10	2.53	1.50707	1.50854	1.51112	1.51314	1.51700	1.52159	1.52540
515547	KF 3	1.70	2.56	1.51008	1.51169	1.51454	1.51678	1.52110	1.52627	1.53060
517522	KF 6	1.61	2.67	1.51274	1.51443	1.51742	1.51978	1.52434	1.52984	1.53446
517642	BK 7	1.00	2.51	1.51289	1.51432	1.51680	1.51872	1.52238	1.52669	1.53024
517643	UBK 7	1.74	2.51	1.51290	1.51433	1.51680	1.51872	1.52237	1.52667	1.53022
518590	K 3	1.64	2.54	1.51404	1.51556	1.51823	1.52032	1.52435	1.52913	1.53311
518602	BaLK N3	1.13	2.61	1.51436	1.51586	1.51849	1.52054	1.52447	1.52914	1.53301

6-digit number	Glass Type	Cost vs BK 7	Glass density	n_r	n_C	n_d	n_e	n_F	n_g	n_h
518651	PK 2	1.50	2.51	1.51434	1.51576	1.51821	1.52011	1.52372	1.52798	1.53148
519574	K 4	5.41	2.63	1.51463	1.51620	1.51895	1.52111	1.52524	1.53017	1.53428
520637	BK 8	?	2.57	1.51620	1.51764	1.52015	1.52210	1.52581	1.53019	1.53380
521697	PK 50	13.54	2.59	1.51690	1.51824	1.52054	1.52232	1.52570	1.52968	1.53294
522595	K 5	0.90	2.59	1.51829	1.51982	1.52249	1.52458	1.52860	1.53338	1.53735
523515	KF 9	1.34	2.71	1.51862	1.52035	1.52341	1.52583	1.53052	1.53616	1.54093
523602	K 50	1.30	2.62	1.51841	1.51992	1.52257	1.52464	1.52861	1.53331	1.53721
523604	UK 50	?	2.62	1.51842	1.51993	1.52257	1.52464	1.52858	1.53327	1.53715
525647	PK 3	2.42	2.59	1.52148	1.52292	1.52542	1.52736	1.53104	1.53538	1.53896
526600	BaLK 1	?	?	1.52223	1.52375	1.52642	1.52851	1.53252	1.53729	1.54125
527511	KzF 6	?	2.54	1.52193	1.52370	1.52682	1.52927	1.53400	1.53969	1.54446
529517	KzF 2	2.37	2.54	1.52457	1.52634	1.52944	1.53188	1.53658	1.54222	1.54695
529770	PK 51	17.53	3.97	1.52528	1.52646	1.52855	1.53019	1.53333	1.53704	1.54010
531511	KF 50	?	2.70	1.52599	1.52776	1.53088	1.53335	1.53814	1.54391	1.54877
531621	BK 6	?	2.69	1.52701	1.52851	1.53113	1.53317	1.53706	1.54166	1.54546
532488	LLF 6	1.39	2.81	1.52661	1.52845	1.53172	1.53431	1.53935	1.54546	1.55064
533460	TiF 2	?	2.51	1.52718	1.52911	1.53256	1.53531	1.54070	1.54734	1.55309
533580	ZK 1	1.35	2.71	1.52877	1.53036	1.53315	1.53534	1.53955	1.54457	1.54875
534553	ZK 5	?	2.76	1.52919	1.53084	1.53375	1.53605	1.54049	1.54579	1.55023
540511	KF 1	?	2.78	1.53544	1.53723	1.54041	1.54293	1.54781	1.55371	1.55869
540597	BaK 2	1.36	2.86	1.53564	1.53721	1.53996	1.54212	1.54625	1.55117	1.55525
541472	LLF 2	1.77	2.87	1.53537	1.53729	1.54072	1.54344	1.54876	1.55521	1.56070
547536	BaLF 5	1.54	2.95	1.54258	1.54432	1.54739	1.54982	1.55452	1.56018	1.56492
548422	TiF 3	?	2.61	1.54169	1.54382	1.54765	1.55072	1.55680	1.56433	1.57091
548458	LLF 1	1.28	2.94	1.54256	1.54457	1.54814	1.55099	1.55655	1.56332	1.56910
549454	LLF 7	?	2.98	1.54320	1.54523	1.54883	1.55170	1.55731	1.56415	1.56998
551497	KzF 1	2.87	2.81	1.54592	1.54781	1.55115	1.55379	1.55891	1.56508	1.57028
552635	PSK 3	1.89	2.91	1.54811	1.54965	1.55232	1.55440	1.55836	1.56303	1.56689
554512	BaLF 8	?	2.99	1.54854	1.55036	1.55361	1.55618	1.56118	1.56721	1.57230
557587	BaK 5	1.81	3.02	1.55219	1.55383	1.55671	1.55897	1.56332	1.56850	1.57280

6-digit number	Glass Type	Cost vs BK 7	Glass density	n_r	n_C	n_d	n_e	n_F	n_g	n_h
558542	KzFS N2	9.55	2.56	1.55339	1.55521	1.55836	1.56082	1.56552	1.57111	1.57578
558673	PSK 50	?	2.93	1.55353	1.55499	1.55753	1.55951	1.56327	1.56771	1.57138
560472	LLF 3	?	2.99	1.55458	1.55658	1.56013	1.56295	1.56845	1.57515	1.58084
560612	SK 20	?	3.03	1.55525	1.55684	1.55963	1.56181	1.56598	1.57093	1.57502
561452	LLF 4	?	3.02	1.55561	1.55768	1.56138	1.56433	1.57009	1.57713	1.58313
564438	LF 8	old	-	1.55846	1.56060	1.56444	1.56750	1.57350	1.58085	1.58713
564608	SK 11	1.90	3.08	1.55940	1.56101	1.56384	1.56605	1.57028	1.57530	1.57946
567428	LF 6	?	3.11	1.56119	1.56339	1.56732	1.57047	1.57663	1.58419	1.59067
568580	BaK 50	2.25	2.93	1.56306	1.56476	1.56774	1.57007	1.57455	1.57987	1.58429
569561	BaK 4	1.02	3.10	1.56402	1.56576	1.56883	1.57125	1.57590	1.58145	1.58609
569631	PSK 2	2.22	3.06	1.56438	1.56597	1.56873	1.57088	1.57498	1.57982	1.58382
571529	BaLF 3	?	3.18	1.56628	1.56810	1.57135	1.57392	1.57890	1.58488	1.58990
573426	LF 1	?	3.16	1.56687	1.56910	1.57309	1.57629	1.58256	1.59025	1.59684
573575	BaK 1	1.29	3.19	1.56778	1.56949	1.57250	1.57487	1.57943	1.58488	1.58940
574521	BaLF 51	?	3.03	1.56875	1.57062	1.57393	1.57656	1.58164	1.58776	1.59290
574564	BaK 6	?	3.10	1.56961	1.57136	1.57444	1.57687	1.58154	1.58713	1.59180
575415	LF 7	1.25	3.20	1.56862	1.57090	1.57501	1.57830	1.58476	1.59271	1.59953
578416	LF 4	?	3.21	1.57203	1.57433	1.57845	1.58175	1.58824	1.59620	1.60305
580537	BaLF 4	1.46	3.17	1.57448	1.57632	1.57957	1.58214	1.58711	1.59307	1.59806
581409	LF 5	1.11	3.22	1.57489	1.57723	1.58144	1.58482	1.59146	1.59964	1.60667
582421	LF 3	?	3.21	1.57576	1.57805	1.58215	1.58544	1.59188	1.59980	1.60660
583465	BaF 3	1.72	3.28	1.57685	1.57893	1.58267	1.58565	1.59147	1.59857	1.60460
583595	SK 12	2.79	3.27	1.57845	1.58015	1.58313	1.58547	1.58996	1.59529	1.59971
584370	TiF 4	?	2.68	1.57690	1.57945	1.58406	1.58779	1.59521	1.60452	1.61274
586610	LgSK 2	18.43	4.15	1.58149	1.58311	1.58599	1.58828	1.59271	1.59801	1.60245
589409	LF 2	?	3.30	1.58259	1.58495	1.58921	1.59263	1.59935	1.60762	1.61473
589485	BaF N6	?	3.17	1.58331	1.58536	1.58900	1.59189	1.59752	1.60435	1.61016
589514	BaLF 50	1.66	3.11	1.58354	1.58549	1.58893	1.59166	1.59695	1.60333	1.60869
589530	BaLF 6	?	3.32	1.58381	1.58569	1.58904	1.59169	1.59681	1.60296	1.60812
589613	SK 5	1.80	3.30	1.58451	1.58619	1.58913	1.59142	1.59581	1.60100	1.60529

6-digit number	Glass Type	Cost vs BK 7	Glass density	n_r	n_C	n_d	n_e	n_F	n_g	n_h
592485	KzFS 6	?	3.03	1.58617	1.58827	1.59197	1.59487	1.60048	1.60723	1.61292
592583	SK 13	2.79	3.37	1.58698	1.58873	1.59181	1.59423	1.59888	1.60442	1.60902
594355	TiF N5	11.92	2.71	1.58598	1.58867	1.59356	1.59751	1.60538	1.61529	1.62405
596392	F 8	1.37	3.38	1.58853	1.59102	1.59551	1.59912	1.60622	1.61500	1.62257
599469	KzFS N9	?	3.01	1.59252	1.59471	1.59856	1.60159	1.60747	1.61456	1.62055
601382	F 14	1.74	3.44	1.59419	1.59676	1.60140	1.60513	1.61249	1.62160	1.62946
603380	F 5	1.17	3.47	1.59615	1.59874	1.60342	1.60718	1.61461	1.62380	1.63174
603425	BaSF 5	?	3.49	1.59669	1.59902	1.60323	1.60660	1.61323	1.62136	1.62833
603606	SK 14	1.86	3.44	1.59834	1.60007	1.60311	1.60548	1.61003	1.61541	1.61988
603654	PSK 52	11.47	3.38	1.59867	1.60028	1.60310	1.60530	1.60950	1.61447	1.61858
604536	SSK 51	2.12	3.28	1.59830	1.60022	1.60361	1.60629	1.61147	1.61770	1.62292
606378	F 15	?	3.48	1.59833	1.60094	1.60565	1.60945	1.61695	1.62623	1.63426
606439	BaF 4	1.27	3.5	1.59925	1.60153	1.60562	1.60889	1.61532	1.62318	1.62990
607494	BaF 5	?	3.54	1.60155	1.60361	1.60729	1.61021	1.61590	1.62279	1.62862
607567	SK 2	1.80	3.55	1.60230	1.60414	1.60738	1.60994	1.61486	1.62073	1.62562
607595	SK 7	1.90	3.51	1.60241	1.60418	1.60729	1.60973	1.61440	1.61995	1.62455
609464	BaF 52	1.43	3.32	1.60251	1.60469	1.60859	1.61170	1.61779	1.62521	1.63154
609589	SK 3	2.24	3.53	1.60388	1.60567	1.60881	1.61127	1.61600	1.62163	1.62630
610567	SK 1	?	3.56	1.60515	1.60699	1.61025	1.61282	1.61775	1.62365	1.62856
611559	SK 8	?	3.57	1.60600	1.60787	1.61117	1.61378	1.61880	1.62479	1.62979
613370	F 3	1.35	3.55	1.60537	1.60806	1.61293	1.61685	1.62461	1.63423	1.64256
613443	KzFS 1	?	3.14	1.60658	1.60894	1.61310	1.61639	1.62276	1.63048	1.63702
613443	KzFS N4	5.01	3.20	1.60689	1.60924	1.61340	1.61669	1.62309	1.63085	1.63745
613574	SK 19	?	3.56	1.60834	1.61018	1.61342	1.61597	1.62087	1.62672	1.63159
613586	SK 4	1.84	3.57	1.60774	1.60954	1.61272	1.61521	1.62000	1.62569	1.63042
614552	SK 9	?	3.60	1.60879	1.61069	1.61405	1.61670	1.62182	1.62795	1.63306
614564	SK 6	2.22	3.60	1.60860	1.61046	1.61375	1.61634	1.62134	1.62731	1.63227
615512	SSK 3	2.54	3.61	1.60920	1.61123	1.61484	1.61770	1.62325	1.62995	1.63559
617310	TiF 6	?	2.79	1.60769	1.61080	1.61650	1.62118	1.63070	1.64308	1.65448
617366	F 4	1.10	3.58	1.60891	1.61164	1.61659	1.62058	1.62848	1.63828	1.64678

6-digit number	Glass Type	Cost vs BK 7	Glass density	n_r	n_C	n_d	n_e	n_F	n_g	n_h
617539	SSK 1	?	3.63	1.61180	1.61375	1.61720	1.61993	1.62520	1.63152	1.63681
618498	SSK N8	2.39	3.33	1.61191	1.61400	1.61772	1.62067	1.62641	1.63336	1.63925
618526	SSK 50	2.79	3.42	1.61243	1.61442	1.61795	1.62075	1.62617	1.63268	1.63816
618551	SSK 4	2.29	3.63	1.61235	1.61427	1.61765	1.62032	1.62547	1.63163	1.63677
620364	F 2	0.86	3.61	1.61227	1.61503	1.62004	1.62408	1.63208	1.64202	1.65063
620381	F 9	1.54	3.56	1.61299	1.61565	1.62045	1.62431	1.63194	1.64140	1.64959
620603	SK 16	1.79	3.58	1.61548	1.61727	1.62041	1.62286	1.62756	1.63312	1.63774
620635	PSK 53	7.48	3.60	1.61548	1.61717	1.62014	1.62247	1.62693	1.63222	1.63660
621362	F N11	1.83	2.66	1.61314	1.61593	1.62096	1.62502	1.63309	1.64320	1.65211
621603	SK 51	6.96	3.52	1.61600	1.61778	1.62090	1.62336	1.62807	1.63368	1.63835
622360	F 13	?	3.62	1.61451	1.61730	1.62237	1.62646	1.63457	1.64465	1.65341
622532	SSK 2	2.38	3.67	1.61679	1.61878	1.62230	1.62509	1.63048	1.63696	1.64240
623569	SK 10	1.84	3.66	1.61761	1.61949	1.62280	1.62541	1.63043	1.63642	1.64141
623581	SK 15	1.79	3.58	1.61788	1.61973	1.62299	1.62555	1.63046	1.63631	1.64118
624470	BaF 8	1.73	3.67	1.61757	1.61978	1.62374	1.62690	1.63305	1.64054	1.64691
625356	F 7	1.27	3.62	1.61737	1.62021	1.62536	1.62953	1.63779	1.64809	1.65704
626357	F 1	1.13	3.65	1.61790	1.62074	1.62588	1.63004	1.63827	1.64851	1.65741
626390	BaSF 1	1.42	3.66	1.61871	1.62133	1.62606	1.62987	1.63740	1.64671	1.65476
636353	F 6	1.29	3.76	1.62818	1.63108	1.63636	1.64062	1.64909	1.65963	1.66879
639452	BaF 12	?	3.60	1.63274	1.63509	1.63930	1.64266	1.64924	1.65728	1.66415
639554	SK N18	1.96	3.64	1.63308	1.63505	1.63854	1.64129	1.64658	1.65290	1.65819
639555	SK 52	?	3.30	1.63307	1.63505	1.63854	1.64128	1.64655	1.65284	1.65808
640346	SF 7	?	3.80	1.63141	1.63439	1.63980	1.64418	1.65288	1.66372	1.67317
640597	LaK L21	6.01	2.97	1.63530	1.63719	1.64048	1.64304	1.64791	1.65367	1.65844
641601	LaK 21	3.62	3.74	1.63538	1.63724	1.64050	1.64304	1.64790	1.65366	1.65844
643480	BaF 9	?	3.85	1.63703	1.63927	1.64328	1.64647	1.65269	1.66023	1.66663
643580	LaK N6	3.67	3.8	1.63722	1.63913	1.64250	1.64514	1.65022	1.65625	1.66127
646341	SF 16	1.32	3.85	1.63751	1.64056	1.64611	1.65061	1.65954	1.67069	1.68043
648338	SF 12	1.23	3.74	1.63963	1.64271	1.64831	1.65285	1.66187	1.67315	1.68301
648339	SF 2	1.17	3.86	1.63902	1.64210	1.64769	1.65222	1.66123	1.67249	1.68233

6-digit number	Glass Type	Cost vs BK 7	Glass density	n_r	n_C	n_d	n_e	n_F	n_g	n_h
650337	SF 17	?	3.87	1.64143	1.64453	1.65017	1.65474	1.66384	1.67521	1.68514
650392	BaSF 10	2.24	3.91	1.64257	1.64527	1.65016	1.65410	1.66188	1.67150	1.67980
651419	BaSF 57	5.44	3.73	1.64431	1.64687	1.65147	1.65516	1.66242	1.67134	1.67901
651559	LaK N22	2.84	3.73	1.64560	1.64760	1.65113	1.65391	1.65925	1.66563	1.67093
652449	BaF 51	1.53	3.42	1.64551	1.64792	1.65224	1.65569	1.66244	1.67067	1.67769
652585	LaK N7	4.51	3.84	1.64628	1.64821	1.65160	1.65426	1.65935	1.66540	1.67042
654337	SF 9	1.76	3.91	1.64565	1.64878	1.65446	1.65907	1.66822	1.67966	1.68965
654396	KzFS N5	9.73	3.46	1.64645	1.64920	1.65412	1.65804	1.66571	1.67512	1.68319
655329	SF 50	?	3.79	1.64575	1.64892	1.65473	1.65944	1.66884	1.68061	1.69093
657367	BaSF 56	?	3.85	1.64901	1.65190	1.65715	1.66139	1.66979	1.68023	1.68931
658509	SSK N5	2.01	3.71	1.65237	1.65456	1.65844	1.66152	1.66750	1.67471	1.68080
658509	SSK 52	?	3.76	1.65236	1.65455	1.65844	1.66152	1.66749	1.67468	1.68072
658573	LaK 11	5.94	3.79	1.65281	1.65480	1.65830	1.66104	1.66630	1.67256	1.67775
660329	SF 51	?	3.81	1.65121	1.65441	1.66025	1.66499	1.67445	1.68630	1.69670
664358	BaSF 2	1.59	3.90	1.65604	1.65903	1.66446	1.66885	1.67757	1.68844	1.69791
667330	SF 19	1.69	4.02	1.65766	1.66090	1.66680	1.67158	1.68110	1.69302	1.70345
667482	BaF 54	?	3.78	1.66026	1.66258	1.66672	1.67001	1.67641	1.68416	1.69072
667484	BaF N11	2.05	3.76	1.66029	1.66260	1.66672	1.67000	1.67637	1.68411	1.69066
668419	BaSF 6	2.30	3.79	1.66023	1.66284	1.66755	1.67133	1.67876	1.68790	1.69577
669450	BaF 13	2.13	3.80	1.66203	1.66450	1.66892	1.67245	1.67937	1.68783	1.69507
669574	LaK 23	?	4.00	1.66326	1.66527	1.66882	1.67159	1.67693	1.68328	1.68855
670392	BaSF 12	?	3.90	1.66216	1.66495	1.66998	1.67403	1.68204	1.69194	1.70052
670471	BAF N10	1.99	3.76	1.66341	1.66579	1.67003	1.67341	1.68001	1.68803	1.69486
670471	BaF 53	?	3.75	1.66340	1.66578	1.67003	1.67341	1.68000	1.68801	1.69482
673289	TiSF 1	?	2.81	1.66300	1.66667	1.67339	1.67889	1.68997	1.70417	1.71698
673322	SF 5	1.51	4.07	1.66328	1.66661	1.67270	1.67764	1.68750	1.69985	1.71068
678552	LaK N12	4.49	4.1	1.67209	1.67419	1.67790	1.68083	1.68647	1.69320	1.69881
681319	SF 62	?	4.15	1.67172	1.67512	1.68134	1.68639	1.69646	1.70909	1.72018
681372	KzFS N7	9.58	3.51	1.67221	1.67523	1.68064	1.68498	1.69353	1.70412	1.71330
682482	LaF 20	3.82	3.87	1.67586	1.67824	1.68248	1.68585	1.69240	1.70033	1.70704

6-digit number	Glass Type	Cost vs BK 7	Glass density	n_r	n_C	n_d	n_e	n_F	n_g	n_h
683445	BaF 50	2.74	3.80	1.67561	1.67816	1.68273	1.68637	1.69350	1.70219	1.70959
689306	SF 52	?	4.10	1.67843	1.68200	1.68852	1.69384	1.70448	1.71789	1.72970
689312	SF 8	1.82	4.22	1.67899	1.68250	1.68893	1.69416	1.70460	1.71772	1.72926
689495	LaF 23	4.04	4.20	1.68250	1.68483	1.68900	1.69232	1.69876	1.70655	1.71311
691547	LaK 9	5.76	3.51	1.68498	1.68716	1.69100	1.69401	1.69979	1.70667	1.71240
693516	LaK 20	old	-	1.68716	1.68944	1.69349	1.69669	1.70289	1.71033	1.71657
694533	LaK N13	5.69	4.24	1.68737	1.68958	1.69350	1.69660	1.70258	1.70975	1.71573
697554	LaK N14	10.17	3.63	1.69078	1.69297	1.69680	1.69980	1.70554	1.71237	1.71804
697564	LaK 31	12.54	3.68	1.69080	1.69296	1.69673	1.69968	1.70531	1.71200	1.71755
698386	BaSF 13	3.07	3.97	1.68935	1.69229	1.69761	1.70190	1.71038	1.72090	1.73004
699301	SF 15	1.90	4.06	1.68853	1.69221	1.69895	1.70445	1.71546	1.72939	1.74174
700347	BaSF 55	?	3.95	1.69066	1.69391	1.69981	1.70459	1.71409	1.72597	1.73641
700350	BaSF 14	?	4.00	1.69061	1.69383	1.69968	1.70442	1.71384	1.72563	1.73596
702410	BaSF 52	4.01	3.96	1.69389	1.69673	1.70181	1.70587	1.71384	1.72362	1.73201
706303	SF N64	2.33	3.00	1.69538	1.69909	1.70585	1.71135	1.72238	1.73637	1.74882
710366	BaSF 50	?	4.07	1.70133	1.70449	1.71020	1.71480	1.72388	1.73515	1.74492
713538	LaK 8	6.68	3.78	1.70668	1.70898	1.71300	1.71616	1.72222	1.72944	1.73545
717295	SF 1	2.10	4.46	1.70647	1.71032	1.71736	1.72311	1.73462	1.74916	1.76199
717480	LaF N3	4.01	4.14	1.71003	1.71253	1.71700	1.72055	1.72747	1.73584	1.74289
720346	KzFS 8	?	4.13	1.71095	1.71434	1.72047	1.72540	1.73516	1.74726	1.75777
720504	LaK 10	7.31	3.81	1.71323	1.71568	1.72000	1.72340	1.72996	1.73784	1.74444
722293	SF 18	2.15	4.49	1.71046	1.71436	1.72151	1.72734	1.73903	1.75380	1.76685
724381	BaSF 51	4.01	4.31	1.71504	1.71813	1.72373	1.72823	1.73712	1.74810	1.75758
728284	SF 10	1.94	4.28	1.71682	1.72085	1.72825	1.73430	1.74648	1.76197	1.77578
728287	SF 53	2.37	4.45	1.71696	1.72096	1.72830	1.73430	1.74635	1.76161	1.77515
734517	LaK N16	10.17	3.95	1.72675	1.72920	1.73350	1.73688	1.74340	1.75119	1.75769
735416	LaF N8	6.51	4.02	1.72702	1.72995	1.73520	1.73940	1.74763	1.75772	1.76638
736322	BaSF 54	?	4.41	1.72597	1.72961	1.73627	1.74169	1.75251	1.76614	1.77817
740282	SF 3	2.69	4.64	1.72829	1.73242	1.74000	1.74620	1.75866	1.77444	1.78844
741276	SF 13	2.49	4.36	1.72884	1.73304	1.74077	1.74710	1.75988	1.77621	1.79084

6-digit number	Glass Type	Cost vs BK 7	Glass density	n_r	n_C	n_d	n_e	n_F	n_g	n_h
741281	SF 54	3.17	4.56	1.72904	1.73318	1.74080	1.74703	1.75955	1.77546	1.78959
744448	LaF N2	5.06	4.34	1.73630	1.73905	1.74400	1.74795	1.75567	1.76507	1.77305
744508	LaK 28	?	4.09	1.73734	1.73985	1.74429	1.74778	1.75451	1.76257	1.76931
746400	LaF 26	?	4.20	1.73736	1.74044	1.74597	1.75040	1.75909	1.76977	1.77897
748277	SF 63	3.77	4.62	1.73637	1.74061	1.74840	1.75477	1.76761	1.78393	1.79845
750350	LaF N7	5.84	4.38	1.73970	1.74319	1.74950	1.75458	1.76464	1.77713	1.78798
751275	SF 61	?	4.64	1.73869	1.74297	1.75084	1.75728	1.77026	1.78677	1.80147
755276	SF 4	2.17	4.79	1.74300	1.74730	1.75520	1.76167	1.77468	1.79121	1.80589
757318	LaF N11	6.31	4.60	1.74616	1.74998	1.75693	1.76256	1.77378	1.78784	1.80016
757478	LaF N24	14.41	4.04	1.74973	1.75242	1.75719	1.76096	1.76826	1.77706	1.78446
762265	SF 14	2.44	4.54	1.74910	1.75358	1.76182	1.76859	1.78229	1.79988	1.81574
762270	SF 55	2.82	4.72	1.74924	1.75366	1.76180	1.76847	1.78193	1.79908	1.81438
773496	LaF N28	12.97	4.24	1.76578	1.76843	1.77314	1.77686	1.78403	1.79264	1.79984
773497	LaF 28	old	-	1.76579	1.76844	1.77314	1.77684	1.78400	1.79257	1.79974
776378	LaF 13	?	4.55	1.76611	1.76946	1.77551	1.78037	1.78996	1.80179	1.81201
782371	LaF 22	9.50	4.21	1.77217	1.77559	1.78179	1.78679	1.79667	1.80895	1.81965
784413	LaF 25	?	4.45	1.77549	1.77863	1.78427	1.78878	1.79762	1.80846	1.81775
784439	LaF N10	13.67	4.16	1.77610	1.77909	1.78443	1.78868	1.79694	1.80699	1.81552
785258	SF 11	2.79	4.74	1.77126	1.77599	1.78472	1.79190	1.80645	1.82518	1.84211
785261	SF 56	2.99	4.92	1.77136	1.77605	1.78470	1.79180	1.80614	1.82448	1.84091
785261	SF L56	6.71	3.28	1.77136	1.77606	1.78470	1.79179	1.80615	1.82461	1.84128
788475	LaF N21	13.27	4.34	1.78050	1.78332	1.78831	1.79226	1.79992	1.80915	1.81693
795284	LaF 9	7.61	4.96	1.78254	1.78695	1.79504	1.80166	1.81495	1.83180	1.84675
802443	LaSF 11	old	-	1.79320	1.79624	1.80166	1.80597	1.81435	1.82452	1.83314
803304	LaSF 32	9.05	3.52	1.79161	1.79581	1.80349	1.80974	1.82225	1.83805	1.85206
803464	LaSF N30	16.41	4.46	1.79506	1.79799	1.80318	1.80730	1.81530	1.82496	1.83309
805254	SF 6	2.40	5.18	1.79117	1.79609	1.80518	1.81265	1.82775	1.84705	1.86436
805254	SF L6	6.71	3.37	1.79116	1.79609	1.80518	1.81265	1.82780	1.84731	1.86497
806342	LaSF 33	9.68	4.48	1.79529	1.79908	1.80596	1.81153	1.82261	1.83646	1.84856
807316	LaSF 8	?	4.53	1.79587	1.79996	1.80741	1.81345	1.82550	1.84062	1.85389

6-digit number	Glass Type	Cost vs BK 7	Glass density	n_r	n_C	n_d	n_e	n_F	n_g	n_h
808408	LaSF N3	19.88	4.21	1.79884	1.80212	1.80801	1.81272	1.82195	1.83326	1.84293
847238	SF 57	4.79	5.51	1.83104	1.83651	1.84666	1.85504	1.87205	1.89391	1.91363
850322	LaSF N9	11.46	4.44	1.83834	1.84256	1.85026	1.85651	1.86899	1.88467	1.89844
855366	LaSF 13	old	-	1.84476	1.84856	1.85544	1.86099	1.87194	1.88547	1.89713
878381	LaSF N15	26.08	4.75	1.86738	1.87118	1.87800	1.88347	1.89424	1.90746	1.91881
878382	LaSF 15	old	-	1.86739	1.87118	1.87800	1.88346	1.89419	1.90735	1.91863
881410	LaSF N31	31.85	5.41	1.87075	1.87429	1.88067	1.88577	1.89576	1.90793	1.91824
913324	LaSF N18	28.55	4.82	1.90068	1.90522	1.91348	1.92016	1.93345	1.95007	1.96462
918215	SF 58	7.13	5.95	1.89900	1.90550	1.91761	1.92765	1.94816	1.97486	1.99923
953204	SF 59	12.37	6.26	1.93221	1.93928	1.95250	1.96349	1.98605	2.01559	2.04279

References

2.1 H. C. King. *The History of the Telescope*. New York: Dover Publications, 1979.

3.1 R. Kingslake. *Lens Design Fundamentals*. New York: Academic Press, 1978.

3.2 W. J. Smith. *Modern Optical Engineering*. New York: McGraw-Hill Book Company, 1966.

3.3 R. S. Longhurst. *Geometrical and Physical Optics*. London: Longman, 1967.

4.1 R. Kingslake. *Lens Design Fundamentals*. New York: Academic Press, 1978.

4.2 W. J. Smith. *Modern Optical Engineering*. New York: McGraw-Hill Book Company, 1966.

4.3 R. S. Longhurst. *Geometrical and Physical Optics*. London: Longman, 1967.

4.4 J. Texereau. *How to Make a Telescope,* Second Edition. Richmond, VA: Willmann-Bell, Inc., 1984.

4.5 M. Simmons. "Optical Configurations for Astronomical Photography." *Sky and Telescope*, July 1980, 64–70.

5.1 R. L. Berry. "Newtonian Telescopes." *Telescope Making* 9 (Fall 1980): 6–13.

5.2 D. Stolzmann. "Newtonian Aberrations." *Telescope Making* 12 (Summer 1981): 4–13.

5.3 W. T. Peters & R. Pike. "The Size of the Newtonian Diagonal." *Sky and Telescope*, March 1977, 220–223.

5.4 D. E. Allen. "Updating Newtonian Collimation." *Telescope Making* 22 (Spring 1984): 4–8.

5.5 R. K. Dakin, "Placing and Aligning the Newtonian Diagonal." *Sky and Telescope*, December 1962, 368-69. (See also corrections in *Sky and Telescope*, February 1963, 114.)

6.1 R. Kingslake. *Lens Design Fundamentals*. New York: Academic Press, 1978.

6.2 W. J. Smith. *Modern Optical Engineering*. New York: McGraw-Hill Book Company, 1966.

6.3 R. S. Longhurst. *Geometrical and Physical Optics*. London: Longman, 1967.

7.1 M. Bottema & R. A. Woodruff. "Third Order Aberrations in Cassegrain-Type Telescopes and Coma Correction in Servo-Stabilized Images." *Applied Optics* 10, (1971): 300–303.

7.2 R. A. Buchroeder. "Cassegrain Optical System." *Telescope Making* 1 (Fall 1978): 1–3, 10–12.

8.1 R. Kingslake. *Lens Design Fundamentals*. New York: Academic Press, 1978.

8.2 D. Hawkins & E. H. Linfoot. "An Improved Type of Schmidt Camera." *Monthly Notices of the R.A.S.* 105 (1945): 334.

8.3 E. H. Linfoot. "On the Optics of the Schmidt Camera." *Monthly Notices of the R.A.S.* 109 (1945): 279–297.

8.4 E. H. Linfoot. *Recent Advances in Optics*. Oxford: Oxford University Press, 1955.

9.1 R. R. Willey. "Eliminating Stray Light in Cassegrain Telescopes." *Sky and Telescope*, April 1963, 232–235.

9.2 R. D. Sigler. "Family of Compact Schmidt Cassegrain Telescope Designs." *Applied Optics* 13 (1974): 1765–1766.

9.3 R. D. Sigler. "Compound Schmidt Cassegrain Telescope Designs with Nonzero Petzval Curvatures." *Applied Optics* 14 (1975): 2302–2305.

9.4 E. H. Linfoot. *Recent Advances in Optics*. Oxford: Oxford University Press, 1955.

9.5 E. H. Linfoot. "The Schmidt Cassegrain Systems and their Application to Astronomical Photography." *Monthly Notices of the R.A.S.* 104 (1944): 48.

9.6 A. S. DeVany. "A Schmidt Cassegrain Optical Design with a Flat Field." *Sky and Telescope*, May 1965, 318–322 and June 1965, 380–381.

9.7 J. G. Baker, *Proc. Amer. Phil. Soc.*, 82 (1940): 323–338.

10.1 D. D. Maksutov. "New Catadioptric Meniscus Systems." *Journal of the Optical Society of America* 34 (May 1944): 270–284.

10.2 A. Bouwers. *Achievements in Optics*. New York: Elsevier, 1946.

10.3 R. E. Cox, (ed). *Bulletin C*. Cambridge, MA: Sky Publishing Corporation, 1963.

10.4 B. Sutherland. "Designing Maksutov Telescopes from Scratch." *Telescope Making* 14 (Winter 1981/1982): 36–42.

10.5 R. Kingslake. *Lens Design Fundamentals*. New York: Academic Press, 1978.

10.6 F. B. Wright. "The Maksutov Lens Applied to Herschelian and Newtonian Telescopes." *Amateur Telescope Making Book III*, A. G. Ingalls, ed. New York: Scientific American, 1953, 574–580

11.1 J. Gregory. "A Cassegrainian-Maksutov Telescope Design for the Amateur." *Sky and Telescope*, March 1957, 236–239.

11.2 R. E. Cox, ed. *Bulletin C*. Cambridge, MA: Sky Publishing Corporation, 1963.

11.3 M. Simmons. "Optical Configurations for Astronomical Photography." *Sky and Telescope*, July 1980, 64–70.

11.4 R. D. Sigler. "A High Performance Maksutov Telescope." *Sky and Telescope*, Sept. 1975, 190–192.

12.1 A. Kutter. "Der Schiefspiegler." *Sterne und Weltraum*, Jan. 1965, 12–16.

12.2 O. R. Knab. "Making a Three-Inch Schiefspiegler." *Telescope Making* 1 (Fall 1978): 4–7.

12.3 R. A. Schmidt. "Constructing a 10–Inch Kutter Schiefspiegler." *Telescope Making* 4 (Summer 1979): 8–13.

12.4 A. Kutter. "A New Three-Mirror Unobstructed Reflector." *Sky and Telescope*, Jan. 1975, 46 and Feb. 1975, 115.

12.5 A. L. Woods. "Building a Kutter 3-Mirror Tri-Schiefspiegler." *Telescope Making* 16 (Summer 1982): 10–17.

12.6 A. S. Leonard. "The Yolo Reflector." *Advanced Telescope Making Techniques Vol. 1, Optics*, A. Mackintosh, ed. Richmond, VA: Willmann-Bell, 1986, 228-231.

12.7 A. S. Leonard. "The Solano Reflector." *Advanced Telescope Making Techniques*. Vol. 1, *Optics*, A. Mackintosh, ed. Richmond, VA: Willmann-Bell, 1986, 231–236.

12.8 R. A. Buchroeder. "A New Three-Mirror Off-Axis Amateur Telescope." *Sky and Telescope*, Dec. 1969, 418–423.

13.1 J. L. Houghton. U.S. Patent 2,350,112. May 30, 1944.

13.2 R. A. Buchroeder. "The Houghton Camera." *Telescope Making* 8 (Summer 1980): 4–7.

13.3 E. Turco. "Making an Aplanatic 4-Inch Telescope." *Sky and Telescope*, Nov. 1979, 473–477.

13.4 R. D. Sigler. "Compound Catadioptric Telescopes with All Spherical Surfaces." *Applied Optics* 17 (1978): 1519–1526.

13.5 R. T. Jones. "A Wide Field Telescope with Spherical Optics." *Sky and Telescope*, Sept. 1957, 548–550.

13.6 B. Brixner. "Barlow Lens Design for a Spherical Primary Mirror." *Sky and Telescope*, Aug. 1966, 103–104.

13.7 T. Bird & A. Bowen. "A Compact All Spherical Catadioptric Newtonian Telescope." *Telescope Making* 3 (Spring 1979): 14–17.

13.8 R. Christen. "An Apochromatic Triplet Objective." *Sky and Telescope*, Oct. 1981, 377–381 and erratum, Apr. 1982, 411–12.

13.9 N. Loveday. "A Folded Newtonian with Dual Focal Length." *Sky and Telescope*, June 1981, 545–548.

13.10 P. A. Valleli. "The Focal Reducer as a Telescope Accessory." *Sky and Telescope*, Dec. 1973, 407.

13.11 A. Bouwers. *Achievements in Optics*. New York: Elsevier, 1946.

13.12 *Zeiss Information* 95 (Feb. 1984). Oberkochen, Germany: Zeiss, 60.

13.13 D. C. Dilworth. "A New Catadioptric Telescope." *Sky and Telescope*, Nov. 1977, 425–432.

13.14 R. D. Sigler. "All Spherical Relay Telescope." *Telescope Making* 2 (Winter 1978): 16–18.

13.15 H. Dall. "A Dall Maksutov Telescope with an Erecting Lens." *Sky and Telescope*, Feb. 1962, 109–110.

13.16 F. B. Wright. "The Maksutov Lens Applied to Herschelian and Newtonian Telescopes." *Amateur Telescope Making Book III*, A. G. Ingalls, ed. New York: Scientific American, 1953, 574–580. In the revised edition (1996) published by Willmann-Bell: *Amateur Telescope Making Book 3,* pp. 429-436.

13.17 E. Everhart. "Making Corrector Plates by Schmidt's Vacuum Method." *Applied Optics* 5, 1966, 713–715.

13.18 R. Gelles. "A New Family of Flat-Field Cameras." *Applied Optics* 2 (1963): 1081–1084.

13.19 R. D. Sigler. "A New Wide-Field All-Spherical Telescope." *Telescope Making* 30 (Summer 1987): 4–9.

13.20 J. Daley. *Amateur Construction of Schupmann Medial Telescopes*. Privately printed, 1984.

13.21 R. A. Buchroeder. "A New Catadioptric Design Suitable for ATMs." *Sky and Telescope*, April 1968, 249–254.

13.22 R. A. Buchroeder. "An Improved Buchroeder Catadioptric Design." *Sky and Telescope*, Nov. 1968, 336–342.

14.1 F. E. Ross. "Lens Systems for Correcting Coma of Mirrors." *Astrophysical Journal* 81 (1935): 156–172.

14.2 D. D. Maksutov. "New Catadioptric Meniscus Systems." *Journal of the Optical Society of America*, 34 (May 1944): 270–284.

14.3 J. Richter. "A Coma Corrector for a Newtonian." *Sky and Telescope*, May 1985, 456–459.

14.4 D. Mikesic. "A Simple Coma Corrector for Newtonians." *Sky and Telescope*, July 1983, 67–69.

15.1 C. R. Hartshorn. "The Barlow Lens." *Amateur Telescope Making Book III*, A. G. Ingalls, ed. New York: Scientific American, 1953, 280. In the revised edition (1996) published by Willmann-Bell: *Amateur Telescope Making 3,* p. 180.

15.2 F. H. Thornton, "Mounting the Barlow Lens." *JBBA* 60, no. 3 (1950): 77.

15.3 P. A. Valleli. "The Focal Reducer as a Telescope Accessory." *Sky and Telescope*, Dec. 1973, 405–410.

16.1 A. König & H. Köhler. *Die Fernrohre und Entfernungsmesser.* Berlin: Springer Verlag, 1959.

16.2 Nagler Eyepiece. Tele Vue. U.S. Patent 4,286,844.

16.3 S. Rosin. "Eyepieces." *Applied Optics and Optical Engineering*,(1996) Vol. III. New York: Academic Press, 1965, 331.

16.4 T. L. Clarke. "Simple Flat Field Eyepiece." *Applied Optics* 22 (1983), 1807–1811.

16.5 T. Bird & A. Bowen. "A Compact All Spherical Catadioptric Newtonian Telescope." *Telescope Making* 3 (Spring 1979): 14–17.

16.6 H. Feijth. "De Jones-Bird Teleskoop." *Zenit*, Jan. 1985, 28. (The author describes his experience with various objective-eyepiece combinations.)

16.7 Multi Purpose Telescope. Tele Vue. U.S. Patent 4,400,065.

16.8 Plössl Eyepiece. Tele Vue. U.S. Patent 4,482,217.

16.9 H. W. Klee and M. W. McDowell. "The Pretoria Eyepiece." *Telescope Making* 29 (Winter 1986/1987): 4–9.

17.1 W. J. Smith. *Modern Optical Engineering*. New York: McGraw-Hill Book Company, 1966.

17.2 R. Kingslake. *Lens Design Fundamentals*. New York: Academic Press, 1978.

17.3 S.B.C. Gascoigne. "Recent Advances in Astronomical Optics." *Applied Optics* 12 (1973): 1419.

17.4 R. Clark. "Cassegrain Telescopes: Limits of Secondary Movements in Secondary Focusing." Applied Optics 15 (1976): 1266.

17.5 R. E. Cox & R. W. Sinnot. "On Focussing a Cassegrain." *Sky and Telescope*, Oct. 1976, 299.

17.6 W. B. Wetherell & M. P. Rimmer. "General Analysis of Aplanatic Cassegrain, Gregorian and Schwarzschild Telescopes." *Applied Optics* 11 (1972): 2817.

18.1 P. M. Duffieux. *L'integrale de Fourier et ses applications à l'optique*. Besançon: Privately printed, 1946.

18.2 K. R. Barnes. *The Optical Transfer Function.* Monographs on Applied Optics. London: Adam Hilgers, 1971

18.3 E. Everhart & J. W. Kantorski. "Diffraction Patterns Produced by Obstructions in Reflecting Telescopes of Modest Size." *Astronomical Journal* 64 (Dec. 1959): 455.

18.4 E. L. O'Neill. "Transfer Function for an Annular Aperture." *Journal of the Optical Society of America* 46 (1956): 285.

18.5 W. Pickering. "Reflectors versus Refractors," *Amateur Telescope Making, Book II*, A. G. Ingalls, ed. New York: Scientific American, 1953. In the revised edition (1996) published by Willmann-Bell: *Amateur Telescope Making 2,* pp. 3–14.

18.6 H. E. Dall. "Diffraction Effects Due to Axial Obstructions in Telescopes." *J.B.A.A.* 48, no. 4 (Feb. 1938): 163.

18.7 W. H. Steel. "Instruments à pupille annulaire." *Revue d'Optique* 32, no. 5 (1953): 272.

18.8 M. A. Danjon. "L'Acuité visuelle et ses variations." *Revue d'Optique* 7, no. 3 (1928): 15.

18.9 H. S. Coleman. "The Influence of Magnification on Resolving Power of Telescopic Systems for Foucault Test Objects of Different Inherent Contrast." *Journal of the Optical Society of America* 39 (1949): 771–774.

18.10 A. Bouwers. *Achievements in Optics*. New York: Elsevier, 1946.

18.11 A. E. Conrady. *Applied Optics and Optical Design*, Part I. London: Oxford University Press, 1929.

18.12 A. Kutter. Personal communication, 1975.

18.13 D. Stoltzmann. "The Perfect Point Spread Function," *Applied Optics and Optical Engineering*, Vol. IX, Academic Press, 1983, 111.

19.1 W. J. Smith. *Modern Optical Engineering*. New York: McGraw-Hill Book Company, 1966.

19.2 R.R. Willey. "Eliminating of Stray Light in Cassegrain Telescopes." *Sky and Telescope*, April 1963, 232–235.

19.3 A. T. Young. "Design of Cassegrain Light Shields." *Applied Optics* 6 (1967): 1063.

19.4 R. Prescott. "Cassegrainian Baffle Design." *Applied Optics* 7 (1968): 479.

19.5 A. W. Greynolds. "Computer Assisted Design of Well-Baffled Axially Symmetric Optical System." *SPIE* 193, Optical Systems Engineering (1979): 129.

19.6 G. Melles. *Optical Guide 3,* Private publication. 1985.

20.1 R. Kingslake. *Lens Design Fundamentals*. New York: Academic Press, 1978.

20.2 W. J. Smith. *Modern Optical Engineering*. New York: McGraw-Hill Book Company, 1966.

20.3 D. Stoltzmann. "Optical Design with a Pocket Calculator," *Sensor Design Using Computer Tools. SPIE*, Vol. 327, 1982.

20.4 J. Gregory. "The H' tan U' Curve," *Telescope Making* 21, Winter 1983.

21.1 R. D. Sigler. "Family of Compact Schmidt Cassegrain Telescope Designs." *Applied Optics* 13 (1974): 1765–1766.

21.2 R. D. Sigler. "Compound Schmidt Cassegrain Telescope Designs with Nonzero Petzval Curvatures." *Applied Optics* 14 (1975): 2302–2305.

21.3 R. D. Sigler. "Compound Catadioptric Telescopes with All Spherical Surfaces." *Applied Optics* 17 (1978): 1519–1526.

21.4 R. Kingslake. *Lens Design Fundamentals*. New York: Academic Press, 1978.

21.5 M. Herzberger. "Colour Correction in Optical Systems and a New Dispersion Formula." *Optica Acta* (London), Vol. 6, (1959): 197.

21.6 W. J. Smith. *Modern Optical Engineering*. New York: McGraw-Hill Book Company, 1966.

21.7 H. H. Selby. "Prism Diagonals, Axial Aberration Effects." *Amateur Telescope Making Book III*, A. G. Ingalls, ed. New York: Scientific American, 1953, 505–509. In the revised edition (1996) published by Willmann-Bell: *Amateur Telescope Making 1,* pp. 157–162.

Index

A

Abbe
  eyepiece 163, 167, 176–178
  number 308
  numbers 34
  sine condition 26, 28
Abbe, Ernst 26
aberration
  aberration-correcting lens 76
  astigmatism 29
  coma 25
  distortion 31
  lateral color 35
accommodation of the Eye 189
achromatic 4
  combinations 160
  Mangin 137
  refractor 305, 306
achromatizing 53
afocal 16
air gap 322
air space, change of 209
air-spaced (Fraunhofer) doublet 55, 56, 317
Airy disk 36, 213
amplification factor, Barlow's 156
angular
  magnification distortion 169
  off-axis distance 43
  resolving power 211
annular baffles 228
aperture stop 235
aplanatic 27, 29
apochromat 58, 59, 322, 323
apochromatic color correction 346
Apoklaas 60
apparent field 163
aspheric optical 111
aspherizing 209
assembly deviations 202
astigmatism 25, 29, 168, 180, 256, 258, 304
  caused by axial shift 204
  negative or overcorrected 30
  Newtonian 48
  obj. eyepiece combinations 188
  positive or undercorrected 30
  role in image sharpness 181
  Schiefspiegler 120
  Schmidt-Newtonian 128
astrocamera 37, 116, 298
astrophotography 38
automatic design 306
axial
  chromatic aberration 35
  deviation, diagram 272
  lateral color caused by tilt 204
  perfornamce
    effect of spherical aberration 222
    effect of surface roughness 222
  shift and aberrations 204, 274

B

back focal length (b.f.l.) 66
baffles 72, 89, 228–234, 335
Baker, James 5
ball spherometer 200
Barlow lens 136, 155, 156, 158
"bending" a lens 55
best focus 22
Bird focal corrector 134
blocking stray light 227
Bouwers, Albert 6, 97, 107
Brachyt 117
brightly illuminated objects 221
Brixner focal corrector 134, 135
Buchroeder-Houghton 131

C

calcium fluorite 58
camera 16
cardinal points 14
Cassegrain 65, 119, 228, 230, 233, 278, 279, 282, 334, 335
Cassegrain, Guillaume 4
catadioptric
  cassegrain 283
  internal reflections 240
  Schiefspiegler 121

two-mirror 334
Wright-Newtonian 144
cemented objectives 55
centering cones 202
central obstruction 89
effects of 222
Chrétien, Henri 6
Christen triplet objective 138, 329
Christen, Roland 329
chromatic aberrations 24, 31, 53, 57
circle
a conic section 245
Clark objective 322
classical
Cassegrain 70, 279
Gregorian 142
close focusing 91
coincidence of tangential and sagittal focal surfaces 189
Coleman, H.S. 223
color
aberration curve 57
aberration in lens 57
corrected concentric system 101
sensitivity curve 57
coma 25, 55, 256, 258, 305
caused by axial shift 204
correcting eyepieces 196, 197
Houghton-derived designs 132
in design of refractor 306
Newtonian 48
reduction, Maksutov method 153
residual caused by glass combinations 314
Schiefspiegler 120
Schmidt-Newtonian 128
Compaan, Klaas 61
Companar 112
compensation of astigmatism by the objective 195
compound systems 11, 127
computer programs 331
FIRST 331
INT 331
LENSDES 331
using the program 341–343
doublet 340, 343
triplet 340
RAYTRACE 331, 347
README 331
TDESIGN 331
available designs 332–340
optimizations 362
Houghton camera 365
Houghton-Cassegrain 366
Maksutov camera 367
Maksutov-Cassegrain 368
Schmidt-Cassegrain 363
Wright 362
using the program 335–340
concave field 147
concentric
distribution 23
meniscus camera 98
Conrady, A.E. 322
contrast 211, 216
for extended objects 215
transfer concept 211, 215, 216
transfer function (CTF) 218
for perfect system 218
convex secondary 65
correcting chromatic aberration 57
corrector
achromatizing 98
focal 127
full-aperture 127, 131
Houghton 131
plate ghost images 240
profiles 77
thickness 113
correctors
full-aperture 131
criteria
Dawes 214
for acceptable image quality 35
for visual use 41
Rayleigh 214
Sparrow 214
critical angle 10
crown glass 54, 308
CTF (contrast transfer function) 218, 219, 220
curvature of field 25, 31, 32, 43, 67, 167, 180, 256, 259
in a Newtonian 48
role in image sharpness 181
curve
color aberration 57
color sensitivity 57

D

Dall relay telescope 144
Dall, Horace 6
Dall-Kirkham 70, 72, 279, 334
Danjon, M.A. 223
Dawes criterion 214

degree of freedom 107
design, automatic 306
DeVany, A.S. 95
deviations 199
diagonal mirror 45
diameters of optical elements
  off-axis light loss and vignetting 239
  why not specified 239
diaphragm, adjustable 236
diffraction
  effects of 24
  for a lens 212
  image diagram 212
  pattern 212
  ring 213
diffuse reflections 227, 228
Dilworth relay telescope 144
direction cosines 263
dispersion 34
dispersion number = Abbe number 308
distant field-flattener 150, 151
distortion 25, 31, 169, 257, 259
  negative, or barrel 31
  positive, or pincushion 31
Dollond, John 4
doublet design 306
  program LENSDES 340, 343
"doughnut" profile = Schmidt shape 247
downscaling 43
drop in illumination, rule of thumb 50
drum micrometer vs dial micrometer 201
Duffieux, P.M. 215, 218

E
eccentricity 246
effective focal length 14, 66
ellipse, conic section 245
entrance pupil 23
Erfle eyepiece 163, 166, 167, 176, 177, 178, 180, 183, 184, 186
exit pupil 18
extended objects, i.e. lunar, planetary surfaces 215
eye
  accommodation 189
  compensating for field curvature 168
  peak sensitivity of 38
  pupil 171
  relief 167
  sensitivity 57
eyepiece 163
  astigmatism 167
  axial color 167
  coma 167
  field diagrams 167
  projection 186

F
fictitious entrance pupil 173
field corrector 147, 284
field curvature 168
  compensated by eye 168
  in obj. eyepiece combinations 188
field flatteners 79, 147, 148, 149
field stop 235
fifth-order term 255
film formats 38
first useful reflector 4
first-order
  calculation 244
  optics 11
  term 255
flat-field
  camera 94
  Schmidt-Cassegrain 84
flint glass 54, 308
fluor-crown glasses 313
fluorite 58, 313
fluorite objective 60, 308
focal
  corrector 127, 133, 135
  distance 11
  extenders 155
  point 11
  ratio 11
  reducer 158, 160
focus tolerance 79
foil spacers 202
Foucault, Leon 4
Fraunhofer doublet 56, 60, 320, 321, 322

G
Gabor, Dennis 6
Galilei, Galileo 3
geometric optics 9
Ghost images, corrector plate 240
"golden mean" 120
Gregorian 142
Gregory Maksutov 107, 110
Gregory, James 3
Gregory, John 6
Greynolds, Alan W. 234

H
H´ tanU´ plot 276
half-apochromats 58
Hall, C.M. 4

Herschel, William 4, 117
Herschelian TCT 144
Herzberger, M. 323
Houghton
  camera design 299, 365
  Cassegrain 132, 289, 290, 335, 366
  corrector 131
  derivatives 131
  Seidel coefficients 290, 299
  system 334
Huygenian eyepiece 120, 163, 165, 166, 176, 177, 178, 179, 180
Huygens, Christiaan 3
hyperbola, a conic section 245

I
image
  aberrations 11, 21
  blur 180
  distance 13
  quality 37
  sharpness 39
image aberrations 24
immersion objective 61
index of refraction 9
intensity distribution 24
internal reflections 240
intrinsic contrast 223
Invar bars 203
invention of the telescope 3
inward curving field 147

J
Jones focal corrector 134, 135
Jones-Bird
  focal corrector 135
  telescope 187

K
Kellner eyepiece 163, 176, 177, 178, 180
kidney-bean effect 170, 182
Kingslake 322
Köhler, H. 255
  formulae 255
König eyepiece 163, 167, 176, 177, 178
Kutter, Anton 6, 117
Kutter's "golden mean" 120

L
LA (longitudinal aberration) 26
lateral color 35, 169, 180, 183
lens
  coatings 241
lens cell, play 204
lens shape recognition procedure diagram 316
lensless Schmidt 75, 82
light loss curves, Schmidt-Cassegrain 239
light rays 9
linear off-axis distance 43
linear resolving power 211
Linfoot, Edward 5
Lippershey, Hans 3
longitudinal
  aberration 26
  beam displacement 18
  chromatic aberration 35, 307
  color aberration 306
  presentation 274
  spherical aberration 26, 275
Lord Rosse 4
Loveday telescope 139, 140
low contrast objects 221
low-astigmatism eyepieces 197
Lurie Houghton 131

M
magnesium fluoride 241
magnification 17, 18
Maksutov 97, 107
  camera design 97, 300, 367
  Cassegrain 108, 116
    baffling 233
    coma correction 335
    design 292, 368
    Seidel coefficient 292, 294
  configuration 334
  corrector 292
  corrector thickness 104
  Gregorian 143
  method of recucing coma 153
  Newtonian 141
  non-compact design diagram 297
Maksutov, Dimitri 6
Mangin 136
Mangin mirror 136, 137
Manufacturing deviations 206
manufacturing tolerance 208
Matching Principle 208
meniscus corrector 97, 113
Meniscus Newtonian corrector diagram 152
meridional
  plane 30
  ray trace 244
Mersenne 141
Mersenne telescope 141

mirror cell, play 204
misalignment 199, 201, 203
modulation transfer function (MTF) 218
monochromatic aberrations 24

N

Nagler eyepiece 166, 167, 181, 182, 183, 185, 186
nebular photography 37
negative lens 13
neutral zone 76
Newton, Isaac 4
Newtonian 129
  aberrations 48
  baffles 228
  field correctors 147, 152
  optical system 45
  use of Barlow 158
non-centered systems 272
non-concentric
  meniscus camera 97
  meniscus corrector 99
  system 101

O

object distance 13
objective 16
objective-eyepiece combinations, performance 187
oblique reflectors 117
obstruction 66
offense against the sine condition (OSC) 27
opaquing 227
optical
  calculation 243
  path difference 276, 322
  transfer function 218
  transfer function (OTF) 218
optimization techniques 90, 303
optimizing predesigns from TDESIGN 362
optimum magnification 211, 223
orthoscopic eyepiece 167
OSC 27

P

P–P diagram 326
P–V diagram 311, 313, 326
parabola 245
paraboloidal mirror 46
paraxial
  calculation 244, 249
  focus 26
partial dispersions, relative 314
peak design 206
peak designs 206
Penning, Karl 6, 97
pentaprism 19, 20
perfect optical system, def. 216
photographic
  criterion 41
  image quality 116
  useful field of a Newtonian 48
Pickering, William 6
plane parallel plates 18
planetary details 224
plano concave lens 147
play 202
Plössl eyepiece 163, 167, 176, 177, 178
point spread function 212
positive lenses 13
predesign, deviations from 362
Pressmann-Camichel 70, 72, 334
  formulae 279
principal
  planes 16
  points of a lens 16, 252
  ray 14
prism 18
  reflecting behaves as a plane-parallel plate 330

R

radius of curvature 13
Ramsden eyepiece 163, 167, 176, 177, 178, 179, 180
ray-trace program, RAYTRACE 347
ray-tracing objective and eyepiece, diagram 172
rectilinear distortion 169
reference
  circle 246
  sphere 246
reflectance of opaquing materials 227
reflection 9, 10, 227
reflectivity and refractive index 241, 341
refraction 9
refractor
  baffles 228
  objective 53
  visual use 320
relative power factor 248
relay telescope 142, 143
rescaling doublet and triplet designs 347
residual aberrations in objective lenses 57
resolution 211

resolving power 37, 211, 215, 221
retouching aspheric surfaces 200
right angle prism 19, 20
ring spherometer 200
Ritchey, George 6
Ritchey-Chrétien 72, 150, 334
  formulae 279
Ross corrector 151
Ross Newtonian corrector diagram 152
Ross, Frank 151
Rumak 110

S
sagitta 200
sagittal
  focal surface 30, 56, 168, 169, 305
  plane 30, 263
scaling optical systems 41
Schiefspiegler 117, 119, 120
Schmidt camera 75, 127, 247, 334
  corrector 75, 204, 298, 300
  design 298
  diagram 80, 297
Schmidt, Bernhard 5, 76
Schmidt-Cassegrain 83, 84, 85, 92, 116, 148, 233, 238, 239, 335, 363
  design 284
  neutral zone shape formulae 302
Schmidt-Newtonian 128, 129, 334
Schott optical glasses 36
Schwarzschild constant 70, 246
Schwerflint Apochromat 58
SCT 83
secondary
  magnification 65, 119
  mirror 49
  spectrum 58, 59, 312
Seidel
  calculation 244
  sums 257
Seidel, Ludwig von 24, 254
sensitivity analysis, difference from tolerance analysis 207
Shapley lens (focal reducers) 158
Sigler relay telescope 144
Sigler, Robert D. 5, 277, 290
sign conventions 249
Simak 38, 39, 150
simple compensation 209
sine condition 55
skew-ray trace 236, 244
sky flooding 72
Slevogt flat-field astrocamera 92
Slevogt, Karl 5
35 mm SLR 150
Smyth lens 182
Snell's law 9, 19
Snell, Willebrord 9
specular reflection 227, 228
spherical aberration 25, 54, 55, 170, 173, 182, 204, 256, 258, 304, 306, 314
spherochromatism 60, 88, 103, 306, 314, 320, 328
spherometer 200
spot diagram 21, 22, 23, 35, 44, 86, 95, 140, 141, 149
  for lenses 38
  made with skew-ray trace 244
star diagonal 19
Steinheil doublet 56, 320, 321, 322
stellar photography 37
stop analysis 234
stray light diagram 240
superachromat 323, 330
surface
  accuracy 199
  coefficients 257

T
tangential focal surface 30, 168, 169, 305
tangential plane 29, 30, 263
tangential surface of a doublet 56
TCTs 117
tertiary mirror 122
Texereau, Jean 224
thermal expansion 203
thick lens 14
thin lens 13
thin lens achromatization formulae 310
third-order
  aberration theory 244
  approximations 303
  calculation 254
  term 255
threshold, of a detector 24
tilt angles 204
tilted
  component telescopes 117
  elements, diagram 273
  surfaces 274
tilted elements, due to weight 203
tolerance 199, 201
  analysis 205, 206, 207
total internal reflection 10
transverse
  aberration 26, 27, 275

presentation 274
shifts 204
triplet corrected for three colors = apochromat 323
triplet design 323, 346
triplet design program, LENSDES 340
Trischiefspiegler 122

U

unobstructed telescopes 222
unsmoothness and CTF curves 218

V

vignetting 89, 229, 232, 234, 236, 238
calculations 350
virtual focal points 13
visual acuity 223
von Liebig, Justus 4

W

wedge error 202
wide-field eyepieces 163
Willey, Jr., Ronald R. 5
Wright camera 128, 129, 142, 297, 334, 362
corrector 298
design 298
Wyld, J. H. 341

Computer Software

In order to compare various telescope designs for *Telescope Optics*, Harrie Rutten wrote programs to run on his home computer. These programs were not "user friendly" since his main goal was only to secure data and spot diagrams. Diane Lucas has adapted these programs for the IBM-PC.

The result of this effort is three easy-to-use programs:

1. TDESIGN, a powerful design program for Cassegrain or catadioptric telescopes, which produces a pre-design based on third-order aberration theory. These designs are preliminary and must be further optimized since they are computed on third-order Seidel theory. In most cases, these predesigns are quite accurate—within 5\% of the final values—and often much better. Final optimization is done by skew ray tracing with RAYTRACE.

2. LENSDES, a powerful design program for doublet and triplet lenses, which produces designs that are complete and require no further optimization.

3. RAYTRACE, a fast and powerful raytracing program that can trace up to 2,800 rays through axially symmetric, tilted, or decentered systems with flat, spherical, conic, or aspherically deformed optical surfaces, with or without vignetting. The output of this program can be in graphical or tabular form.

The programs were written in Borland TURBO BASIC. They are provided as compiled, executable files and are protected under United States Copyright Law but are not copy protected. You can make archival copies. Each diskette is serialized. The source code is not provided. Software is available only from Willmann-Bell, Inc., which will maintain a list of software users eligible for updates or enhancements at reduced rates.

TDESIGN and LENSDES will run on any IBM PC compatible and most work-alikes. RAYTRACE requires CGA, EGA, or Hercules compatible monitors for graphic display of the spot diagrams. However, raytrace data can be displayed in tabular form, in which case any monitor will work. The raytrace program can direct spot diagrams to a dot matrix printer capable of graphics.

In addition to the three "core" programs, startup files and a selection of "canned" designs are provided to aid you in learning the programs. The programs are not sold without the purchase of the book *Telescope Optics*. The serial number for this copy of *Telescope Optics* and the order form to order software is on the following page. Photocopies of the order form will be acceptable.

TELESCOPE OPTICS
SOFTWARE ORDER FORM

☐ Yes, send me Telescope Optics for the IBM-PC at $24.95 plus $1.00 handling for delivery in the United States.[1] I want a ☐ 3.5-inch 720k or ☐ 5.25-inch 360k diskette.

I wish to pay with:

☐ **Check** ☐ **Money Order**
☐**Visa** ☐ **MasterCard** ☐ **American Express**

Card No. ______________________________

Card expiration date ______________________________

Signature ______________________________

Name (Please Print) ______________________________

Street ______________________________

City, State, ZIP ______________________________

Willmann–Bell, Inc.

P.O. Box 35025
Richmond, Virginia, 23235
Voice (804) 320-7016 FAX (804) 272-5920

SERIAL NUMBER OF THIS BOOK 038664

Prices and Specifications Subject to Change

[1] **Foreign Orders:** Shipments to foreign countries (including Canada) should include an additional $3.20 for postage. We do not guarantee delivery to foreign countries unless the shipment is made by registered mail. If you have experienced lost mail in the past, we suggest that you include an additional $4.40, and we will ship by registered mail. $3.20 + $4.40 = $7.60. Payment must be in United States funds, either an International Postal Money Order, Visa (EuroCard), American Express or Master Card, or a check drawn on a United States branch of your bank.